OECD
Environmental Outlook

OECD
ORGANISATION FOR ECONOMIC CO-OPERATION AND DEVELOPMENT

ORGANISATION FOR ECONOMIC CO-OPERATION AND DEVELOPMENT

Pursuant to Article 1 of the Convention signed in Paris on 14th December 1960, and which came into force on 30th September 1961, the Organisation for Economic Co-operation and Development (OECD) shall promote policies designed:

– to achieve the highest sustainable economic growth and employment and a rising standard of living in Member countries, while maintaining financial stability, and thus to contribute to the development of the world economy;

– to contribute to sound economic expansion in Member as well as non-member countries in the process of economic development; and

– to contribute to the expansion of world trade on a multilateral, non-discriminatory basis in accordance with international obligations.

The original Member countries of the OECD are Austria, Belgium, Canada, Denmark, France, Germany, Greece, Iceland, Ireland, Italy, Luxembourg, the Netherlands, Norway, Portugal, Spain, Sweden, Switzerland, Turkey, the United Kingdom and the United States. The following countries became Members subsequently through accession at the dates indicated hereafter: Japan (28th April 1964), Finland (28th January 1969), Australia (7th June 1971), New Zealand (29th May 1973), Mexico (18th May 1994), the Czech Republic (21st December 1995), Hungary (7th May 1996), Poland (22nd November 1996), Korea (12th December 1996) and the Slovak Republic (14th December 2000). The Commission of the European Communities takes part in the work of the OECD (Article 13 of the OECD Convention).

Publié en français sous le titre :
LES PERSPECTIVES DE L'ENVIRONNEMENT DE L'OCDE

Photo Credit:
Rothko, Mark
Untitled, 1949 (detail). National Gallery of Art, Washington
Gift of The Mark Rothko Foundation, Inc.
Photograph © Board of Trustees, National Gallery of Art, Washington

Preface

Just over three years ago, in 1997, OECD published a study entitled "The World in 2020 – Towards a New Global Age". It gave a vision of the world economy in the year 2020, where governments and societies seize the challenge of realising global prosperity. In my preface to that Report I wrote that economic expansion of the kind envisaged by the New Global Age would bring the foreseen benefits only if pursued within the context of environmental sustainability. I pointed out that there was an urgent need to deal effectively with issues such as greenhouse gas emissions, hazardous waste production, megacity evolution, intensification of agriculture, timber and fisheries exploitation, and demands on freshwater. Following from this, it was only logical that Environment Ministers would ask the OECD for a forward-looking environmental strategy when they met in April 1998. And the OECD Environment Policy Committee wisely considered that such a strategy would only be credible if underpinned by a thorough analysis of how our use of natural resources and the quality of the environment may develop in the medium- to long-term future.

This OECD *Environmental Outlook* to 2020 provides the requested analysis. It presents an economy-based assessment of environmental pressures and conditions to 2020, indicates practical policy options to change the outlook in a more environmentally friendly direction, and analyses the potential economic and environmental consequences of implementing such policies. In preparing the *Environmental Outlook*, data collected throughout the OECD were used. The economic model was based on the one used in the 1997 study, and was combined with other models to provide projections on environmental pressures and conditions. Many OECD Directorates and affiliated agencies have contributed their experience and knowledge to the *Environmental Outlook,* ensuring that it is another good example of the unique ability of the OECD to provide high-quality cross-sectoral policy advice.

The *Outlook* also constitutes an important analytical background for the environmental pillar of the OECD-wide Initiative on Sustainable Development, and it has provided a substantial contribution to the Analytical Report and the Policy Report on Sustainable Development. The results of the *Outlook* are presented in such a way that a wide audience can find this study a worthwhile basis for thinking about the future of our environment.

In preparing for a New Global Age, it is becoming clearer that we don't own this world, but rather have borrowed it from our children and the generations that will follow them. We have responsibility to pass it on in a state which will allow them to fulfil ambitions similar to ours. Ambitions not only with respect to economic aspirations, but also in terms of living in good health and in a clean environment, without natural resources becoming scarce. The OECD *Environmental Outlook* provides a vision and a means for achieving these ambitions. However, the road to the future has more red and yellow lights than green ones. Were the late philosopher Lewis Mumford to review this *Outlook* he would likely repeat his famous observation of being "...optimistic about the possibilities; pessimistic about the probabilities". I hope this *Outlook* will help to ring alarm bells in capitals around the world and inspire concrete policy implementation which will make us optimistic about both.

Donald J. Johnston

Secretary-General of the OECD

Acknowledgements

This OECD *Environmental Outlook* was prepared by a horizontal task force in the Environment Directorate of the OECD. Joke Waller-Hunter, Director of the Environmental Directorate, initiated and assumed an oversight role in the preparation of the report. It was written and edited by Lars Fogh Mortensen and Helen Mountford – together with Nils Axel Braathen, Hoe-Seog Cheong, Nis Christensen, Elaine Geyer-Allely, Tom Jones, Sanna Keijzers, Niels Schenk, Dian Turnheim and Ponciano Villafuerte-Zavala – under the supervision of Rob Visser. General guidance was provided by Michel Potier and Kenneth G. Ruffing. The modelling framework for the outlooks was provided by Nils Axel Braathen, Helen Mountford and Niels Schenk, in co-operation with Erik Kemp-Benedict of the Stockholm Environment Institute – Boston.

The report was prepared in close co-operation with, and based on background documents prepared by, or under the responsibility of, experts in the Environment Directorate, including Jean-Philippe Barde, Noreen Beg, Carla Bertuzzi, Dan Biller, Gerard Bonnis, Chris Chung, Jan Corfee-Morlot, Jane Ellis, Nicky Grandy, Henrik Harjula, Nick Johnstone, Jan Keppler, Myriam Linster, Gene McGlynn, Voula Mega, Laurie Michaelis, Grzegorz Peszko, Richard Sigman, Soizick de Tilly, Heino Von Meyer, Peter Wiederkehr and Frédérique Zegel. Management support was provided by Mark Baldock, Katherine Kraig-Ernandes, Patricia Lachaud, Amy Plantin, Bonnie Schaefer and Stéphanie Simonin-Edwards. Fatiha Caid-Lecland, Kristin Eggleston, Wendy Fages, Jennifer Griffin and Barbara Ladeuille gave secretarial assistance. The report was edited for language by Beatrix de Koster. OECD's Public Affairs and Communication Directorate – in particular Catherine Candea, Jean-Michel Lebrun, Anne-Lise Prigent and Mubeccel Valtat-Gevher – was responsible for its publication. The Translation Division of the OECD was in charge of the translation into French.

The OECD *Environmental Outlook* benefited from the extensive knowledge and experience of members of the OECD Advisory Panel on the Environmental Outlook – consisting of leading experts on environmental outlooks and assessments – who provided valuable guidance throughout the process. Members of the Advisory Panel were Dr. Allen L. Hammond, World Resources Institute (WRI); Mr. Domingo Jimenez-Beltran and Dr. David Stanners, European Environment Agency (EEA); Mr. Fred Langeweg, Dutch National Institute of Public Health and the Environment (RIVM); Dr. Nebojsa Nakicenovic, International Institute of Applied Systems Analysis (IIASA); Dr. Shuzo Nishioka, Keio University; Mr. Nicholas C. Sonntag and Dr. Paul Raskin, Stockholm Environment Institute (SEI); and Dr. Veerle Vandeweerd, United Nations Environment Programme (UNEP).

The report – and its background documents – benefited from comments from colleagues in the Environment Directorate and in other OECD Directorates and affiliated organisations, including the General Secretariat, the Economics Department, the Directorate for Food, Agriculture and Fisheries, the Directorate for Education, Employment, Labour and Social Affairs, the Directorate for Science, Technology and Industry, the Trade Directorate, the Directorate for Financial, Fiscal and Enterprise Affairs, the Territorial Development Service, the Development Co-operation Directorate, the International Energy Agency (IEA), and the European Conference of Ministers of Transport (ECMT).

Valuable comments and guidance were also received from experts in Member countries through various OECD committees and working groups, including: the Environment Policy Committee (EPOC), the Working Party on Pollution Prevention and Control, the Working Party on Economic and Environmental Policy Integration, the Working Group on the State of the Environment, the Chemicals Committee, the Joint Working Party on Agriculture and the Environment, the Working Group on Transport, the Committee on Fisheries, and the Steel Committee.

In the process of preparing the *Outlook*, two stakeholder consultations were organised, and the report benefited from comments from representatives from business, trade unions and environmental citizens organisations. Furthermore, valuable comments were received at mini-workshops and consultations held for a number of specific sectors examined in the report. A workshop on forestry was hosted by Poland, and consultations were held for the chapters on the chemicals industry, the pulp and paper industry and the steel industry.

The preparation of the OECD *Environmental Outlook* was made possible by extra-budgetary contributions from Austria, Denmark, the European Commission, Japan, Korea, Mexico, the Netherlands, Norway, Poland, Switzerland and the United States.

Table of Contents

Section IV

ENERGY, CLIMATE CHANGE, TRANSPORT AND AIR QUALITY

Section V

HOUSEHOLDS, SELECTED INDUSTRIES AND WASTE

Section VI

SELECTED CROSS-CUTTING ISSUES

Section VII

INSTITUTIONAL FRAMEWORKS AND POLICY PACKAGES FOR ADDRESSING ENVIRONMENTAL PROBLEMS

List of Boxes

List of Tables

List of Figures

Section I

EXECUTIVE SUMMARY, CONTEXT AND STRUCTURE

Executive Summary

Challenges for the future

Developing an OECD Strategy to tackle the main environmental problems…

At the beginning of the 21st century, OECD countries are taking stock of their natural resources, the damage that is being done to the environment, and what actions they can take to ensure a clean, healthy and productive environment to pass on to future generations. To support the development of the OECD Environmental Strategy for the First Decade of the 21st Century, the OECD was asked to prepare an environmental outlook report to analyse the likely environmental changes to 2020, and to evaluate policy options for addressing the main problems.

… requires an **Environmental Outlook** *examining recent trends and outlooks to 2020.*

This OECD *Environmental Outlook* provides an analysis of the forces driving environmental change, the recent and projected pressures on the environment, and the resulting changes in the state of the environment to 2020. The *Outlook* focuses on selected economic sectors and key environmental issues. The main drivers of environmental change include economic drivers (economic growth and development, trade and investment liberalisation), social drivers (demographic and labour force developments, and consumption patterns), and technological innovation. When analysing recent trends and the projected outlook for the driving forces and the sectoral pressures that will affect the state of the environment to 2020, the extensive work of the OECD on economic and social developments was used. OECD data and analysis of environmental change and the OECD Pressure-State-Response (PSR) framework form the backbone of the analysis of environmental changes, and of possible policy responses to address the environmental problems.

Traffic lights are used to signal key findings.

Traffic lights are used to signal key findings of the report. The "green light" signals pressures that are decreasing or environmental conditions for which the outlook to 2020 is positive. It is also used to signal societal responses that have proved to help alleviate the problems identified. The "yellow light" signals areas of uncertainty or potential problems. Finally, the "red light" signals pressures on the environment or environmental conditions for which recent trends have been negative and are expected to continue to be so to 2020, or for which recent trends have been more stable, but are expected to worsen.

Green lights: proceed with caution

For many years, OECD countries have been trying to tackle environmental problems they face. For some of these problems, major improvements have been registered or are expected by 2020, such as decreasing emissions of some air pollutants, reversal of deforestation trends in OECD regions, and reductions in point source pollution from industry (see Table 1).

There are some "green lights" indicating decreasing pressures or improving conditions…

OECD countries have virtually eliminated lead emissions from petrol and emissions of ozone depleting CFCs, and significantly decreased emissions of sulphur oxides, carbon monoxide and some particulate matter. Implementation of already agreed emission reduction targets in many OECD countries is expected to lead to further decreases in the emission of these pollutants to 2020. The three industries examined in this report – steel, pulp and paper, and chemicals – are examples of industries where increased efficiency and improved production methods are significantly reducing the intensity of resource use and the pollution and waste emitted per unit of product. After centuries of deforestation, forested area and forest volume have begun to increase in most OECD countries over the last two decades, and are expected to remain stable in the OECD area to 2020. "Green" purchasing of environmentally friendly goods and products is gaining ground in many OECD countries. The share of waste that is diverted from landfilling to recycling is continuing to increase, and is expected to account for 33% of municipal waste generated in OECD countries in 2020, compared with the current 18%. At the same time, organic agriculture and other environmentally friendly farm practises are spreading rapidly. For these "green lights", OECD countries should "proceed with caution".

Yellow lights: require further investigation or additional action

... and "yellow lights" signalling uncertainty or potential problems.

Several environmental pressures and problems fall under "yellow lights", which signal uncertainty or potential problems. One such issue is water use in OECD countries, which is continuing to increase in total, but has been declining per person in over half of the OECD countries since 1980 and is expected to continue to decline per unit of GDP to 2020 (see Figure 1). Similarly, while there have been significant improvements in surface water quality in OECD regions, with many of the worst polluted water bodies having been cleaned up, few OECD countries satisfactorily meet basic water quality objectives. Also, although CFC emissions have decreased significantly in recent years due to the implementation of the Montreal Protocol, the ozone layer continues to become thinner as past emissions gradually reach the stratosphere.

For many environmental concerns, available data or scientific understanding are inadequate. Effects on human health and ecosystem functions due to toxic emissions from industry are still uncertain and poorly understood. Data on hazardous waste generation in OECD countries have not been reliably collected, so recent trends are difficult to discern, although there are some indications that hazardous waste generation has been increasing. Similarly, with respect to many aspects of modern biotechnology, the potential effects on both human and ecological health are still poorly understood.

While both aquaculture and plantation forests can help to alleviate pressures from increased fish and forest product demand on natural resources (*e.g.* marine capture fisheries and old growth forests) through intensive production, both have potentially negative effects on local ecosystem quality.

Red lights: need to be addressed urgently

"Red lights" need to be addressed urgently.

A large number of environmental pressures and conditions have been placed under a "red light", signalling recent trends which have been negative and are expected to continue to 2020, or recent trends which have been more stable, but are expected to worsen. These are the major problem areas, and need to be addressed urgently.

These include global environmental problems: biodiversity loss, tropical deforestation,...

Many of the "red lights" relate to global issues – the state of environmental resources or sinks of global significance, and for which OECD countries are only some of the users or polluters. Over-fishing is a clear example of a "red light" issue of global importance: one-quarter of the world's marine fisheries are already either exhausted, over-fished, or recovering from over-fishing. All increases in the demand for fish to 2020 will need to be supplied through aquaculture, since marine capture fisheries show no signs of increasing yield; instead, their yields are likely to remain stable or even decline. Global deforestation is another serious problem, as is biodiversity loss. Efforts in OECD countries to improve conditions at home – such as through aforestation programmes and the expansion of protected areas – are steps in the right direction, but insufficient to dominate the global trends, with non-OECD regions expected to lose almost a further 10% of their forested area by 2020.

... climate change and urban air pollution...

Climate change as a result of greenhouse gas emissions is, arguably, one of the most important "red light" issues faced by OECD countries. Despite commitments by Annex I countries[1] to reduce emissions, the release of total greenhouse gases in OECD countries is expected to continue to increase to 2020, with emissions from non-OECD countries also growing rapidly. Under current policies, OECD countries are likely to increase CO_2 emissions by a further one-third to 2020, far from the overall Kyoto Protocol target for Annex I countries of a 5% reduction of greenhouse gas emissions from 1990 levels to 2008-2012. Stronger policies to tackle this problem are urgently required if the worst effects of climate change are to be averted. Energy use and transportation are the main contributors to greenhouse gas emissions, as well as to various air pollutants that lead to urban air pollution. If current policy patterns continue, the impacts on climate change of these activities are likely to continue in OECD countries and worldwide to 2020, with motor vehicle kilometres travelled in OECD countries expected to increase by 40% from 1997 to 2020, and passenger air kilometres expected to triple. Meanwhile, energy use in OECD regions is expected to increase by 35% to 2020.

... to which energy use and transport are the main contributors.

Groundwater pollution and chemicals in the environment are also expected to be "red lights".

While groundwater pollution has not been a major concern for OECD countries in the past, it is increasingly becoming one now. The main source of groundwater contamination is agricultural pollution. As human populations draw more and more upon groundwater sources for drinking water and other uses, the build-up of nitrates and other pollutants in these sources will pose a growing problem. To 2020, nitrogen loading to waterways from agriculture in OECD countries is expected to increase by more than one-quarter, and persistent and toxic chemicals are expected to continue being widespread in the environment over the next 20 years, causing serious effects on human health.

1. Under the UN Framework Convention on Climate Change (UNFCCC), Annex I (industrialised) countries have agreed to aim to limit their greenhouse gas emissions. Annex B of the Kyoto Protocol established legally binding greenhouse gas limitation commitments for almost all Annex I countries.

Table 1. **Signals of the OECD *Environmental Outlook***

	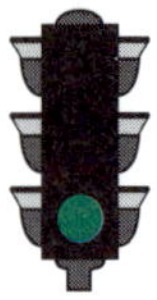	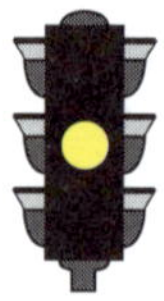	
PRESSURES ON THE ENVIRONMENT	• Industrial point source pollution • Some air pollutants (lead, CFCs, CO, SO_x)	• Water use • Toxic emissions from industry • Hazardous waste generation • Energy production and use	• Agricultural pollution • Over-fishing • Greenhouse gas emissions • Motor vehicle and aviation air pollution emissions • Municipal waste generation
STATE OF THE ENVIRONMENT	• Forest coverage in OECD regions	• Surface water quality • Forest quality in OECD regions • Ozone layer integrity	• Biodiversity • Tropical forest coverage • Fish stocks • Groundwater quality • Urban air quality • Climate change • Chemicals in the environment
RESPONSES	• "Green" purchasing • "Green" agriculture • Protected areas • Resource efficiency • Energy efficiency	• Biotechnology • Forest plantations • Aquaculture • Energy and transport technologies • Waste management	

Integrity of ecosystems

The integrity of ecosystems is threatened.

Many renewable resources are used in unsustainable ways, threatening the integrity of ecosystems. As indicated by the "red lights", unsustainable marine fisheries are now causing concern that current fishing practices and levels are not sustainable in the long run, and that the diversity and catch of the world's oceans may decrease significantly to 2020. Human induced climate change is evident, and the effects on ecosystems are expected to be severe, including rising temperatures and rising levels of the world's oceans. Biodiversity is threatened by economic development in OECD and non-OECD countries, and tropical forests are continuing to diminish. Finally, freshwater scarcity is becoming a problem in some regions and desertification, land degradation, and soil erosion are also threatening the health of ecosystems.

With the world's population expected to increase by one-quarter to 2020, and economic growth and globalisation continuing steadily, human pressures on the environment are not expected to ease unless strong policy actions are taken to protect ecosystems and maintain the essential services they provide. In order to ensure ecosystem integrity over the long term, policies will be needed to secure detoxification of the substances released to the environment, decarbonisation of energy, conservation of biological diversity and the sustainable use of renewable natural resources.

Technological change

New technologies can help to alleviate some pressures...

Technological responses have lessened a number of environmental pressures in the past, and they can be expected to alleviate many – though certainly not all – of the expected future pressures. Most of the technological developments have been in the form of efficiency improvements, practices or technologies to increase the level of renewable resources production (*e.g.* intensive forestry and agriculture, use of biotechnologies), and to reduce or eliminate the release of pollutants to environmental media.

In many fields, there have been new technological breakthroughs recently, and others are expected to occur in the period to 2020. These include new energy technologies, such as fuel cell technology, that can contribute to improving energy efficiency and reducing pressures on the environment.

... but may also lead to new risks.

Modern biotechnology, including genetically modified organisms (GMOs), have the potential to reduce the amount of damaging inputs (pesticides, fertilisers) used in natural resource sectors (agriculture, forestry, fisheries), and increase production levels to meet human needs. However, there may be negative effects on human health and on global ecosystems; coping with these potential threats will require more research and improvements in risk management regimes.

De-coupling environmental degradation from cconomic growth

Some de-coupling of environmental degradation from economic growth is expected, but...

In general, environmental degradation has increased at a slightly slower rate than economic growth. The use of energy and other resources, like agricultural raw materials, water and metals, now appears to be increasing at a slower rate than GDP in many OECD countries, and the pollution intensity of output is growing even more slowly. These trends suggest a potential for de-coupling economic growth from environmental degradation. In some cases, reductions in resource intensity have been large enough to lead to absolute, rather than just relative, environmental improvements by offsetting the overall effects of growth in per capita incomes and population.

... will not be enough to counter the volume effects of increases in pressures.

However, despite improvements in resource efficiency, overall environmental degradation has persisted in most areas as the volume effects of total increases in production and consumption have outweighed the resource efficiency gains per unit of product. Following recent trends, OECD countries are expected to reduce the energy intensity of their economies by 20% to 2020, while increasing total energy use by 35% (see Figure 1). Even with the use of new, more efficient, energy and transport technologies, it is unlikely that total emissions from these sources will decrease much over the next two decades. OECD countries will need to achieve more significant changes in the fuel mix than are currently foreseen, with greater substitution of the more polluting fossil fuels with renewable resources and cleaner fuels.

Figure 1. **Resource and material intensity of OECD economies, total use and intensity of use relative to GDP, 1980-2020**

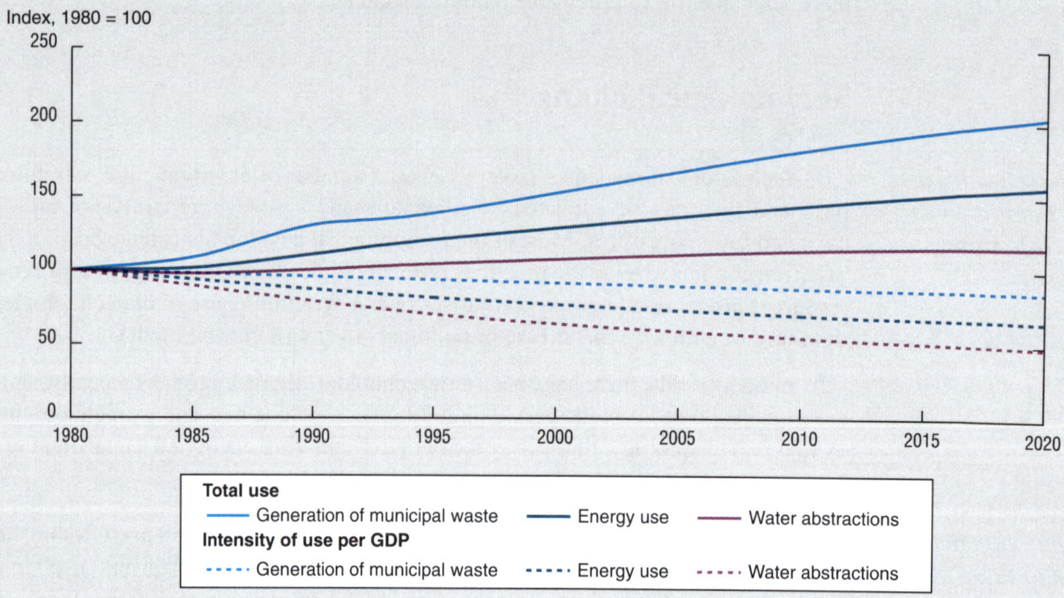

Sources: OECD (1999) and Reference Scenario.

Policy action: learning from the past

We can learn from our past successes in the use of a variety of policies, including economic instruments,...

What can policy-makers do to tackle these environmental problems? For a start, they need to look at examples of where improvements have already taken place or are starting to happen. Improvements have often been linked to pricing incentives or regulatory intervention. Recent reductions in water use have been most pronounced in countries that have removed subsidies for water use and applied charges which better reflect the marginal costs of water. Similarly, the main reductions in the energy intensity of OECD economies, while not driven by environmental policy, occurred during the major oil price shocks of the 1970s when energy prices increased rapidly.

... regulations and restrictions, and...

Government regulations and restrictions have been particularly successful in reducing industrial pollution, cleaning up the worst polluted surface waters, and reducing the levels of some air pollutants, for example by phasing out the use of CFCs. Direct intervention by governments has also been successful in increasing the size and number of natural areas in OECD countries which are conserved or where only restricted use is allowed, protecting ecosystems and biodiversity. Again, while municipal waste generation continues to increase almost as rapidly as GDP, waste management policies in OECD countries have nevertheless led to a greater share of waste that is recycled or reused, reducing its impacts on the environment.

... encouraging changes in consumption and production patterns,...

In other cases, government policies can facilitate environmentally beneficial changes in consumption patterns. This is true for the development of organic agriculture and other environmentally sustainable agricultural practices, the rapid growth of which in OECD countries is partly due to greater consumer demand, and partly to government support. Similarly, the use of eco-labels and certification schemes can help to increase consumer awareness of the environmental and health impacts of their consumption choices, and direct demand towards less environmentally-damaging products and services. In general, growing public access to environmental information and the policy-making process can help to inform individual consumption choices and increase support for environmental policies.

Addressing the social and environmental interface

... while bearing in mind the social effects.

Perhaps some of the most significant effects of environmental degradation from a social and economic perspective are the effects on human health. It is estimated that environment-related health effects – such as increased asthma and respiratory diseases from urban smog, skin cancer from a thinner ozone layer, and poisoning through chemicals in the environment – may amount to as much as 2-6% of total health-related expenditures in OECD countries per year. Furthermore, it is often the poorest regions of the world and the poor communities in society that are particularly exposed to these effects, and that are the least able to access the benefits, resources and services that the environment provides.

Policies that are developed to protect the environment and manage natural resources should consider the distributive effects of environmental quality and access to natural resources and, of course, ensure that they do not further exacerbate these effects. Some strategies for addressing distributive issues in environmental policy design are discussed in this report, including measures to reduce any potentially regressive effects of environmental taxes or charges (*i.e.* where lower-income groups spend a larger proportion of their incomes on the tax or charge than higher-income groups).

Policy packages

Because of the increasing complexity of many environmental concerns, and their often global nature...

To some extent, the most tractable environmental problems have already been dealt with. The problems of the future are likely to be more complex, and their resolution will require more difficult trade-offs and greater international co-operation. Thus, many of the issues identified as "red lights" are problems of sharing the use of global resources or sinks, such as biodiversity loss, over-fishing and the release of greenhouse gases. While unilateral action can help to alleviate some of the pressures on these resources, strong internationally co-ordinated policies will be required. A number of the other "red lights" are the result of diffuse sources of pollution – *e.g.* poor urban air quality, the dispersion of chemicals in the environment, and groundwater pollution – which are difficult to tackle because they come from large numbers of dispersed or mobile sources (*e.g.* cars, farms, consumers). In many cases it is difficult to identify the particular source of pollution, and thus to design regulations or policies to directly address the problem. Increasing the availability of and access to environmental information, as well as increased public participation in environmental decision-making processes, can help to improve understanding of environmental issues and increase support for environmental policy development.

... there is a need for packages of policy instruments to successfully address them.

For each of the "red light" issues examined in the *Outlook*, examples of appropriate policy instruments for addressing the problem are identified, and – where possible – their potential effects are quantitatively assessed. This report outlines a "policy package" or combination of instruments – regulatory, economic, and others – that can be used to tackle many of the most pressing environmental problems. It is often difficult to design a single policy instrument that will successfully provide the right incentives for a total reduction in resource use or in pollution and waste generation. Instead, it will generally be necessary to employ a mix of policy instruments. The policy mix suggested here involves the combination of a robust regulatory framework with a variety of other instruments, such as stronger pricing mechanisms to influence the behaviour of consumers and producers, voluntary agreements, tradable permits, eco-labels and information-based incentives, land use regulation and infrastructure provision. In particular, the *Outlook* recommends the removal of environmentally harmful subsidies and a more systematic use of environmental taxes, charges and other economic instruments to get the prices right.

The results will be both beneficial for the environment and economically efficient.

Adopting this policy package could deliver significant environmental benefits at relatively low economic costs in OECD countries. A policy simulation was undertaken to examine the potential effects of some of the key elements of the combined policy package: namely, the removal of all the subsidies identified in OECD countries, the application of an energy tax linked to the carbon content of fuels, and a tax on all chemical use. The environmental benefits from this policy mix would be substantial (see Figure 2). As a result of

Figure 2. **Effects in 2020 of removing subsidies, applying a fuel tax and a chemical use tax in OECD regions**

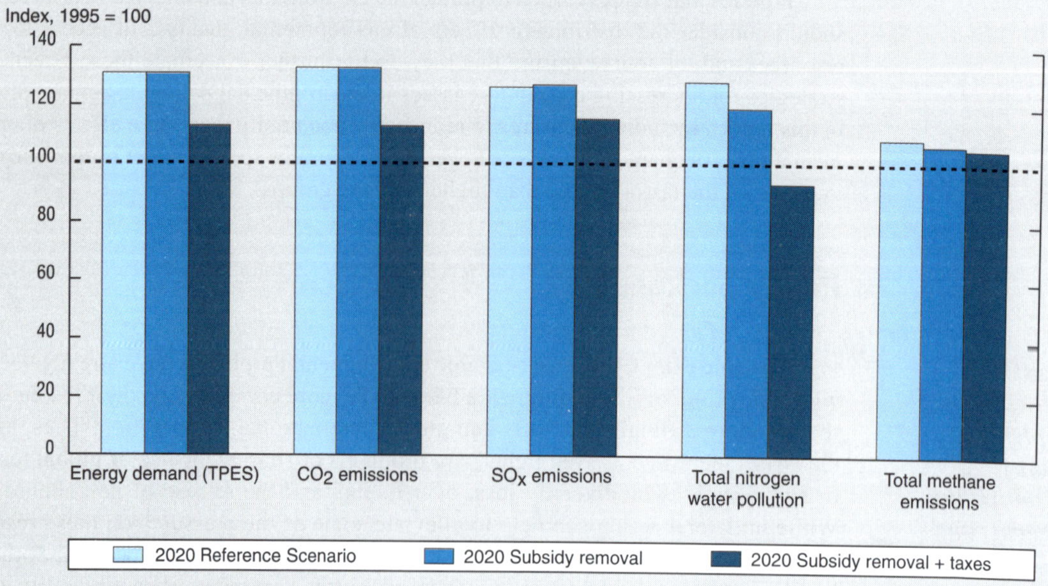

Source: Reference Scenario and Policy Simulations.

implementing the policy mix, CO_2 emissions from OECD countries would be 15% lower in 2020 compared with the Reference Scenario, SO_x emissions would be 9% lower, and methane emissions 3% lower. Largely because of the effect of the chemical tax on fertiliser use in agriculture, nitrogen loading to waterways would be almost 30% lower in 2020 compared with the Reference Scenario. With this policy package, the economic costs of achieving these environmental benefits were estimated to be quite low – less than a 1% decrease in GDP for OECD regions overall in 2020 compared with the Reference Scenario. Thus, implementing such a policy package would be cost-effective and lead to significant environmental improvements by 2020.

1 Context and Structure of the Report

1.1. Objectives and context of the report

In April 1998, OECD Environment Ministers adopted a set of Shared Goals for Action. This included an invitation to OECD "to develop a new environmental strategy for the next decade... to ensure excellence in the OECD's contribution to the implementation of sustainable development in the next century". The Ministers requested that a draft of a new environmental strategy be presented for consideration at their meeting in 2001. In order to support this exercise, the OECD Environment Policy Committee (EPOC) decided in November 1998 that the Strategy should be underpinned by an economy-based *Environmental Outlook* to 2020. To directly support the identification of objectives to be outlined in the Environmental Strategy, the *Outlook* would also include an analysis of the policy options and potential policy packages that could be used to address the identified environmental problems and meet the objectives to be outlined in the Environmental Strategy. The *Outlook* provides an overview of developments in the economy to 2020, and examines how these will affect the environment, which policies can be used to address the environmental problems identified, and how the implementation of these policies would affect the economy.

This *Environmental Outlook* is also a response to the 1997 report of the High-Level Advisory Group on the Environment to the Secretary-General of the OECD, which recommended that "the OECD should, as a matter of urgency, develop into the key intergovernmental organisation providing the industrialised nations with the analytical and comparative framework of policy necessary for their economies to make the transition to sustainable development." By providing an environmental perspective on the policy challenges for the future, the *Environmental Outlook* reinforces the environmental "pillar" of the OECD-wide Initiative on Sustainable Development.

While some environmental outlooks have been produced by other international organisations (such as UNEP, 1997 and 2000; EEA, 2000; and UNCSD, 1997), the OECD *Environmental Outlook* is distinct in that it provides a detailed economy-based environmental outlook to 2020. It draws on the long experience and analysis by the Organisation of economic and sectoral developments, and focuses on change in OECD regions, recognising that the OECD countries have a special responsibility for taking action on environmental problems and for achieving progress towards sustainable development. The broader global situation is also considered, particularly how developments in OECD countries can affect other countries, and *vice versa*.

The *Outlook* draws on OECD data and projections for economic and sectoral developments that affect the environment to provide projections of the changes that may occur in the state of the environment to 2020. Based on in-house analysis of environmental policies, the report examines potential policy options to tackle the most significant environmental problems for OECD countries, and analyses the potential environmental effects, economic costs and distributive impacts of these policies. The *Outlook* does not exhaustively address all sectors and environmental issues of importance to OECD countries, but examines selected pressures and issues, and focuses on those of highest policy relevance and for which information was available, both inside and outside the OECD.

1.2. Policy context

The *Environmental Outlook* starts from an economy-based vision of environmental issues. It recognises that there are underlying economic, social and technological factors that drive the pressures on the environment and that most of these pressures arise from activities in specific economic sectors. Global and regional economic developments are central to determining future pressures on the environment, and these pressures may either be supported or alleviated by the policies implemented by countries. In conjunction with increasing (but slowing) population pressures, the most significant drivers of environmental change over the next two decades will be continued economic growth (particularly in non-OECD countries) further globalisation of trade and investment, the use of new technologies (including information and communication technology) and changes in consumption patterns. Changes in these drivers could have both negative and positive effects on the environment, and policies will be needed to address the negative effects.

The broader context in which environmental policies operate has been changing in recent decades, and is expected to continue to evolve (see Table 1.1). Environmental problems that caused the greatest concern 20-30 years ago related to point-source pollution, with political interest being driven mainly by short-term public health concerns. The need for decisive political action was clear, and there tended to be more winners than losers from such policy actions. There was also relatively wide agreement on the policies that were needed (generally a ban on the offending pollutant or its containment), and progress was rapid. In effect, environmental policies were good by definition. While business did face additional costs in some instances, it was able to keep a certain "distance" from the environmental movement by respecting the new technical constraints that resulted from the environmental policies. The media coverage of environmental problems was extensive, but usually supportive, given the public health dimension of many of the underlying issues. There was also a general perception that existing institutions were capable of solving these problems. During this period, many of the most tractable environmental problems were "cherry-picked", in the sense that the easiest issues were largely dealt with, leaving behind the more difficult (and more costly) questions to be faced in the future.

Table 1.1. **The evolving context of environmental policy-making**

	Last 25 years	Next 25 years
Issues	• Air and water quality • Noise and nuisances • Contaminated land • Wastes and recycling • Toxic chemicals • Radioactivity • Endangered species	• Climate change • Food security (topsoil) • Fisheries • Forests • Biodiversity • Water • Biotechnology, genetically modified organisms
Politics	• Need is clear • More obvious winners than losers • Many "win-win" options	• Need is less clear (long-term; more diffuse issues) • More obvious losers than winners • Fewer "win-win" options, more difficult trade-offs required
Policy tools	• Available	• New ones needed
Public driver	• Immediate threats to public health	• Threats to strategic natural resources and "common" resources
Resolution mode	• Confrontation	• Collaboration
Status	• Significant progress • Tactical gains	• Little progress • Strategic trade-offs
Impact on business	• Additional cost of market entry	• Market structure changes
Implications for business	• Environment is mainly a *technical* issue • Managing "objectivities"	• Environment is mainly a *business* issue • Managing "subjectivities"
Institutions	• Adequate	• More substantial changes needed
Media	• High Profile	• Lower profile

Source: Adapted from Burke (1998).

Attention has thus shifted from point-sources of pollution to non-point, or diffuse, sources, such as agriculture and other sectors. The number of actors with an interest in a given environmental problem has increased, complicating the policy task considerably. Awareness has also grown that environmental problems are very much part of a broader "system", and cannot be viewed in isolation from that system. For example, deforestation is now seen as not just a forest management problem – it also increases erosion, alters flood risks, degrades water quality, and ruins fish habitat, as well as causing "rebound" effects on the economy by destroying the resource base. The diverse ecological services provided by the environment (as "sinks" for air and water pollutants, as habitats for commercial fisheries and wildlife, and as a fundamental source of life itself) have therefore gradually come to be recognised. Problems related to resource management (air, water, land, wildlife) are increasingly competing for attention with problems related to pollution control. Because of these more complex interactions, the urgency of some environmental problems seems less evident than in the past. After all, few countries appear to be running out of water or land, and a key characteristic of many of the resource management problems is that they can be put off until later ("out of sight, out of mind"). As a result, there has often been a lack of a strong political constituency to lobby on behalf of these "new" environmental issues. Even where potential for lobbying exists, the issues involved are often so complex that they are difficult to explain to the general public. As a consequence, media attention has declined in many countries.

The scale of environmental problems has also been changing rapidly. Whereas many environmental externalities were previously confined to local sources, these externalities have begun to spread to include entire countries, to cross borders and, increasingly, to affect the entire world. Climate change, biodiversity losses, acid rain and reductions in the thickness of the ozone layer all indicate that the world might be overexploiting the assimilative capacity of some of its natural resources. As these issues become broader and more complex, touching on every aspect of social and economic activity, it is becoming clear that environmental policies alone may not be sufficient to resolve them. Sectoral and economic affairs ministries, in particular, need to be actively engaged in the process of protecting the environment, together with the business community and environmental groups. In short, the focus has shifted away from simply "the environment" to "the environment in the context of sustainable development".

Another reason why economic affairs ministries and others are showing a greater interest in environmental policies is their growing cost. Whereas environmental policies were "obviously" needed in the early days, environmentalists are now increasingly asked to justify their calls for new policies. The result has been a new emphasis on "valuing" the environment (and its services) in economic terms. Efforts to quantify environmental externalities across a wide range of issues have intensified, although these efforts are met with suspicion by many for which the very idea of placing an economic "value" on environmental services is inappropriate. Nevertheless, tools such as contingent valuation, cost-benefit analysis, risk assessment, regulatory impact assessment and cost-effectiveness analysis are now finding their way into environmental policy decisions.

As we look toward the future, environmental issues are likely to continue to become increasingly complex. The long-term nature of many environmental problems will become more apparent, as evidence grows of the accumulation of pollutants in ecosystems and of the gradual degradation of renewable resource stocks and qualities. The role of these ecosystems in underpinning economic and social activity will become clearer, as will the need to examine the impacts of economic activities on ecosystems. Environmental "science" will therefore take on greater importance. On the other hand, the ability of science to deliver credible solutions to these problems is also likely to be called into question more often, and debates about the need for "precaution" in dealing with environmental matters will probably intensify.

We can also anticipate that decisions concerning environmental policy will increasingly have to be integrated with social and economic policy decisions in a sustainable development context. Two key messages will in particular need to be delivered to economic and social policy-makers. The first is the idea that the environment is a vital base upon which all economic and social activity ultimately depends. If this base is jeopardised, then these "derivative activities" will also be jeopardised. Second, it is no longer possible to completely separate economic and social policies from environmental needs, as if "someone else" were looking after the environment. In order to achieve full "policy integration", policy-makers in the economic and social spheres will have to accept more environmental responsibilities (and *vice versa*).

1.3. Sources of information and modelling

This report examines the recent and projected future state of the environment to 2020 and the pressures exerted upon it. The projections to 2020 are developed using an OECD global, dynamic general equilibrium model (JOBS) to project economic developments in 12 regions and 26 different sectors, in combination with the Stockholm Environment Institute's PoleStar framework to generate environmental projections. A Reference Scenario has been developed and is used throughout the report. Results of selected policy simulations are compared to the Reference Scenario.

Demand, production and prices in all sectors and regions are determined simultaneously in JOBS. The assumed household income elasticities are among the important "drivers" in the model. They reflect how much household demand for a given category of products or services will change when incomes change. The assumed substitution elasticities between various production factors are also important in determining simulation results. These elasticities tell how much the composition of factor use will change when the relative price between factors alters. The assumptions used are described in greater detail in Annex 2 and in the individual chapters of the report.

The results from the JOBS model are fed into the PoleStar framework, with the macroeconomic variables setting the scale of activity within the sectoral modules. Once the economic and demographic variables have been entered, projections for environmental and resource pressures are developed. PoleStar is essentially an accounting framework for combining economic, resource and environmental information to examine alternative development scenarios. The module algorithms and scenarios rely on an update of the Global Scenario Group's *Bending the Curve* scenarios (Raskin *et al.*, 1998; Heaps *et al.*, 1998).

Figure 1.1. **Regions used in the OECD Environmental Outlook**

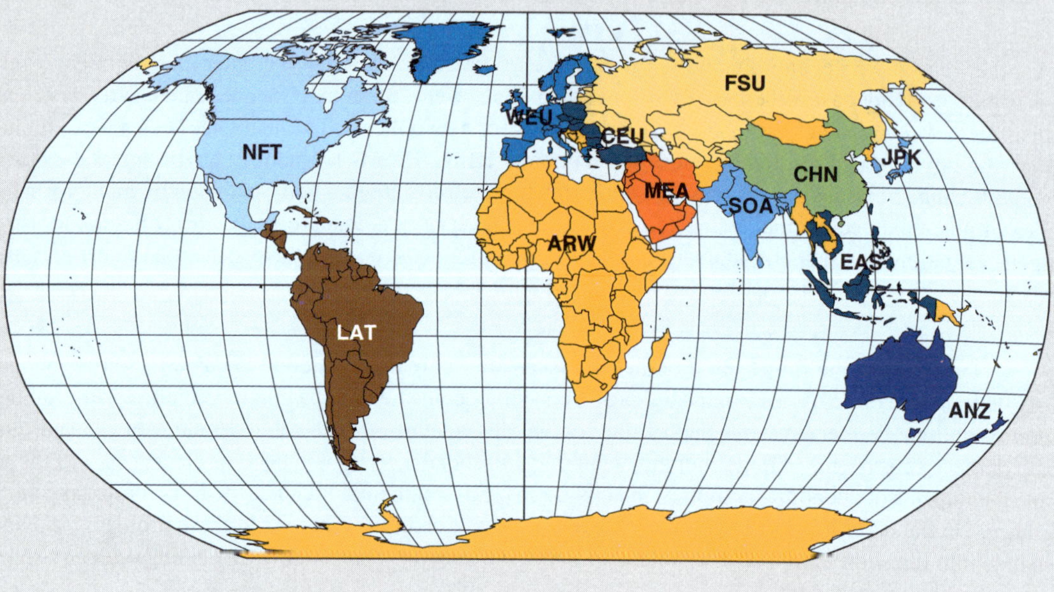

Note: OECD regions: ANZ = Australia & New Zealand, CEU = Central & Eastern Europe, JPK = Japan & Korea, NFT = Canada, Mexico & United States, and WEU = Western Europe.
Non-OECD regions: ARW = Africa & the Rest of the World, CHN = China, EAS = East Asia, FSU = Former Soviet Union, LAT = Latin America, MEA = Middle East, and SOA = South Asia.

The quantitative data regarding recent past trends and the current state of the environment are primarily drawn from OECD environmental data and indicators, particularly the country data collected in the *OECD Environmental Data: Compendium 1999* (OECD, 1999). Data from outside sources (such as various UN bodies) are also used where in-house data are not available. The 1995 base year data in the Reference Scenario is a combination of

OECD and some non-OECD data for environmental issues. The base year data used in the JOBS model were mostly taken from the GTAP (Global Trade Analysis Project) database developed at Purdue University. In addition to the base year data, assumptions are made in the Reference Scenario concerning:

- total GDP developments (based on OECD Economics Department projections);
- population growth (based on UN median fertility estimations);
- labour supply (based on OECD Economics Department estimates and UN population data); and
- supply and productivity of certain agricultural inputs (based on OECD Agricultural Directorate analyses).

The Reference Scenario is based on current activities and recent trends. It does not take into account the adoption or implementation of new policies.

A more detailed sectoral model for transport (MOVE II) was used to develop projections of specific developments and pressures arising from the transport sector that were not well covered in the Reference Scenario. The underlying assumptions of the MOVE II model are broadly consistent with the other models used in the Reference Scenario.

In addition to its application in developing the Reference Scenario, the modelling framework has also been used to examine the effects of particular environmental policies on economic and environmental conditions. Drawing on previous OECD analysis of policy options available for addressing environmental pressures and issues, the impacts of potential "policy shocks" were first simulated in the JOBS model where, for instance, effects on input use and real value-added for each sector were quantified. The changes compared to the Reference Scenario were then fed into the PoleStar system, which simulated effects of the policy shocks on environmental conditions. Because of the nature of the modelling exercise and the models available, the policy shocks examined quantitatively were primarily of the following types:

- the removal of subsidies to the production or consumption of particular products; and
- the introduction of taxes or charges on the production or consumption of given products.

Sensitivity analyses on other exogenous parameters, such as changes in the assumptions regarding autonomous energy efficiency improvement or transport and trade margins, were also performed.

While policy shocks were developed and analysed for particular sectoral pressures or environmental issues, and are described throughout the report, a "package of instruments" was also developed that combines some of the key policies for addressing the main environmental concerns of OECD countries in the next few decades. Those elements of the package of instruments that could be modelled in a meaningful way have been put together in a single policy shock to examine the net effects of a combination of policies – both in terms of synergies and in terms of potential policy conflicts.

Consistency has been sought with other OECD modelling exercises, such as the modelling work of the Economics Department on greenhouse gas emissions using the GREEN dynamic general equilibrium model, and the work of the International Energy Agency (IEA) using the partial equilibrium World Energy Model. Consistency in the basic assumptions was sought, but because *inter alia* of the different structures of the models, as well as assumptions about the future value of exogenous variables, the resulting projections necessarily differ to some extent.

1.4. Identifying the most pressing environmental concerns

Based on recent trends and the future outlooks described in the chapters, the report identifies the most pressing environmental concerns facing OECD countries over the next two decades, as well as some of the issues and pressures that are currently being relatively well-addressed. The overall results of the analysis can be summarised using the symbols of traffic lights. These serve to indicate and simplify the major results, but are to no extent a detailed summary of all the conclusions of the analysis. More specific findings, including recent trends and projections to 2020 for economic sectors and environmental issues, are discussed in the chapters themselves.

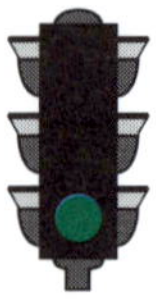

The "green light" signals pressures on the environment or environmental issues for which recent trends have been positive and are expected to continue in the future, or for which the recent trends have not been so positive, but are expected to improve. Green light responses have also been identified, signalling societal responses that help to alleviate environmental problems. For the "green lights", current policy approaches in OECD countries thus seem adequate, but in most cases could be improved.

The "yellow light" signals areas of uncertainty or potential problems. These relate primarily to environmental pressures and environmental issues for which current understanding is inadequate. "Yellow lights" are also attributed to societal responses that may help to alleviate the pressures on the environment, but for which there is uncertainty or concern regarding their effects (*e.g.* on human health or ecosystems). For the "yellow lights", efforts should focus on better understanding the pressures and monitoring the state of the environment, and precaution is needed when addressing these issues or pressures.

The "red light" signals pressures or environmental conditions for which the recent trends have been negative and are expected to continue, or for which the recent trends have been stable, but are expected to worsen in the future. The "red lights" need to be urgently addressed by OECD countries.

1.5. Structure and contents of the report

The conceptual framework underlying the analysis and structure in the *Environmental Outlook* is the OECD Pressure-State-Response framework (PSR). In this framework, "pressure" indicates pressures on the environment from human activities, "state" indicates changes in the state of the environment and its natural resources resulting from these pressures, and "response" indicates societal response to changes in the state of the environment (see Figure 1.2).

Figure 1.2. **The OECD Pressure – State – Response (PSR) framework**

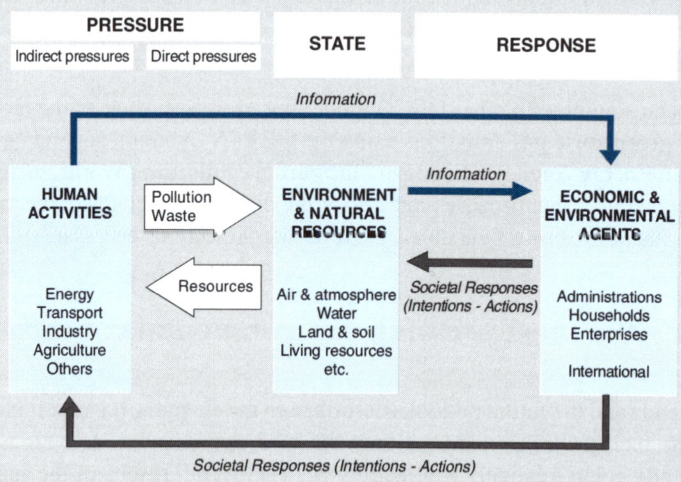

Source: OECD (1999).

The underlying drivers of the pressures and of environmental change – population growth, economic development, trade and investment liberalisation, technological innovation, and consumption patterns – are discussed in Section II of the *Environmental Outlook*. Sections III-V focus on the recent and projected future developments in selected economic sectors and for selected environmental issues. Given the resources and time available, it was decided to focus on those sectors or environmental variables for which there was high policy relevance and sufficient material (both from within and without the OECD). Among the sectors and issues not examined in detail in this *Outlook* are the service sector, tourism and soil quality. Section VI examines selected cross-cutting issues relating to the social and economic aspects of environmental change, including human health and the environment, the social and environmental interface, and resource efficiency across the economy. The final section, Section VII, describes past and possible future developments in institutional frameworks for the environment and analyses the potential effects of packages of instruments that could be adopted by OECD countries to address the most pressing environmental problems (the "red lights") identified in the *Outlook*.

The *Outlook* includes the following sections:

Section I: Executive Summary, Context and Structure

Section II: Economic, Social and Technological Drivers of Environmental Change

Section III: Primary Sectors and Natural Resources

Section IV: Energy, Climate Change, Transport and Air Quality

Section V: Households, Selected Industries and Waste

Section VI: Selected Cross-cutting Issues

Section VII: Institutional Frameworks and Policy Packages for Addressing Environmental Problems

Using the Pressure-State-Response framework to structure the individual chapters, the majority of the 25 chapters of the *Outlook* follow one of three types of structures, depending on whether the chapter describes a driver of environmental change, a sectoral pressure, or an environmental issue.

The chapters examining social, economic or technological drivers of environmental change first briefly outline the links between the drivers and the environment, then describe the recent trends and projected outlook for changes in the driver, followed by an analysis of how these changes might affect environmental conditions. Where relevant, some policy issues regarding the driver and its effect on the environment are discussed.

The chapters examining the selected sectors present the trends and projected outlook for major developments in the sector, and analyses how these developments might affect the environment.

For the environmental issue chapters, there is first an analysis of the sectoral and other pressures affecting the environmental issue, and this is followed by a discussion of the recent trends and projected outlook for changes in this environmental media or natural resource.

Both the sectoral chapters and environmental issue chapters conclude with an examination of some of the main policy options in use or under consideration in OECD countries for addressing environmental problems. This includes, where feasible, an analysis of the potential effects that implementing these policies may have on the environment, the economy and particular sectors. Such an analysis is made possible by comparing the "policy shocks" with the Reference Scenario. The main categories of policy options examined include technological development and diffusion, regulatory instruments, economic instruments, voluntary agreements and information and other instruments (see Table 1.2).

Table 1.2. **The array of environmental policy options used in the *Outlook***

Category	Description	Examples
Technological development and diffusion	Policies that encourage the development and use of environmentally benign technologies, and limit the use of those that are environmentally damaging. While direct measures can be taken to guide technological development and use, many of the policy instruments listed below also provide incentives that encourage "environmental" technological development and diffusion.	*Direct:* Technology information provision Technology demonstration projects Support to environmental research and development *Indirect:* Regulatory standards Environmental taxes and charges Voluntary approaches
Regulatory instruments	Legal instruments to enforce or restrict certain activities, or the conditions under which they are carried out. Can target particular activities, processes or products and set exact environmental results for achievement. Are widely used but can be economically inefficient as the costs of additional abatement will tend to vary between polluters. Furthermore, some regulations can lock in the use of specific technologies or practices, limiting the development of more environmentally friendly alternatives.	Licenses/permits Ambient quality standards Emission standards Process standards Product standards
Economic instruments	Economic instruments typically involve a financial transfer (*e.g.* taxes, charges, subsidies) and/ or the creation of markets (tradable permits and quotas). They can maximise economic efficiency by making the costs of additional abatement equal between polluters, and by giving polluters a continued incentive to develop low-cost abatement options. Economic instruments also provide revenue (taxes and auctioned permits) that can be set aside for environmental protection (earmarking) or to reduce economic distortions (*e.g.* distortionary taxes). However, they depend on the measurability of the relevant activities or emissions, and when used in practice, their economic and environmental efficiency is often reduced through a large number of exemptions.	Charges Taxes Tradable permits Environmental subsidies Deposit-refund systems Resource pricing
Voluntary approaches	Approaches whereby firms make commitments to improve their environmental performance beyond legal requirements, including self-regulatory arrangements. Can also involve other stakeholders, like trade unions and NGOs, and are often used as a supplement to regulatory and economic instruments. The extent to which they induce abatement beyond "business as usual" is not clear, and they are often not economically efficient instruments.	Unilateral commitments Public voluntary schemes Negotiated agreements
Information and education	Makes environmental information and training available to decision-makers and stakeholders, enabling informed decisions and participation in policy discussion. Includes the collection of relative data and the development of indicators of environmental quality and performance.	Education campaign for the general public Data collection and monitoring Development and use of indicators Diffusion of technical information Publicity of sanctions for non-compliance. Eco-labelling

REFERENCES

EEA (European Environment Agency) (2000), *Environmental Signals* 2000, European Environment Agency, Copenhagen.

Heaps, C., E. Kemp-Benedict, and P. Raskin (1998), *Conventional Worlds: Technical Description of* "Bending the Curve" *Scenarios*, PoleStar Series Report No. 9, Stockholm Environment Institute, Boston.

Burke, T. (1998), "Globalisation, the State and the Environment", in OECD, *Globalisation and the Environment – Perspectives from* OECD *and Dynamic Non-Member Economies*, OECD, Paris.

OECD (1999), OECD *Environmental Data Compendium* 1999, OECD, Paris.

Raskin, P., G. Gallopin, P. Gutman, A. Hammond, and R. Swart (1998), *Bending the Curve: Towards Global Sustainability*, PoleStar Series report No. 8, Stockholm Environment Institute, Boston.

UNCSD (United Nations Commission on Sustainable Development) (1997), *Critical Trends: Global Change and Sustainable Development*, United Nations, New York.

UNEP (United Nations Environment Programme) (1997), *Global Environment Outlook* 1997, United Nations Environment Programme, Nairobi.

UNEP (2000), *Global Environment Outlook* 2000, United Nations Environment Programme, Nairobi.

Section II

ECONOMIC, SOCIAL AND TECHNOLOGICAL DRIVERS OF ENVIRONMENTAL CHANGE

This section examines underlying drivers of the pressures on the environment and environmental change. These include economic drivers (economic development; globalisation, trade and investment liberalisation), social drivers (demographic and labour force developments; and consumption patterns), and technological change. The chapters in this section briefly outline the links between the driver and the environment, then describe the recent trends and projected outlook for changes in the driver, followed by an analysis of how these changes might affect the impact of the driver on environmental conditions. Where relevant, some policy issues regarding the driver and its effect on the environment are discussed.

2
Demographic and Labour Force Developments

KEY SIGNALS

- Human populations have an impact on the environment through how and what they produce and consume, the waste and pollution they generate, the natural resources they deplete, and the natural habitats they destroy. These impacts vary according to demographic variables such as population size, structure and density, and labour force participation.
- Population increases are expected to put further stress on the environment. World population is projected to grow by one-quarter, from the current 6 billion to 7.5 billion in 2020; the population in OECD countries is projected to increase by less, from 1.1 billion in 1995 to 1.2 billion in 2020.
- The share of people over 60 years old in OECD countries is expected to increase from just over 17% in 2000 to almost 24% in 2020.
- Urban areas are expected to continue to grow in the next two decades especially through urban sprawl, but with lower growth rates than during the last two decades.
- The growth rates for the working age population (15-64 years old) in OECD regions are expected to continue to decline to 2020 and, for some OECD countries, the growth rates may become negative.

2.1. Introduction

The environmental impacts of changes in population size and spatial distribution are a concern in both OECD and non-OECD countries. These impacts are linked to production levels in economic sectors, land use patterns, natural resource use, and waste and pollution generation. Qualitative aspects of populations – such as education levels, affluence, workforce participation, age structure, etc. – and how these influence consumption and production patterns are also potentially significant drivers of economic and environmental change.

The main demographic trends in OECD countries are quite different from those in developing countries. While population levels are still slowly increasing in most OECD regions, there is a tendency towards stabilisation in the period to 2020. In non-OECD regions population levels are continuing to grow, though at a slower rate than in the past. Population sizes in OECD countries are significantly influenced by migration levels, with many of the arriving migrants of working age. The ageing of the population in OECD countries will affect demographic pressures on the environment – particularly through changes in consumption patterns and the proportion of the total population that is active in the labour force. Many developing countries, on the other hand, are characterised by very large shares of young people, due to both high birth rates and reduced child mortality.

Population density and changes in settlement patterns also have important implications for the environment. Many countries are continuing to experience an expansion of urbanisation – both in terms of a growing percentage of the population living in urban areas and in terms of the physical expansion of urban areas into rural zones (*i.e.* urban sprawl). As far as the population density of urban areas is concerned, the environmental effects are not entirely clear. On the one hand, high population density leads to high local pressures on natural resources, such as water and air quality, as well as natural habitats. On the other hand, urban areas are more compact than suburban or rural areas, with less land used per capita, more effective distribution of electricity and water, less travelling, and more centralised waste collection. In terms of the physical expansion of urban zones through urban sprawl, the environmental effects are more clearly negative. Urban sprawl in most cases causes disproportionally high levels of land use, energy consumption and resource use compared with more central high-density zones in cities.

The physical expansion of urban areas (urban sprawl) leads to increasing pressures on the environment.

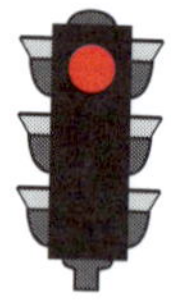

To some extent, the size of the population will have less influence on the environment than the number of households, with the related demand for space, heating, equipment, etc. (see Chapter 16). The decreasing number of persons per household being seen in many OECD countries, coupled with a trend towards urban sprawl, may partially offset the beneficial environmental effects of a relative stabilisation of their total population levels.

2.2. Population size

The process of economic development in the 20th century has significantly affected the size of both the world population and the population in OECD countries. Before the industrial revolution, most OECD countries experienced high birth rates and high mortality rates resulting in slow or no population growth. In the era of industrialisation, a demographic transition took place with death rates falling drastically while birth rates remained high, resulting in very rapid population growth. This was largely the result of improved medical care and higher incomes, with less exposure to health risks such as poor diets and poor living conditions. A further transition has taken place, that of declining birth rates. This has primarily been driven by factors such as the greater participation of women in the workforce, a dramatic reduction in infant mortality, the introduction of social security and pension systems, and the availability of reliable birth control.

Overall, as a result of these changes, the population in OECD countries and the rest of the world has grown rapidly since the 1950s, but with slower growth in the 1990s. It is projected to exhibit even lower growth rates in the period to 2020, accompanied by an ageing of the population in OECD regions.

The population in OECD countries increased from about 1 billion in 1980 to 1.1 billion in 2000 (average annual increase: 0.8%) and is projected to increase to 1.2 billion in 2020 (average annual increase: 0.4%) (Figure 2.1). Some OECD regions are experiencing very low (less than 1%) natural increases, and the natural balance (births minus deaths) has even been negative in a number of OECD countries. Western Europe, as well as Japan & Korea, are projected to face a decrease in total population after 2010.

Figure 2.1 also shows projected total percentage changes in population levels in the period from 1995 to 2020. The growth rates of total population have declined and are projected to continue to decline in virtually all regions of the world over the period up to 2020. The total world population was 4.4 billion in 1980, 6.1 billion in 2000 (average annual increase: 1.6%) and is estimated to be 7.5 billion in 2020 (average annual increase: 1.1%).[1] Thus, the world population is still increasing, but more slowly than before.

Besides natural increases, the populations of many countries also increase due to net migration. Since the mid-1980s, international migration has rapidly gained in importance as a component of population change. The world total of international migrants increased from 75 million in 1965 to 119 million in 1990, with an annual growth rate that peaked at 2.6% in 1985-1990. Among OECD regions, Western Europe and North America have accepted the highest number of international migrants (with total immigrant populations of 24 and 25 million respectively in 1990), while Japan and Korea have experienced low immigration (300 000 and 60 000 migrants respectively in the early 1990s). Net migration is the source of 75% of the population growth in EU countries and will most likely be their main source of population increase in the short and medium-term.

1. This follows the UN medium fertility variant (UN, 2000*a*).

Figure 2.1. **Total population, 1980-2020**

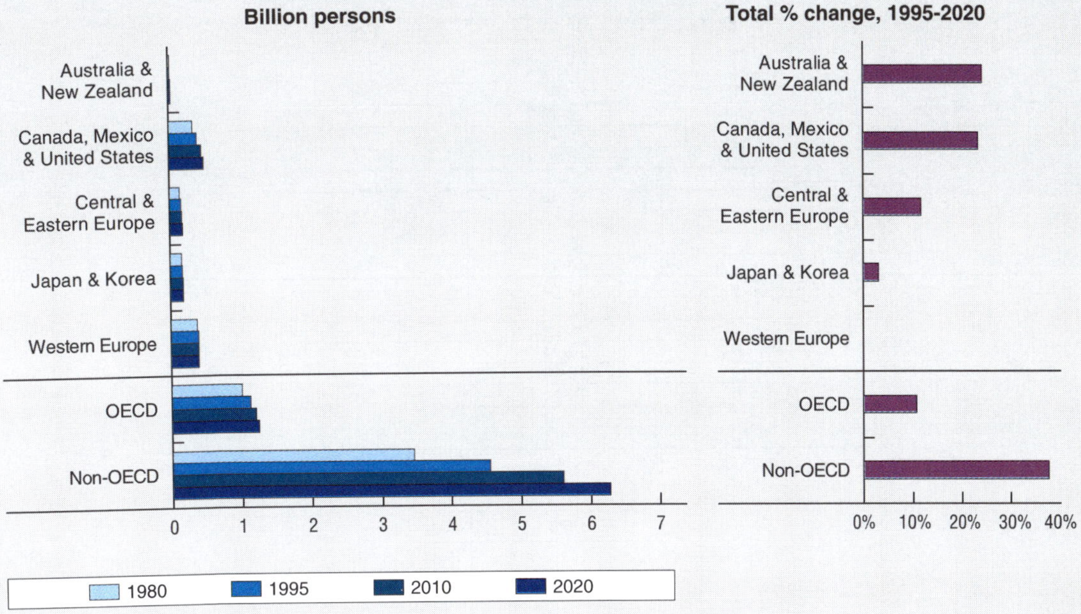

Source: UN (2000*a*).

2.3. Population structure

The structure of the population also influences the environment, particularly through changing consumption patterns and the workforce participation rates of different age groups, of men and women, etc. In OECD regions, a significant change in population structure over the next twenty years is expected to be the ageing of the population. The median age – which divides the number of younger and older people into two equal parts – passed from 28.6 years in OECD countries in 1950 to 36.8 in 1998, and is projected to exceed 40 by 2020. The percentage of the population over 60 years of age in OECD countries is expected to increase from over 17% in 1995 to almost 24% in 2020, with OECD Europe experiencing the most dramatic ageing of the population.

Changing age pyramids will also have important implications for the ability of society to generate sufficient income to support dependants. Under current circumstances, it is likely that the proportion of dependants in society (*i.e.* those not actively participating in the workforce) will grow in OECD countries as the ageing of society results in a significant increase in retired people (Figure 2.2). A number of countries are already considering policies to address this, including raising the retirement age to augment the labour force.

The effects of ageing populations on the environment are not well understood.

One way in which ageing populations in OECD countries are likely to affect environmental pressures is through changing consumption patterns (see also Chapter 5). There is still substantial speculation as to the net effects that will result from these developments. More leisure time and the better health of retired persons may lead to higher consumption levels (*e.g.* through increased tourism travel), or to relatively moderate or low consumption effects (*e.g.* through staying at home or lower income level). Furthermore, decisions on how to reduce or increase working time (daily, weekly, annually, early retirement, etc.) have major implications on spatial settlement patterns and thus on the environment.

Figure 2.2. **OECD age pyramids, 1995 and 2020**

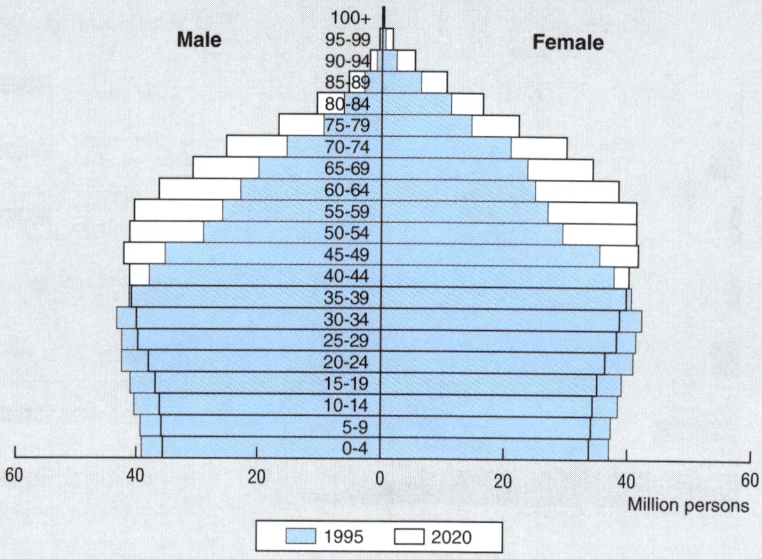

Source: UN (2000*a*).

2.4. Labour force

As a result of the tendencies described above, OECD countries are experiencing changes in the labour force supply. One example is female participation in the labour force, which has increased in most OECD countries. After a period of high unemployment, unemployment rates are now declining and slowly approaching 1980 levels. The increase in the total population of working age, however, has led to an increase in the absolute number of unemployed in OECD regions, such that it is now 57% higher than in 1980.

A long list of economic and social factors affect labour force participation rates, including the participation of women in the workforce (which currently varies significantly across OECD countries and worldwide), education levels, changes in retirement age, etc. For example, higher education levels have mixed impacts on labour force participation rates. Higher education tends to decrease employment levels among persons under 30 years of age, as university attendance is prolonged, but may increase participation of older age groups if it leads to greater job satisfaction. Because of the inherent difficulty in projecting the net impacts of various conflicting trends on labour force participation rates, the Reference Scenario for this *Outlook* assumes that a constant share of the population aged between 15 and 64 participates in the labour force.

The growth in the number of working-age people varies considerably across OECD regions with Canada, Mexico & United States showing the highest projected increase, and the regions of Japan & Korea and Western Europe showing a projected decrease (Figure 2.3). In OECD countries as a whole, the total number of persons in this age group is expected to increase by 0.6% annually from 1995 to 2010 – *i.e.* from 745 to 810 million – and then stabilise between 2010 and 2020. For non-OECD regions, the number of people of working age is expected to increase by 1.9% annually from 1995 to 2010, and 0.8% annually over the period 2010-2020.

Recent labour market developments and future prospects highlight the increase in part-time working and more varied and flexible working lives (*e.g.* working from home, shift work, etc.). The effects of more part-time workers and generally shorter working weeks on the environment are not clear. While they may lead to extra time available for consumption and travel, they may also lead to less disposable income for such activities. Similarly, the environmental impacts of the noticeable trend toward an increase in tele-commuting (working from home using new information technologies) are not obvious although a number of observers expect that this may lead to reduced transportation use, with an easing in pressures on the environment from commuting.

Figure 2.3. **Persons aged between 15 and 64 years, 1980-2020**

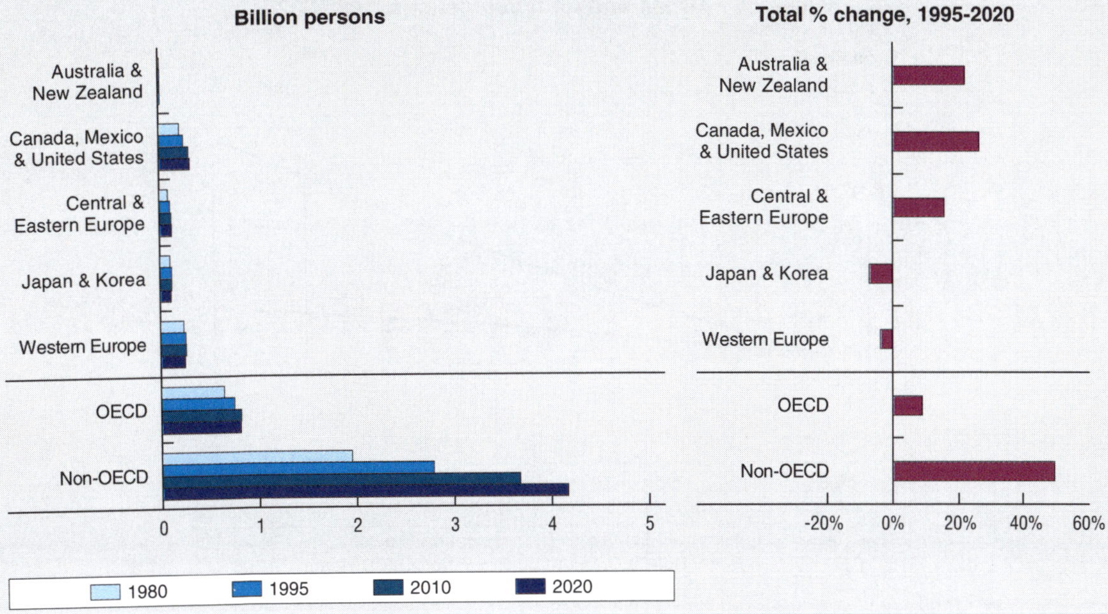

Source: UN (2000a).

2.5. Population distribution

The rapid population increases of the last century, and continued economic development, have had a major impact on rural and urban settlements. The declining relative importance of the primary sectors and the growing economic significance of the industrial and service sectors in OECD countries have led to larger and economically more important cities. While the average population density of OECD regions lies below the median world population density, it has grown slowly from 27.2 inhabitants per km^2 in 1980 to 31.2 inhabitants per km^2 in 1995, representing an average annual increase of 0.9%. It is expected to increase only slightly in most OECD regions over the next twenty years, following population levels, which show a tendency toward stabilisation at the end of this period. Population density differs strongly across OECD countries and within them, with most of the population concentrated in coastal zones and river valleys, and remote areas, like mountains and deserts, often being sparsely populated.

The same tendencies of urbanisation can be observed when using the OECD regional typology of sub-national territorial units (OECD, 1994; OECD, 1996).[2] Over the past decade, the share of the rural population has dimin-

2. The percentage of the regional population that lives in rural local communities – which correspond to local level administrative or statistical units and where the population density does not exceed 150 inhabitants per square kilometre (500 for Japan) – is used to distinguish between three types of sub-national regions:
1) Predominantly rural – more than 50%;
2) Intermediate – between 15 and 50%; and
3) Predominantly urban – less than 15%.

ished in the majority of OECD countries. High population growth is concentrated in major metropolitan regions, while many rural regions experience population loss. The trends have been for large urban areas to grow faster than the rest of the country, driven by better employment opportunities and higher incomes. The leading metropolitan regions are still recording rapid growth, although some other intermediate and rural regions appear to be catching up. Overall, it is unlikely that these trends will change the regional population balance.

Figure 2.4. **Urban and rural populations, 1980-2020**

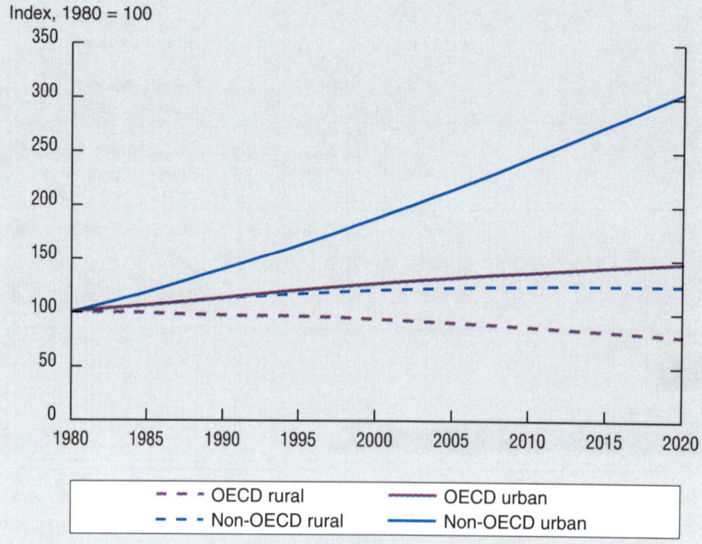

Source: UN (2000*b*).

From an environmental perspective, the total population density of a country is perhaps not as important as the density of populations at the local level within each country. Populations have been concentrating more and more in OECD regions, with OECD countries generally being more urbanised than less developed ones. The gap in urbanisation levels increased between 1950 and 1975, but has narrowed since 1975, and is expected to narrow further by 2020, essentially because of higher urbanisation rates in developing countries. The world rate of urbanisation was 2.5% in 1980-2000 and is expected to be 2% in 2000-2020. The global urban population is projected to increase, from 2.6 billion in 1995 to 4.3 billion in 2020. In OECD regions, urban dwellers, according to the UN definition, represented 50% of the total population in 1950, 77% in 2000 and are projected to reach 83% in 2020. The annual rate of urbanisation in OECD regions is projected to decrease from 1.2% in 1980-2000 to 0.7% in 2000-2020.

2.6. Policy issues

Total world population is projected to increase by almost one-quarter in the period from 2000 to 2020, with intensified pressures on the environment as a result. In OECD countries in general, population levels are still increasing but at a lower rate than before. Natural increases in population levels (not including migration) in some OECD countries have been negative in recent years, and the trend seems likely to continue. The migration levels assumed in the Reference Scenario (based on UN projections) will not be enough to fully offset declines in both total population and the potential workforce in some OECD regions. There is a possibility that some OECD countries may find it necessary to attract more immigrants in the future to bridge this shortcoming, and so may adjust current migration policies accordingly.

Another option is to encourage greater labour force participation. This could be done, for example, by raising the retirement age, a policy solution already widely discussed in OECD countries. Another possibility may be to encourage an increase in the number of women in the workforce. In some OECD countries, the rate of women working is still relatively low, and stimulating their workforce participation could potentially balance out the higher proportion of elderly people.

The link between demography and the environment is however, still not well understood. The example of ageing populations is symptomatic of the complexities involved in examining how demographic factors are likely to affect the environment. Increased leisure time and longer lives may lead to higher consumption levels and a higher share of services in total consumption, but reduced income levels can also be a constraining factor on actual spending power. It thus remains to be seen how an ageing population influences pressures on the environment. What is clear, though, is that demographic developments are important drivers of environmental change because of both the direct and indirect influences they have on consumption and production patterns (see Chapters 5 and 16), waste (see Chapter 20) and pollution generation, and on the destruction of natural habitats, etc.

How the population is distributed – *i.e.* urban versus rural population – is another important demographic development in relation to the environment. Urbanisation offers some environmental advantages. Greater population density in inner cities, for example, can lead to less travelling and more efficient water and electricity distribution. Population density is also strongly related to biodiversity loss. It has been estimated that population growth in areas adjacent to the so-called "biodiversity hotspots" will be substantially higher than both global population growth and population growth in developing countries (Cincotta *et al.*, 2000). The physical expansion of urban areas – *i.e.* urban sprawl – into nearby rural areas that is occurring in many OECD countries tends to have detrimental effects on the environment, particularly through the negative environmental impacts from providing sufficient public transport facilities to such dispersed housing. These pressures will need to be addressed in many OECD regions in the future, possibly through better urban planning and public infrastructure development.

REFERENCES

Cincotta, R.P., J. Wisnewski, and R. Engelman (2000), "Human Population in Biodiversity Hotspots", *Nature*, 404: 990-992.

Mega, V. (2000), "Demography and Employment, Urbanisation and Land Use", Background Paper for the OECD *Environmental Outlook*, OECD, Paris.

OECD (1994), *Creating Rural Indicators for Shaping Territorial Policy*, OECD, Paris.

OECD (1996), *Territorial Indicators of Employment* , OECD, Paris.

OECD (1999), OECD *in Figures*, OECD, Paris.

UN (United Nations) (2000*a*), "World Population Prospects – the 1998 Revision", UN, New York.

UN (2000*b*): UNFAOSTAT Database Gateway – Urban and Rural Populations *http://apps.fao.org/lim500/nph-wrap.pl?*Population.LTS&Domain=SUA.

3
Globalisation, Trade and Investment

KEY SIGNALS

- Since 1945, the global economy has experienced a rapid expansion in international trade and cross-border investment. This process accelerated following the completion of the last GATT trade round and the establishment of the World Trade Organisation (WTO).
- In the Reference Scenario, world exports are projected to grow by 90% from 1995 to 2020.
- The environmental implications of globalisation can be both positive and negative. Globalisation can improve efficiency in natural resource use and management, but it can also pose threats to the health of the world's natural ecosystems. The net effect of globalisation depends on the pace and direction of economic growth, and the existence and effectiveness of institutions and policy frameworks to maximise the positive and minimise the negative environmental impacts.
- Well designed policies in the areas of trade and investment, as well as in the environment field, are needed.

3.1. Introduction

Globalisation can be thought of as a process in which economic markets, technologies, and communications gradually come to exhibit more "global" characteristics and less "national" or "local" ones. Viewed at the level of the global economy, globalisation can contribute to the efficiency of economic activities. International trade and international investment are highly visible parts of the economic globalisation process. Liberalisation of international trade regimes, financial market deregulation, intensified competition, as well as rapid technological changes, particularly in information, communication, and transportation technologies, are the main drivers of this process. These trends accelerated in the 1990s, with the improvement in the investment climate and rapid expansion of trade liberalisation.

Trade and investment policies are very powerful economic levers that can contribute to the achievement of environmental goals. Economic globalisation contributes to income growth and higher standards of living through a more efficient allocation of resources. However, it is also true that the global environment has been placed under increasing strain by the resulting intensification of human economic activities. In addition, even if the effects of globalisation on the environment and society are on the whole positive, the distribution of environmental pressures will vary depending, *inter alia*, on the assimilative capacity and resource endowment and the degree of development in the different areas. The challenge for governments will be to find ways to maximise the positive environmental consequences, while limiting the negative ones.

3.2. Developments in globalisation, trade and investment

Between 1970 and 1998, world exports of goods and services, measured in current US$, increased 17-fold and foreign direct investment (FDI) increased 45 times (Figure 3.1). Over the same period, global GDP increased by a

Figure 3.1. **World exports, foreign direct investment and GDP, current US$, 1970-1998**

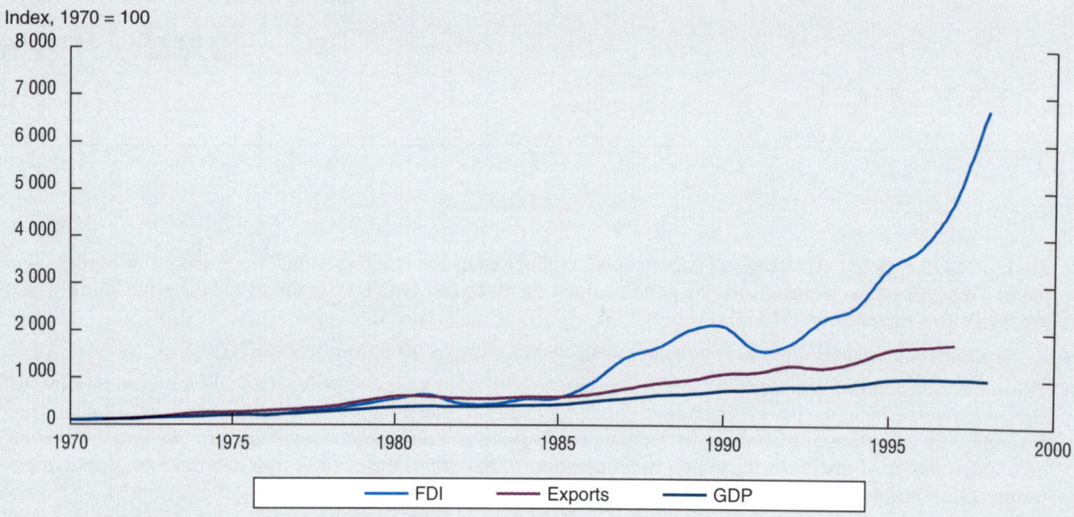

Source: World Bank (2000).

factor of 10. Several underlying and mutually reinforcing reasons for this strong growth in trade and investment can be identified (OECD, 1997; WTO, 1999):

- the steady expansion of the multilateral trading system founded under the General Agreement on Tariffs and Trade (GATT) and now overseen by the World Trade Organisation (WTO);
- the creation of regional trading blocs which have removed barriers to trade among groups of countries, such as the European Union (EU), the North American Free Trade Agreement (NAFTA), the Common Market of the South (Mercosur), the Association of South East Asian Nations (ASEAN), and the Asia-Pacific Economic Cooperation (APEC);
- the evolution of global corporations operating in many countries simultaneously (internal trade between subsidiaries of such companies accounts for about 30% of total world trade);
- rapid income growth in almost all countries, leading to increases in demand for capital and consumer goods and new opportunities for foreign investment and trade;
- the explosive expansion of new means of communication and transportation that has helped to create a global culture and fuel consumer demand for products not necessarily produced locally;[1]
- the collapse of Soviet-style communism in the late 1980s and 1990s, and the opening up of the economies of the former Soviet Union and Central & Eastern Europe to the world economy.

Trade

Over the past two decades, the ratio of world merchandise exports to global production has nearly doubled, from 8% to 15%. As indicated above, this process has accelerated in recent years, following the completion of the

1. Communication and information technologies can significantly widen the scope for trade, allowing for the geographical unbundling of once-integrated operations, including research and development, legal services, and customer services such as insurance claim processing.

last GATT trade round, the Uruguay Round, in 1994. World trade growth doubled from 4% per annum between 1980 and 1993 to 8% in 1994 to 1996, outpacing world output growth by a widening margin. Between 1980 and 1995, trade in services grew even more rapidly, at double the rate of merchandise trade (8.7% compared to 4.5%). Although a far lower proportion of services than goods is traded at present, trade in commercial services accounted nonetheless for 23% of total world trade in merchandise goods and services in 1995.[2]

For OECD countries, trade has grown rapidly in recent years. Between 1990 and 1993, trade among OECD countries grew at half the rate of world trade (2.2% compared to 4.4%), while their trade with non-OECD countries grew much faster, at 7.5%. Trade between non-OECD countries also grew rapidly, rising more than 15% per annum in the mid-1990s, until the sharp fall in commodity prices (which started in 1997) and the Asian financial crisis of 1998. The product composition of world trade has also changed markedly over the past two decades. Between 1985 and 1995, the share of primary products in exports from non-OECD regions to OECD countries fell by half (to 29%). Exports from OECD countries, however, remained dominant in a number of high-tech and medium-tech sectors (see Chapter 6). In terms of commercial services, OECD countries and a few Asian non-OECD countries, dominate world trade, with the EU accounting for about 45% of exports, Asia just over 20%, and North America just under 20%.

With further reductions in communication and transport costs, and with high GDP growth in non-OECD economies stimulating demand, it is expected that international trade will continue to grow strongly (Figure 3.2). Under

Figure 3.2. **World exports, 1995-2020**

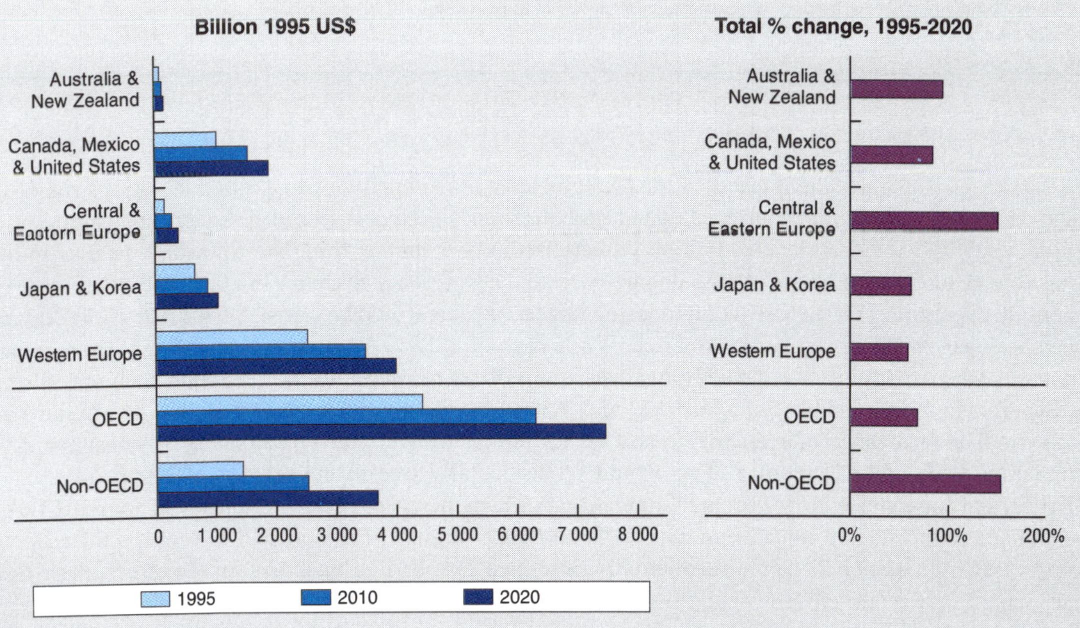

Source: Reference Scenario.

2. Trade in services is in fact considerably underestimated by current balance-of-trade statistics, which fail to reflect transactions such as the cross-border intra-firm exchange of technical and financial advice, remote data processing and transmission, and revenue from services offered in the host country by foreign affiliates of multinationals (Brack, 2001).

the Reference Scenario, global exports are projected to grow by 90% from 1995 to 2020, measured in 1995 US$.[3] Exports are expected to increase by 2.8% annually between 1995 and 2010 and by 2.3% between 2010 and 2020. Corresponding export growth rates for OECD countries are 2.4% in 1995 to 2010 and 1.7% in 2010 to 2020. They will surpass projected GDP growth rates throughout the projection period. The highest export growth rates are expected in non-OECD regions, such as China and East Asia.

Five simulations were carried out to obtain a quantitative impression of how the global export projections might change if economic globalisation was to grow even faster than the Reference Scenario suggests (see Box 3.1). In all cases, the effects on real GDP in both OECD countries and the world were found to be positive, reflecting increases in productivity resulting from improved allocative efficiency. Under Simulation 5 (combining all of the various shocks representing increased globalisation), GDP is expected to be 0.5% higher in 2020 compared with the Reference Scenario in OECD, and 1.1% higher worldwide.

Box 3.1. **Model simulations of effects of accelerated globalisation**

- Simulation 1: Export taxes and export subsidies are cut by 25% annually in all regions, thus promoting "undistorted trade"
- Simulation 2: Simulation 1 plus a 25% annual reduction in import tariffs in all regions, further promoting "undistorted trade"
- Simulation 3: Simulation 2 plus trade and transport margins in all regions decreased by 2% annually, instead of 1% annually as foreseen in the Reference Scenario, making trade between regions cheaper
- Simulation 4: Simulation 3 plus a 5% annual increase in Armington elasticities for all products in all regions, making domestic and imported products more substitutible among consumers
- Simulation 5: Simulation 4 plus the autonomous energy efficiency improvement in non-OECD increased by 50% more per year than in the Reference Scenario, thus significantly increasing energy efficiency outside OECD regions; possibly reflecting increased technology transfers

Figure 3.3 illustrates the impacts of the five globalisation simulations on exports in various regions of the world. In general, the impacts of accelerated globalisation on exports would be much higher in non-OECD regions, as OECD regions are already highly liberalised. For example, Simulation 5 would lead to an increase in the volume of OECD exports of 14% compared with the Reference Scenario in 2020, and an increase of 22% in non-OECD regions. Of the OECD regions, the largest impacts would be felt in Australia & New Zealand, followed by Japan & Korea and North America. While each of the five simulations generates positive impacts on exports, the degree of impact and the regional variations differ considerably. For example, the impacts of reducing export taxes and subsidies (simulation 1) and transportation improvement (simulation 3) would be significantly smaller than those of reducing import tariffs (simulation 2) and globalisation of consumer demands (simulation 4), for OECD regions. This would be particularly true in the regions of North America, Japan & Korea, and Australia & New Zealand. Simulation 5, which increases energy efficiency in non-OECD regions, is estimated to have small impacts on exports from OECD regions. For non-OECD regions, the reduction of import tariffs (simulation 2) would substantially raise their exposure to trade. As for the effects these developments would have on the environment, these would to a large extent be materialised through the impacts on specific economic sectors (see Box 3.2).

3. The assumptions relating to trade in the Reference Scenario are 1) export taxes and subsidies and tariffs will remain unchanged over the simulation period; 2) trade and transport margins in all regions decrease by 1% annually. See Annex 2 for further details of the model assumptions.

Figure 3.3. **Change in exports in 2020 under globalisation policy scenarios, compared to the Reference Scenario**

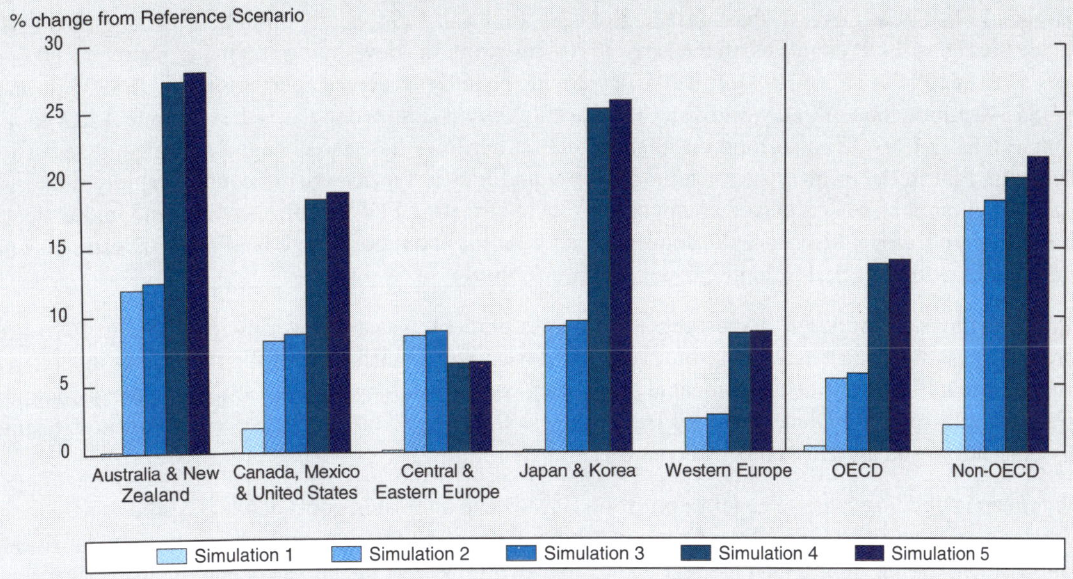

Sources: Reference Scenario and Policy Simulations.

Box 3.2. Sectoral effects of increased globalisation – forestry example

Depending on circumstances, the aggregate impacts of increased globalisation on the environment (both local and global) may be positive or negative. Although not enough is known at present about their precise nature, the specific effects of globalisation and trade liberalisation will be most noticeable at the sectoral level. One example is forests.

The world's forests and wooded land decreased by 2% in the 1980s. While the forest area in industrialised countries remained largely unchanged, the natural forest cover in developing regions was reduced by 8% (see Chapter 10). The value of global trade in forest products has risen steadily over the last few decades, climbing from US$47 billion in 1970 to US$139 billion in 1998. Exports as a share of total world production also increased significantly over this period (French, 2000). These trends are expected to continue. The Reference Scenario projects that globally forest sector production will increase by 87% from 1995 to 2020, with highest growth in non-OECD regions. The policy simulations show that with accelerated globalisation, forest production in non-OECD regions is likely to increase compared with the Reference Scenario, while in OECD regions it is expected to remain unchanged. As such, greater attention to sustainable forest management will be needed in the developing world.

International investments

If growth in world trade is the most obvious indicator of expanding economic integration, then rapid growth in cross-border investment suggests the deepening of such integration. Reductions in transport and communications costs, the opening of capital accounts, deregulation in financial markets and privatisation of state enterprises have all created an attractive environment for international financial flows.

The volume of global foreign direct investment (FDI) has grown almost 47-fold since 1970, reaching US$644 billion in 1998, and the number of transnational corporations worldwide grew from 7 000 in 1970 to some 60 000 in the late 1990s (OECD, 1997; OECD, 1998; World Bank, 2000). FDI outflows from OECD economies have risen at 15% annually over the past decade. Between 1988 and 1994, nearly three-quarters of global FDI flows took place between OECD countries. In the early 1990s, however, the developing countries' share of FDI outflows rose, from 7% in 1991 to 17% in 1994. In 1995, developing countries received approximately US$90 billion (38%) of the US$240 billion total of FDI worldwide. The vast majority is destined towards the dynamic Asian economies and China (OECD,1997). The sectoral destination of FDI outflows has also changed significantly over the last decades, with FDI in the primary sector falling steeply and FDI in services rising equally rapidly. Privatisation, especially of infrastructure, has played an important role in attracting FDI to Latin America and, more recently, to Central & Eastern Europe. Mergers and acquisitions account for about 60% of global FDI, with estimates of 80% in the US, and low figures in developing countries (Brack, 2001).

Portfolio flows[4] comprise the largest segment of cross-border transactions worldwide – 55% in 1996 – though they are much less significant as a source of finance for developing countries. Portfolio flows have tended to be far more volatile than FDI. Portfolio investment in developing countries fell by 50% in 1998 due to the economic crisis in Asian countries, while FDI declined by less than one-sixth. As with FDI, however, portfolio investment is heavily concentrated in a small number of countries.

Commercial lending can also generate portfolio flows when interest in pools of loans is sold as debt securities in secondary markets. In 1996, bank loans accounted for 23% of all international private investment flows. The main motivation driving commercial lending is the credit-worthiness of the borrower and the likely rate of return. As with portfolio investments, this can contribute to a high degree of volatility. Between 1996 and 1998, for example, annual commercial lending to developing countries crashed from over US$100 billion to almost zero.

While the Reference Scenario does not provide explicit projections for international investment flows, it can be assumed that further liberalisation of trade and foreign investment will increase the attractiveness of – and reduce barriers to – such financial flows. FDI will most likely show continued strong growth, and spread more to non-OECD countries. With increasing stability and effective regulation of financial markets, new investment opportunities may be created that would attract portfolio investors as well.

3.3. Environmental implications of globalisation

Globalisation changes the scale of economy, the economic structure, available technologies, the composition of production and consumption activities, and the design and implementation of regulations. Its implications for the environment can be either positive or negative, depending on the rate and direction of economic growth and the presence and effectiveness of institutions and policy frameworks. The effects of globalisation on the environment will depend on the aggregate outcome of a number of impacts from trade and investment liberalisation. The effects are mainly from an expansion of world economic output (scale effects), structural adjustment within and between economies (structural effects), technological development and diffusion stimulated by expanding trade and investment (technology effects), changes in the mix of products made and consumed within an economy (product effects), and changes in the regulatory framework (regulatory effects).

Scale effects

Scale effects occur when the liberalisation and expansion of trade and investment drive and accelerate economic growth and resource use. This in itself can have both positive and negative environmental impacts. Positive scale effects occur partly from the reduction in poverty-driven environmental degradation and partly from the increased attention countries pay to environmental quality and regulation as income rises. Furthermore, following income

4. The term "portfolio" refers to the combined holdings of various financial assets by individuals and institutions. The most common types are equity securities of less than 50% of a firm's total equity and debt securities.

growth, environmental quality may become a political priority for citizens and governments with more resources to devote to the issue. Some studies have shown that the scope of environmental regulation increases steadily with growth in per capita income (Dasgupta *et al.*, 1995) and that many indicators of environmental quality also improve after an initial decline at low income levels – reflected in "environmental Kuznet curves" (see also Chapter 4) (Grossman and Krueger, 1995). Negative impacts are most likely to occur when the expansion of production and consumption coincide with market failures (such as ill-defined property rights and a failure to incorporate environmental externalities) or intervention failures, and through increased pollution resulting from transporting more goods.

Scale effects of economic growth on the environment, spurred by further globalisation, are likely to be on balance negative.

Scale effects of economic growth caused by further globalisation are likely to be on balance negative, largely because this contributes to increased production with the ensuing increases in resource and energy use and waste production worldwide. However, the effects will still depend on the particular environmental factor under investigation. Some local environmental effects – particularly water quality – may well improve in many countries, as growth delivers the income necessary for abatement actions. Effects on local air quality, on the other hand, will be determined by the balance of increasing expenditure on pollution prevention and clean-up, and increasing emissions from expanding industry and the increased distances travelled by motor vehicles. The most worrying scale effect will be the expected increases in emissions of greenhouse gases in OECD countries, and increases in local air pollution in non-OECD countries.

Structural effects

Structural effects on the environment are likely in general to be positive, as comprehensive liberalisation of trade and investment improves resource efficiency.

Structural effects relate to the adjustments within and between economies that occurs when the pattern of resource use shifts. To the extent that trade promotes allocative efficiency among nations, globalisation will have positive structural effects in ensuring that goods are produced with lower input and capital per unit of output worldwide. In general, newly industrialising countries have experienced significant positive effects on environmental quality from structural changes as they have moved from primary commodity production to resource processing and then to light manufacturing and services. This last step in particular has resulted in environmentally beneficial effects.

Structural effects are in general positive, as the comprehensive liberalisation of trade and investment reduces or eliminates subsidies in areas such as agriculture, energy production, forestry and fisheries, thus improving resource efficiency. Evidence suggests that the more open an economy is to international trade, the less pollution-intensive it will become (Lucas *et al.*, 1992). However, the correct pricing of environmental assets and externalities continues to be important, since, in its absence, trade could have negative structural effects, shifting activity to more environmentally damaging sectors.

Technology effects

The technology effects of trade and investment liberalisation derive from the more rapid rate of diffusion of new technologies which, in general, tend to use fewer resources and produce less pollution than their predecessors – thereby improving eco- and resource efficiency in countries where trade and investment are being liberalised. Trade liberalisation will also accelerate and promote the growing international market for environmental goods and services – pollution clean-up equipment, cleaner production techniques, capacity-building assistance, and so on. The liberalisation of investment, particularly of FDI, is important in the area of technology diffusion because foreign investors often bring with them modern technologies that represent environmental improvements over equipment available domestically (see Chapter 6).

Technology effects on environmental performance are generally expected to be positive or neutral, but can occasionally also be negative. There have been instances of "technology dumping", where equipment banned in environmentally strict countries (because of its poor environmental performance) is sold to or "dumped" in countries with less demanding environmental standards. Environmentally beneficial technologies can be promoted by removing subsidies for polluting technologies, thereby stimulating the development of more energy efficient alternatives. Even some of the negative scale effects may accelerate the use of environmentally beneficial technologies since regulations to protect the environment will call for low- or zero- emission vehicles, energy efficient consumer durables, water-saving appliances and more resource-efficient agricultural techniques.

Technology effects of globalisation on the environment are generally expected to be positive or neutral, but may be negative in some circumstances.

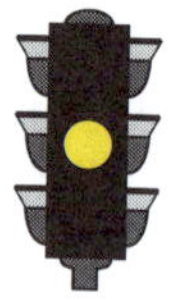

Product effects

The product effects of trade and investment liberalisation refer to changes in the mix of products made and consumed within an economy. Some may be environmentally less damaging than the goods they replace (*e.g.* organic foods, recyclable materials, low-sulphur coal, etc.); some may be more environmentally damaging (*e.g.* high-performance vehicles). Trade liberalisation also makes outsourcing – for example for the manufacture of components – more feasible, which may increase the transport-intensity of products.

In aggregate, product effects can be either positive or negative. To some extent, trade and investment liberalisation encourages environmentally beneficial changes in consumer behaviour, such as using low-emission vehicles, increasing investment in energy efficiency and using telecommunication to reduce commuting. But, at the same time, rising incomes, especially in less developed countries, may lead to growing demand for environmentally damaging products such as air conditioning.

Regulatory effects

Regulatory effects refer to the legal and policy effects of trade and FDI liberalisation on the design and implementation of environmental regulations, standards and other measures, such as extended producer responsibility and voluntary approaches. These effects can be positive when a country's ability to issue and implement effective environmental policies and regulation is enhanced as a result of the provisions of specific trade and investment agreements. For example, WTO agreements, to some extent, incite governments to move away from policies with negative environmental impacts, such as subsides for environmentally damaging activities. Negative regulatory effects occur when the ability of governments to enact and implement appropriate environmental regulations is undermined by the provisions of trade and investment agreements. In general, however, it seems likely that most of the provisions of trade and investment agreements have no direct impact on the framework of environmental policy in any particular country.

3.4. Policy issues

Considering the projected developments of trade and investment liberalisation, their overall impacts on the environment are likely to be significant. Ensuring that environmental policy instruments are put in place to maximise the positive and minimise the negative effects will be a great challenge to the globalised world. Governments will find that longer-term environmental goals will often conflict with shorter-term economic competitiveness goals. Where the environmental effects are negative, this will often have to do with insufficient environmental policies, rather than poorly designed economic policies. Globalisation is thus likely to intensify the need for stronger and better environmental policies, rather than to lessen it. With the globalisation of environmental issues, and the economic competitiveness problems associated with these issues, a stronger environmental regime will also inevitably require more international co-operation.

Global trade agreements can have positive environmental effects by discouraging governmental policies with negative environmental impacts such as, for example, subsidies to intensive agriculture, fisheries and fossil fuel production and use. Removing restrictions on imports of environmentally sound technologies can also contribute to better environmental performance. Nonetheless, much of the discussion has focused on the potential negative environmental effects of the GATT and WTO agreements. One of the issues debated is the ability of governments to take into account environmental concerns related to process and production methods (PPMs), as opposed to product standards, for the goods they import while remaining in compliance with the disciplines of trade agreements.

There are two ways in which some PPMs can have a negative impact on the environment. They can affect the characteristics of the product, so that the product itself may pollute or degrade the environment when it is consumed or used (product-related PPMs), or they may have an environmental effect at the production stage (non-product related PPMs). This distinction is important in the context of trade rules. WTO rules apply to "products" since like products are to be accorded "like treatment". When PPMs affect the characteristics of products, existing trade rules permit the use of PPM criteria for imports, subject to agreed disciplines. However, WTO rules make no provision for import restrictions based on the characteristics that are not physically embodied in the imported products. Although recent interpretation of the relevant GATT rules seems to allow measures based on non-product related PPMs, the application of such measures is subject to strict conditions (including the demonstration of a "nexus" between the country imposing the restriction and the environmental resource to be protected). Other environmental measures with potential trade implications include labelling, taxation, and trade bans and quotas.

Trade measures incorporated in Multilateral Environmental Agreements (MEAs) are another potential source of trade conflict. At present, around 200 MEAs are in place, about twenty of which incorporate trade measures. A number of them deal explicitly with the regulation of trade, *e.g.* the Convention on International Trade in Endangered Species (CITES) and the amended Montreal Protocol on Substances that Deplete the Ozone Layer (OECD, 1999*a*). Although no complaint has ever arisen under the GATT or WTO agreements with respect to trade measures taken in pursuit of MEAs, the relationship remains uncertain, particularly in the event of a conflict between countries which are party to an MEA and those which are not. The WTO's Committee on Trade and Environment was established *inter alia* to help resolve these potential conflicts. The Committee has not yet reached any agreement on possible modifications to the WTO system, but it has carried out extensive analytical work.

Regional trade agreements are paying increasing attention to environmental issues. The European Union has an important environmental policy component to its work, and environmental protection is one of the objectives set out in the Treaty of Rome, subsequently strengthened through various amendments (see Chapter 24). The North American Free Trade Agreement (NAFTA) is another example of linkages to environmental policy. It is accompanied by an environmental side agreement, the North American Agreement on Environmental Co-operation, designed to safeguard the environment from any negative effects of NAFTA-stimulated trade expansion. Other examples of promoting co-operation in environmental policy formation and implementation include the Common Market of the South America (Mercosur), the Association of South East Asian Nations (ASEAN), and the Asia-Pacific Economic Cooperation (APEC). With growing concern over the regional environment, these trends are likely to spread and intensify in the near future.

The Uruguay Round included the adoption of the WTO Agreement on Trade-Related Investment Measures (TRIMS Agreement), which is fairly limited in scope and application. OECD Member countries, as well as some non-members, started to negotiate a much wider agreement, the Multilateral Agreement on Investment (MAI),[5] but this initiative was eventually abandoned in late 1998. The interaction between FDI and the environment has been the subject of considerable debate. Does FDI seek out destinations where environmental regulations are lacking or poorly enforced ("pollution havens")? Alternatively, does FDI contribute to rising environmental standards or performances in developing countries through superior technology and management methods ("pollution halos")? So far, there is not much evidence of pollution havens or pollution halos (see Leonard, 1988; French, 1993; USTR, 1992; Zarsky, 1999; Gentry, 1999). Similarly, environmental factors seem to have little impact on portfolio investors' decisions (OECD, 1999*b*). Nevertheless, portfolio investment may have an effect on the location of invest-

5. The MAI aimed to apply "national treatment and most favoured nation" principles to investments and investors, and provide a high standard of protection to foreign investors and investments when operating overseas.

ments and thereby on environmental quality. Also, the volatility of portfolio flows needs to be mitigated by government action: stable financial markets, for example, created by effective financial regulatory structures can do much to bring long-term environmental benefits.

FDI and portfolio investments can be both vehicles for raising environmental standards and a transfer mechanism for environmentally sound technologies. The greater the number of environmental factors that are integrated into the framework of private investment, the more effective this process will be. Thus, host country governments need to set clear rules about minimum environmental standards for investment. In addition, future negotiations on international investment agreements need to ensure that FDI and portfolio investments do not undermine countries' efforts to maintain high environmental standards.

REFERENCES

Brack, D. (2001), "The Environmental Implications of Trade and Investment Liberalisation", Background document for the OECD *Environmental Outlook*, OECD, Paris.

Dasgupta, S. *et al.*, (1995) *Environmental Regulation and Development: A Cross-Country Empirical Analysis*, World Bank, Washington DC.

French, H. (1993), *Costly Tradeoffs; Reconciling Trade and the Environment*, Worldwatch Institute, Washington DC.

French, H. (2000), "Coping with Ecological Globalisation", *State of the World* 2000, Worldwatch Institute, Washington DC.

Gentry, B. S. (1999), "Foreign Direct Investment and the Environment: Boon or Bane?" in OECD, *Foreign Direct Investment and the Environment*, OECD, Paris.

Grossman, G. M. and A. Krueger (1995), "Economic Growth and the Environment", *Quarterly Journal of Economics* 112: 353-377.

Leonard, J. (1988), *Pollution and the Struggle for the World Product*, Cambridge University Press, Cambridge.

Lucas, R.E.B., D. Wheeler and H. Hettige (1992), "Economic Development, Environmental Regulation and the International Migration of Toxic Industrial Pollution 1960-88", Low, P. (ed.), *International Trade and the Environment*, World Bank, Washington DC.

OECD (1997), *The World in 2020: Towards a New Global Age*, OECD, Paris.

OECD (1998), *Open Markets Matter: The Benefits of Trade and Investment Liberalisation*, OECD, Paris

OECD (1999*a*), *Trade Measures in Multilateral Environmental Agreements*, OECD, Paris.

OECD (1999*b*), *The Environmental Effects of International Portfolio Flows*, OECD, Paris.

USTR (1992), *Review of US-Mexico Environmental Issues*, USTR, Washington DC.

World Bank (2000), *World Development Indicators*, World Bank, Washington DC.

WTO (1999), Press Release, "World Trade Growth", 22 April, WTO, Geneva.

Zarsky, L. (1999), "Havens, Halos and Spaghetti: Untangling the Evidence about FDI and the Environment", in OECD *Foreign Direct Investment and the Environment*, OECD, Paris.

4

Economic Development

KEY SIGNALS

- The *Outlook* uses a Reference Scenario, which is largely based on a continuation of recent trends and present policies. It represents a likely economic development path, but it should not be seen as a forecast of future developments.
- The largest absolute increases in GDP are assumed to take place in the regions of North America, Western Europe and Japan & Korea. Relative growth is expected to be highest in Central & Eastern Europe and in regions outside the OECD.
- It is assumed that GDP growth in OECD regions and worldwide will decrease gradually. In non-OECD regions, however, growth is expected to accelerate.
- The Reference Scenario assumes a continuation of present globalisation processes, reflected in an annual decrease in the costs of buying products from a foreign supplier.
- Energy efficiency is assumed to increase 0.75% per year autonomously in all sectors and regions, while the efficiency of resource use – as a technical assumption – remains largely unchanged.

4.1. Introduction

Economic development is a major underlying factor driving environmental change. In many cases, environmental pressures stemming from a given sector are directly related to the level of activity in that sector. On the other hand, the higher income levels which result from economic growth generally increase the willingness – and ability – to implement policies and measures to protect the environment.

In the Reference Scenario, main economic developments to 2020 in the different regions of the world are included either as assumptions or as results from simulations in the general equilibrium model – JOBS – that has been used in the preparation of this *Environmental Outlook* (see Annex 2 for a description of the model and a number of technical assumptions used in the simulations).

4.2. Assumed developments in GDP

The *Outlook* is based on assumptions of the OECD's Economics Department concerning long-term developments in total GDP by region, which in turn are in part based on World Bank projections. The assumptions used here are, as far as possible, the same as those used in the climate change policy analyses done by the OECD on the GREEN model (Burniaux, 2000). The assumed growth rates are meant to reflect likely developments in a case where no changes have been made to present policies; they should be seen as neither a forecast of future GDP growth nor as representing an "optimal" development path.

It is not always obvious what represents "no change" in policy. For example, an underlying assumption built into the Reference Scenario is that the globalisation process, in particular the trend of gradually increasing market integration will continue. This is reflected in an assumed annual 1% decrease in the extra costs associated with purchasing a given product from abroad, compared to buying it from a local producer.

Another underlying assumption incorporated in the Reference Scenario is that energy efficiency will improve by 0.75% autonomously every year. Autonomous energy efficiency improvements mean that a given amount of output can be produced with less and less energy inputs. This effect comes on top of any changes in energy requirements that result from changes in the composition of demand and production. In general terms, the assumed rate of energy efficiency improvement is based on recent historical trends. Also, lacking any firm foundation for differentiating among sectors or regions, uniform rates of efficiency improvement have been assumed in the Reference Scenario, while simulations described in later chapters illustrate impacts of some alternative assumptions.

A further underlying assumption included in the Reference Scenario is that non-energy inputs in production are used in fixed proportions throughout the simulation period. This should be seen as a technical simplification used in the model simulations, while in practice resource efficiency will actually be affected by a number of trends, as discussed in Chapter 23.

Figure 4.1 shows real GDP levels in 1980 and 1995, and the assumed levels for 2010 and 2020 in each region in 1995 prices. It is evident that the regions of Canada, Mexico & United States, Western Europe and Japan & Korea are the dominant ones in terms of GDP.

Figure 4.1. **Real GDP, 1980-2020**

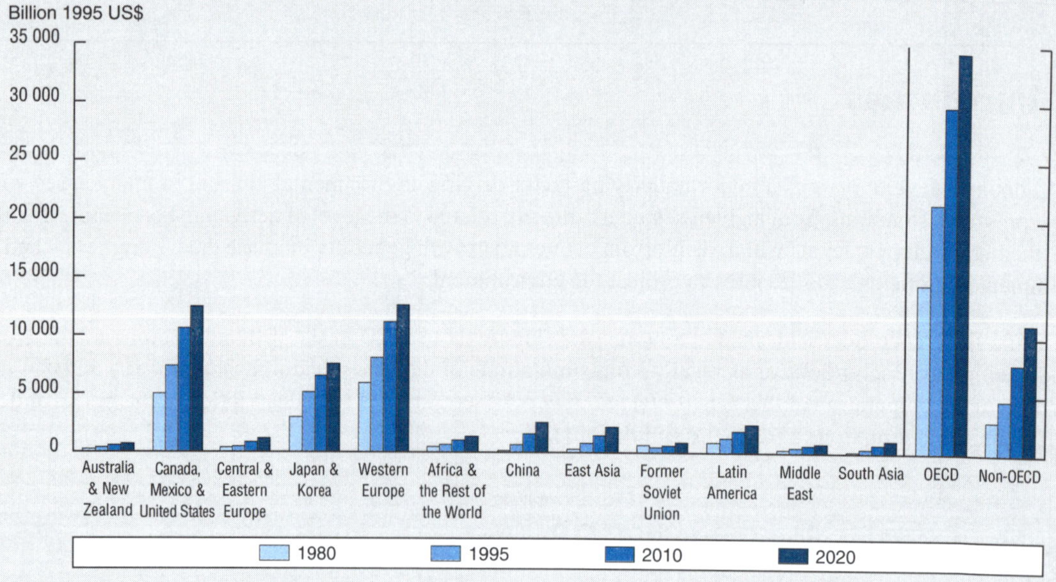

Note: The 1980 numbers were adjusted to reflect differences in 1995 levels in the IMF and GTAP databases.
Sources: IMF (2000*a*) and IEA (2000) for 1980 numbers; GTAP database for 1995 numbers; and Reference Scenario for 2010 and 2020 numbers.

It should be noted that GDP here is valued at market exchange rates. Thus, no correction is made for purchasing power differences reflecting, for example, the lower prices of many goods and services in non-OECD regions due to lower wage levels. The following comparison is instructive, while non-OECD regions accounted for only 16% and 18% of global GDP in 1980 and 1995 respectively with GDP measured at market exchange rate, their share of global GDP would be 35% and 39% in the same years if GDP were measured in terms of purchasing power parities (IMF, 2000*b*).

Figure 4.2. **Average annual change in real GDP, 1980-2020**

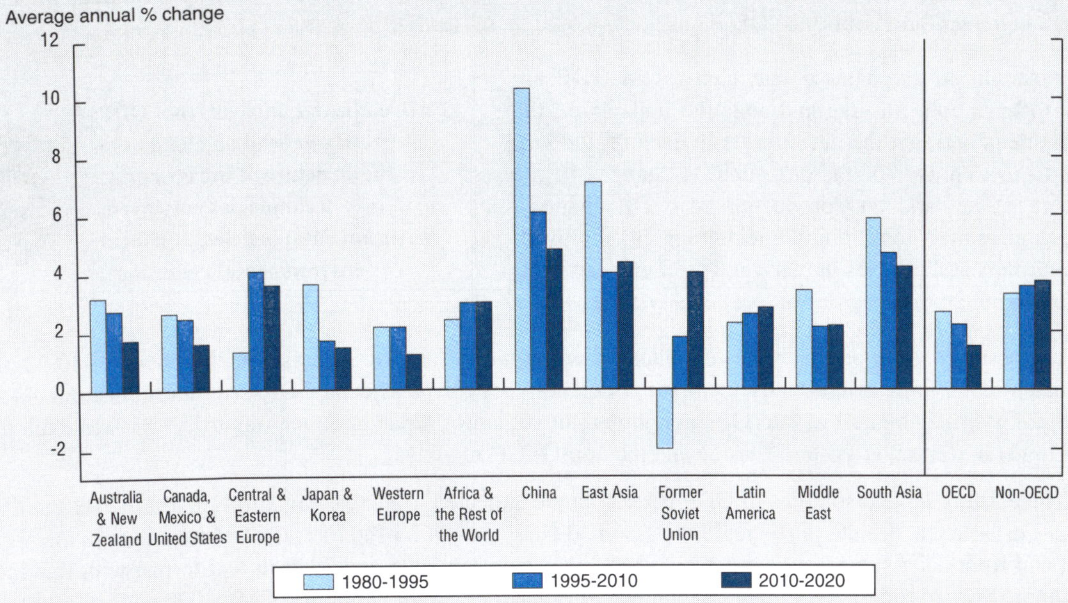

Note: The 1980 numbers were adjusted to reflect differences in 1995 levels in the IMF and GTAP databases.
Sources: IMF (2000*a*) and IEA (2000) for 1980 numbers; GTAP database for 1995 numbers; and Reference Scenario for 2010 and 2020 numbers.

The increasing share of non-OECD regions in global GDP is to a large extent explained by the rapid growth assumed in China. While China accounted for 1% of global production in 1980, its share will have risen to 6% in 2020 under the Reference Scenario.[1]

Figure 4.2 shows the annual percentage changes in GDP between 1980 and 1995, and the assumed annual growth rates for the periods 1995-2010 and 2010-2020 for each region. It is clear that some of the non-OECD regions (China, East Asia and South Asia in particular) are assumed to experience the strongest growth rates in the decades to come. Nevertheless, GDP is still expected to be larger in both the regions of Western Europe and Canada, Mexico & United States individually than the combined GDP of all non-OECD regions in 2020.

GDP growth is assumed to decrease significantly in a number of regions. For the world as a whole, the decrease is from an annual growth of 2.8% in 1980 to 1995, to 2.5% annual growth between 1995 and 2010, and 2% in the decade to 2020. This decline in global growth rates will to a large extent be driven by decreasing growth in OECD regions, from 2.7% annually between 1980 and 1995, to 2.2% annually between 1995 and 2010, and 1.5% annually from 2010 to 2020.

For the non-OECD regions seen together, growth is assumed to accelerate from 3.3% annually between 1980 and 1995, to 3.5% and 3.7% in the periods 1995 to 2010 and 2010 to 2020 respectively. Average annual growth over the 1980-1995 period was kept down by a 2.2% annual decrease in GDP in the Former Soviet Union as a result of the collapse of the communist system, while annual GDP increases in China of more than 10% pushed the average growth rate upwards. For the coming decades, more stable growth paths have been assumed for both of these regions.

1. Measured in purchasing power parities, China accounted for 3.5% of global GDP in 1980 and about 10% in 1995.

The decrease in the labour force assumed for several regions, as described in Chapter 2, played a major role when assumptions concerning GDP growth were made, in particular the relatively low growth rates assumed for OECD regions between 2010 and 2020. A higher labour supply than what is currently assumed, for example through increased immigration to OECD regions, would increase GDP growth.

It should be emphasised that changes in GDP are likely to have both positive and negative impacts on the environment, and that the net impacts in each region will depend on a number of factors, such as changes in the structure of demand and production as GDP changes; changes in capital stocks and the technologies employed; the environmental policies in place and their enforcement; and the assimilative capacities of local eco-systems.

Environmental impacts from GDP growth depend on changes in the structure of the economy, technologies employed, environmental policies, and local environmental conditions.

The assumed GDP per capita developments used in the Reference Scenario for the period 1980-2020 were obtained by combining assumptions regarding population (see Chapter 2) and GDP (see Figures 4.3 and 4.4). GDP per capita is clearly highest in four OECD regions (the region of Japan & Korea in particular), while the level is much lower in Central & Eastern Europe and the non-OECD regions.

The existing large disparities in GDP per capita are expected to increase in absolute terms over the *Outlook* period. For instance, despite fairly moderate assumed GDP growth for the future, GDP per capita in the region of Japan and Korea is assumed to increase by US$13 000 between 1995 and 2020 in real terms, with this increase being more than 20 times larger than the total level of GDP per capita in South Asia in 2020.

Figure 4.3. **Real GDP per capita, 1980-2020**

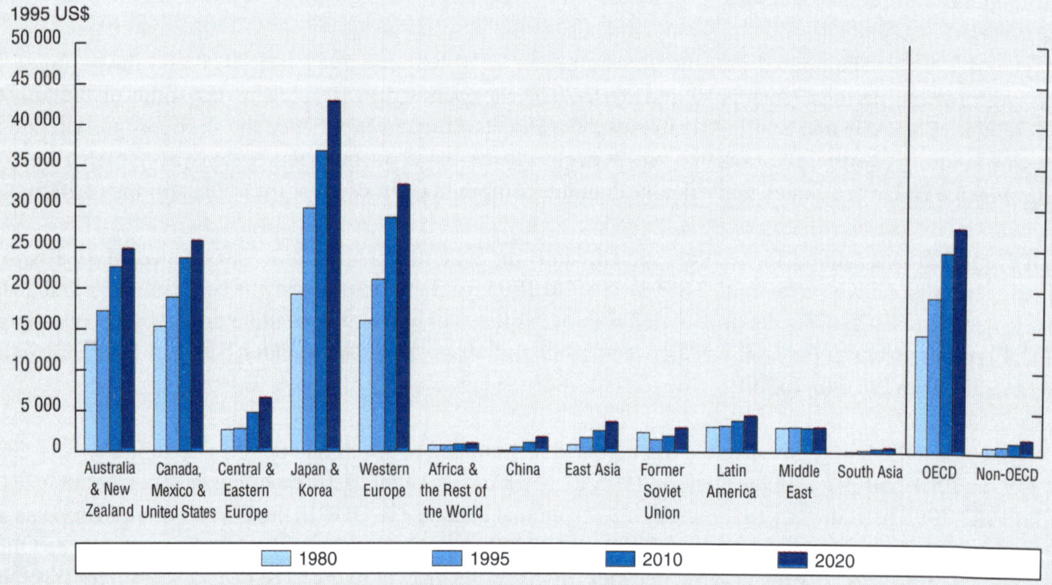

Note: The 1980 numbers were adjusted to reflect differences in 1995 levels in the IMF and GTAP databases.
Sources: IMF (2000*a*) and IEA (2000) for 1980 numbers; GTAP database for 1995 numbers; and Reference Scenario for 2010 and 2020 numbers.

Figure 4.4. **Average annual changes in real GDP per capita, 1980-2020**

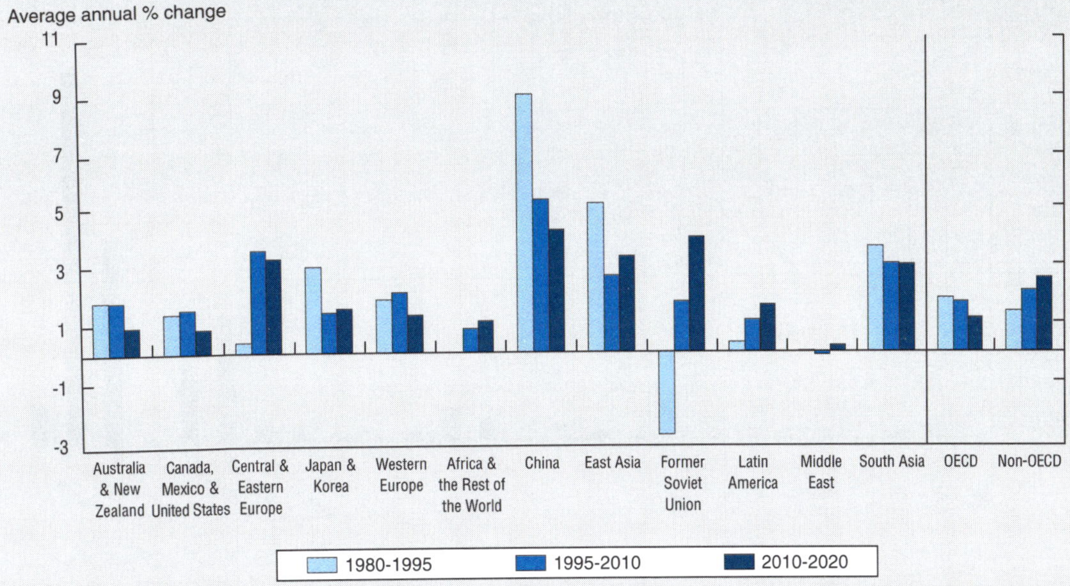

Note: The 1980 numbers were adjusted to reflect differences in 1995 levels in the IMF and GTAP databases.
Sources: IMF (2000*a*) and IEA (2000) for 1980 numbers; GTAP database for 1995 numbers; and Reference Scenario for 2010 and 2020 numbers.

However, as can be seen in Figure 4.4, the percentage of growth in GDP per capita is assumed to be higher outside OECD regions than within. While GDP growth per capita is assumed to decrease on average in OECD regions, it is assumed to accelerate on average outside OECD. Between 2010 and 2020, GDP per capita is in fact assumed to grow more than twice as fast outside OECD regions as within.

4.3. Changes in the composition of production

Whereas the developments in GDP and GDP per capita are entered as assumptions in the Reference Scenario, changes in the composition of demand and production are simulated by the modelling exercise. These changes are to a large extent driven by the assumed income elasticities in each region,[2] which reflect how much household demand for a certain category of goods and services is expected to increase when income increases by 1%.

Figures 4.5 and 4.6 illustrate the composition of real value added in the OECD and non-OECD regions in 1995, 2010 and 2020 respectively, as calculated in the Reference Scenario. There are some clear differences in the composition of value added between the two regions: the service sector exhibits a much larger share within OECD regions than outside (around 37% vs. less than 25% respectively), while the agriculture sector (rice, other crops and livestock) and the natural resource-based sectors (fisheries, forestry, minerals, coal, natural gas and, in particular, crude oil) constitute significantly larger shares of value added outside OECD regions than within. Both in OECD and in non-OECD regions the shares of the agriculture sector and the natural resource-based sectors in total value added are in most cases estimated to decrease to 2020.

2. The income elasticities used in the simulations are taken from the (GTAP) database.

Figure 4.5. **Composition of value added in OECD regions, 1995-2020**

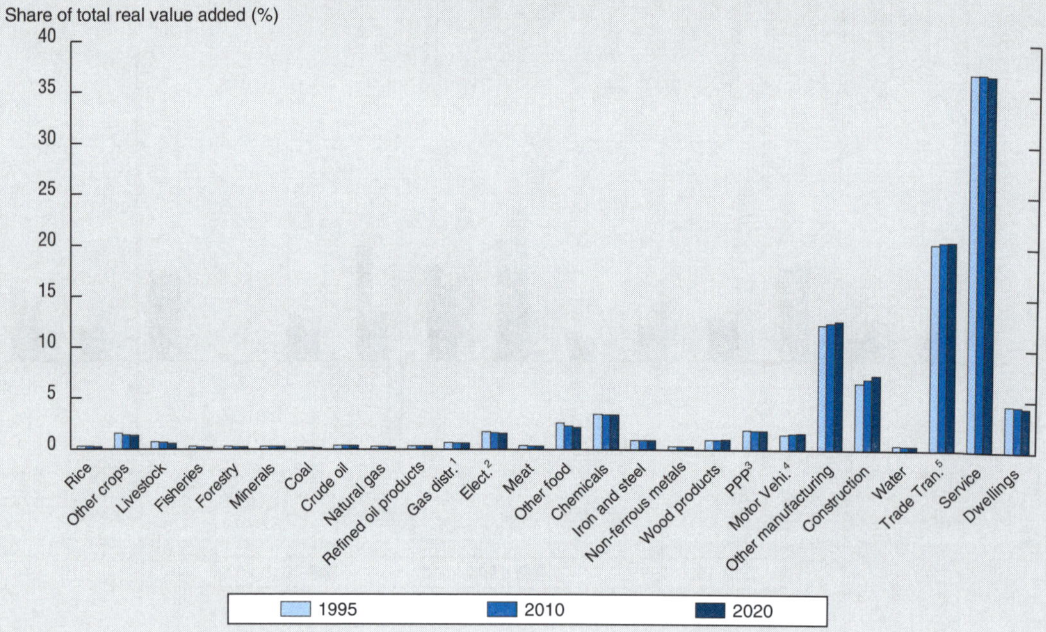

Notes:
1. Gas manufacture and distribution.
2. Electricity generation and distribution.
3. Pulp, paper and publishing.
4. Motor vehicle manufacturing.
5. Trade and transport services.

Source: Reference Scenario.

One might have expected a shift in the composition of value added over time towards the service sector, and that the production structure in non-OECD regions in 2020 would converge towards that of OECD regions in 1995. However, the figures reflect the net impacts of various trends and assumptions, such as:

– a shift towards a larger share of the service sector in production in many, but not all, regions;

– a shift in total GDP – both in OECD and non-OECD areas – towards regions where the service sector at the outset plays a smaller role in production;

– whereas the composition of household demand is estimated to shift towards services (see Chapter 5), this sector needs inputs from "material" sectors to produce the services that households demand;

– the technical simplification referred to in Section 4.2 causing non-energy inputs in production to be used in fixed proportions;

– demand parameters are re-calibrated from year to year to make income elasticities more or less constant over time.[3] Hence, over the simulation period, a given income increase will lead to a fixed percentage increase in demand for each demand category (but this percentage increase will differ between different demand categories);

– the graph illustrates developments in real terms, *i.e.* using constant 1995 prices. In many cases, the service sector is estimated to show increasing relative prices, thus its share in nominal value added is slightly higher in 2020 than its share in real value added, both in OECD and non-OECD regions.

3. With the so-called "Extended Linear Expenditure (demand) System" used in JOBS, the income elasticities would otherwise in the long term all converge towards 1, meaning that a 1% income increase would cause the demand for all demand categories to increase 1%. Such a result would be wrong for most goods, in particular for food, clothing, etc.

Figure 4.6. **Composition of value added in non-OECD regions, 1995-2020**

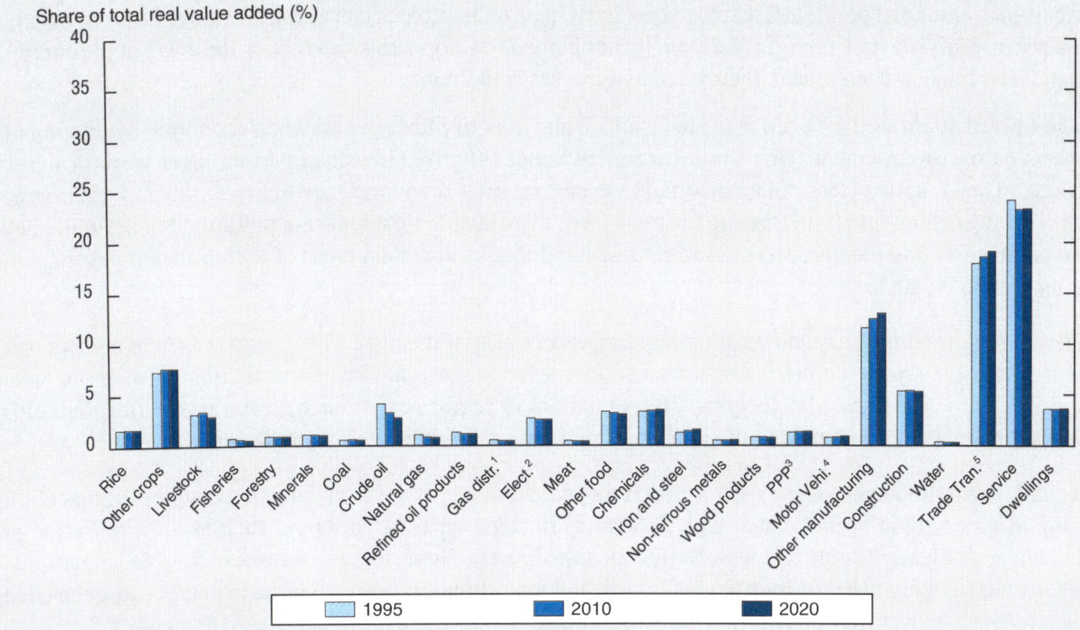

Notes:

1. Gas manufacture and distribution.
2. Electricity generation and distribution.
3. Pulp, paper and publishing.
4. Motor vehicle manufacturing.
5. Trade and transport services.

Source: Reference Scenario.

4.4. Economic development and environmental pressures

The proceeding sections have illustrated some of the major economic developments in the Reference Scenario of this Outlook. These developments impact on the environment in a number of ways. The increased scale of economic activity will – providing other factors remain unchanged – tend to increase negative environmental externalities through larger volumes of emissions of pollutants to air and water, increased (over-) use of renewable or non-renewable natural resources, etc. However, most other things will normally not remain unchanged when the scale of economic activity changes, and in order to estimate net impacts on the environment a closer look at the changes in demand and supply is necessary.

One important effect in this context is that environmental quality is believed to be a "normal good" in economic terms. This means that the demand for environmental quality rises as household incomes rise. Thus, the willingness to pay for improvements in environmental quality increases.

It is also possible that the composition of demand for other (non-environmental) goods and services will change when the scale of economic activity changes. This can have both positive and negative impacts on the environment – consumers might demand less or more of the goods and services that are particularly environmentally harmful when they are produced, used or disposed of as waste. In general, it is expected that the demand for services will rise more than the demand for goods as incomes increase (see Figure 5.2). Many – but not all – services cause fewer negative environmental impacts than goods from other sectors.[4]

4. See Rosenbaum, Horvath and Hendrickson (2000) for a study on environmental implications of service industries.

Furthermore, it is important to determine the "cause" of the increased scale of economic activity. It may be an indication that more productive (and newer) plants are displacing other less productive, and often more polluting, plants. In many cases new technologies will be "cleaner" than older technologies (*e.g.* if they are developed as a response to environmental problems). On the other hand, new technologies can also give rise to new environmental problems, or magnify existing ones. In addition, technological developments can reduce the costs of producing polluting products, causing decreases in their price and increases in their use.

A number of attempts have been made to establish and quantify linkages between economic development and net impacts on the environment. Two studies from 1993 and 1995 by Grossman and Krueger were of particular importance.[5] They found evidence of an inverted U-curve[6] relating economic growth to a number of environmental problems: at low income-levels, increasing incomes were associated with increasing pollution levels, while beyond a certain point increasing incomes were associated with reductions in certain types of pollution. However a number of caveats need to be made.

First, the studies do not explain what causes the perceived relationships. There is no reason to believe that the process is automatic. On the contrary, Grossman and Krueger suggest that the strongest link between income and pollution is in fact via an induced policy response: as nations or regions experience greater prosperity, their citizens demand that more attention be paid to the non-economic aspects of their living conditions.[7]

Second, environmental problems differ in drivers and character, as described in other chapters of this *Outlook*. Increasing incomes might be associated with reductions in some types of pollution, such as local pollution problems that are more clearly visible and which affect people directly. However, any threshold for CO_2-emissions, for example, would be likely to be too high to stop human-induced climate changes. This is in part because the changes in public opinion which cause policy responses to mounting problems are more likely to happen when they concern relatively local environmental problems having immediate or near term consequences than for global and longer-term problems.

5. Bradford, Schlieckert and Shore (2000) have proposed a re-specification of the latter study. Magnani (2000) suggests that the income distribution within richer countries is important for determining the emergence of a downward sloping segment of the inverted U-curve. For a recent comprehensive discussion of the relationships between economic growth and the environment, see de Bruyn (2000).
6. Such a curve has also been labelled an "environmental Kuznets curve", an analogy referring to the "inverted U" relationship between the level of economic development and the degree of income inequality suggested by Simon Kuznets (1955). Most recent empirical studies, however, reject Kuznets' proposition.
7. A study by Komen, Gerking and Folmer (1996) indicates, for instance, that the income elasticity of public research and development expenditures for environmental protection is positive, but less than unity.

REFERENCES

Bradford, D. F., R. Schlieckert and S.H. Shore (2000), "The Environmental Kuznets Curve: Exploring a Fresh Specification", *Working Paper No.* 8001, National Bureau of Economic Research – (NBER) Website, *www.nber.org/papers/w8001*.

de Bruyn, S. M. (2000), *Economic Growth and the Environment*, Kluwer Academic Publishers, Dordrecht, the Netherlands.

Burniaux, J.-M. (2000), "A Multi-gas Assessment of the Kyoto Protocol", Working Papers No. 270 by the Economics Department, OECD, Paris.

Grossman, G. M. and A. Krueger (1993), Environmental Impacts of a North American Free Trade Agreement" in Peter Garber, (ed.), *The Mexico-US Free Trade Agreement*, Cambridge, Massachusetts. MIT Press: 13-56.

Grossman, G. M. and A. Krueger (1995), "Economic Growth and the Environment", *Quarterly Journal of Economics*, 112: 353-377.

IEA (International Energy Agency) (2000), Database for Energy Statistics, IEA, Paris.

IMF (International Monetary Fund) (2000*a*), *World Economic Outlook*, IMF, Washington DC.

IMF (2000*b*), "The World Economic Outlook Database, September 2000", IMF, Washington DC.

Rien, K., S. Gerking and H. Folmer (1996), "Income and Environmental Protection: Empirical Evidence From OECD Countries", Mimeo.

Kuznets, S. (1955), "Economic Growth and Income Inequality", *American Economic Review* (XLV-1): 1-28.

Magnani, E. (2000), "The Environmental Kuznets Curve, environmental protection policy and income distribution", *Ecological Economics* 32: 431-443.

Rosenbaum, J., A. Horvath and C. Hendrickson (2000), "Environmental Implications of Service Industries" *Environmental Science and Technology*, Nov. 15, 2000: 4669-4676.

5
Consumption Patterns

KEY SIGNALS

- Private and public consumption in OECD countries leads to a wide range of impacts on resource and material flows, air and water quality, land use and wildlife habitat, climate change and waste generation.
- The main drivers of consumption patterns are current and expected income, the relative prices of goods and services, the social and cultural context, available technologies and infrastructure, information provision, and institutional arrangements.
- Per capita private consumption has increased steadily in OECD countries over the last two decades, and is expected to continue to follow GDP growth in the period to 2020. In general, households are spending less on food, clothing, furniture and household operation and relatively more on rent, fuel and power, health, services and recreation.
- Environmental pressures from consumption in OECD countries intensified in the second half of the 20th century and are likely to increase in many areas over the next twenty years – driven by increasing per capita incomes and resource- and pollution-intensive lifestyles.
- The most important future challenges for addressing the environmental effects of consumption will be to draw clearer links between environmental quality objectives and consumption patterns; provide stronger signals for technology and product innovation and for infrastructure investment that reflect the environmental costs of products and services; and stimulate public awareness and action.

5.1. Introduction

OECD economies are the world's largest consumers of energy, materials and natural resources. With roughly one-fifth of the world's population and one-fourth of its land area, OECD countries account for 80% of global GDP and 80% of world trade. In the late 1990s, OECD consumers consumed or used 45% of all meat and fish, 58% of total energy, 68% of all fixed telephone connections, 82% of all mobile telephone networks, 84% of all paper, and 87% of all cars (UNDP, 1998; OECD 1999*a*). Forty-five to eighty-five tons of natural resources are required per person every year to supply the products and services that OECD consumers, governments and businesses use – the equivalent of 300 shopping bags per person per week (WRI, 1997).

Consumption patterns in OECD countries have wide environmental impacts on air and water quality, land use and wildlife habitat, climate change and waste generation. Product and technological innovations have helped reduce the energy and material intensity of many consumer goods. A small but growing number of consumers are also making lifestyle changes to lessen the environmental impact of their consumption patterns. However, the increasing volumes of goods used and discarded, and the structure of consumer demand in key areas such as energy and transport, have outweighed many of these gains. As a result, environmental pressures from consumption intensified in the second half of the 20th century and are likely to grow in many areas over the next twenty years driven by increasing per capita incomes and resource- and pollution-intensive lifestyles. This Chapter focuses on the environmental impacts in OECD countries of OECD consumption patterns, but it is important to note that OECD consumption patterns also lead to environmental impacts (*e.g.* on deforestation, biodiversity loss, material flows, fish stocks) in other regions of the world.

Widely defined, "consumption patterns" include not only household purchases and use of goods and services, but also public sector consumption and intermediate consumption in the private sector. This Chapter outlines the broad context and policy challenges related to private household and public sector consumption patterns. Further details on environmental impacts from household actors are provided in Chapter 16.

5.2. General consumption drivers and trends

Household consumption

Income and relative prices

Consumption patterns are influenced by a number of different drivers. Two primary determinants are current and expected income and the relative prices of goods and services. Net disposable income per capita expanded rapidly in OECD countries in the 20th century, doubling and even tripling in several countries between 1985 and 1997/8 alone (OECD, 2000*a*), and is projected to continue to rise to 2020. At the same time, consumers in OECD countries enjoy a steadily expanding range of low-priced, mass-produced goods and access to a progressively more global marketplace.

Private consumption per capita has increased steadily in OECD countries over the last two decades, rising by approximately 40% from US$8 000 per capita in 1980 to US$11 000 per capita in 1998. Household consumption is projected to continue to increase along with growth in GDP in OECD countries in the period to 2020 (Figure 5.1). Government consumption per capita in OECD countries, on the other hand, has only increased by

Figure 5.1. **Private and government consumption per capita, 1980-2020**

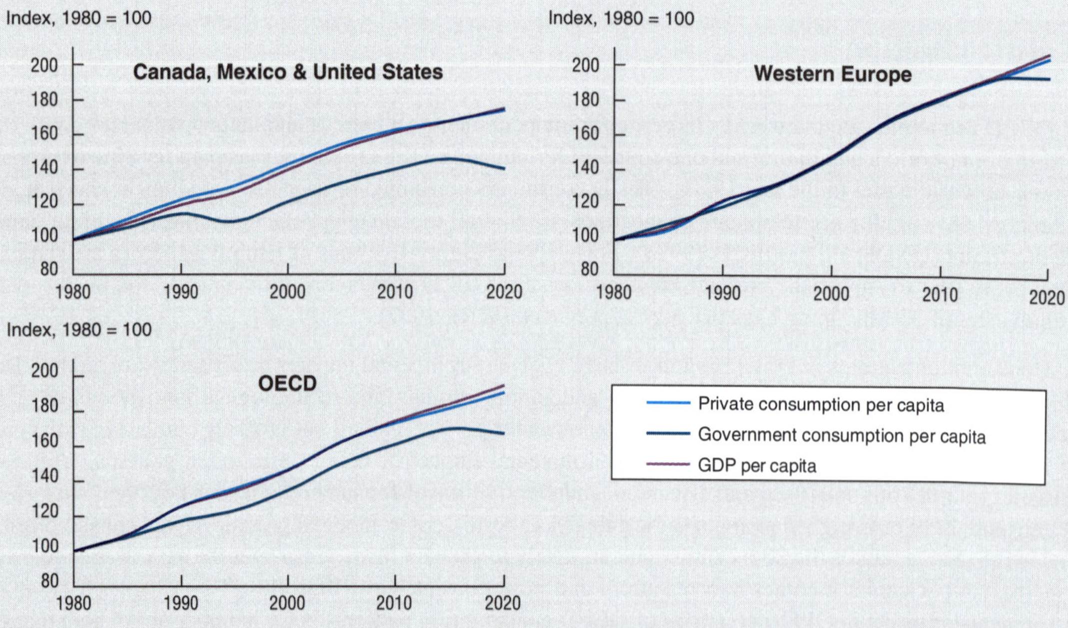

Source: OECD (2000*a*) and Reference Scenario.

approximately 30% in the same period, largely due to actual decreases in North America in the 1990s. The share of total private expenditure[1] in GDP has been almost constant at 60% in Western Europe and Japan over the last twenty-five years, although this share varies significantly among OECD countries (range: 46% to 74%).

The composition of household consumption has also changed. Trends are not uniform across OECD countries but, in general, households are spending less than in the past on food, clothing, furniture and household operation, and relatively more on rent, fuel and power, health, services and recreation, including restaurants, cafés and hotels (OECD, 2000*b*; EEA, 1999). There are large differences among countries in some consumption patterns (Internet connections, paper, air conditioning, water), but there are signs of convergence in others (televisions, refrigerators, clothes washers, personal automobiles). It is significant, however, that nearly all OECD countries share a trend towards larger houses, more household electrical appliances and greater travel distances, particularly for tourism. In the period to 2020, the composition of household consumption in all OECD regions is projected to continue to change towards higher shares of consumption of energy (including fuel and electricity), transport (including motor vehicles), and services, while the share of food consumption is expected to decrease significantly (Figure 5.2).

Figure 5.2. **Change in household expenditure, 1995-2020**

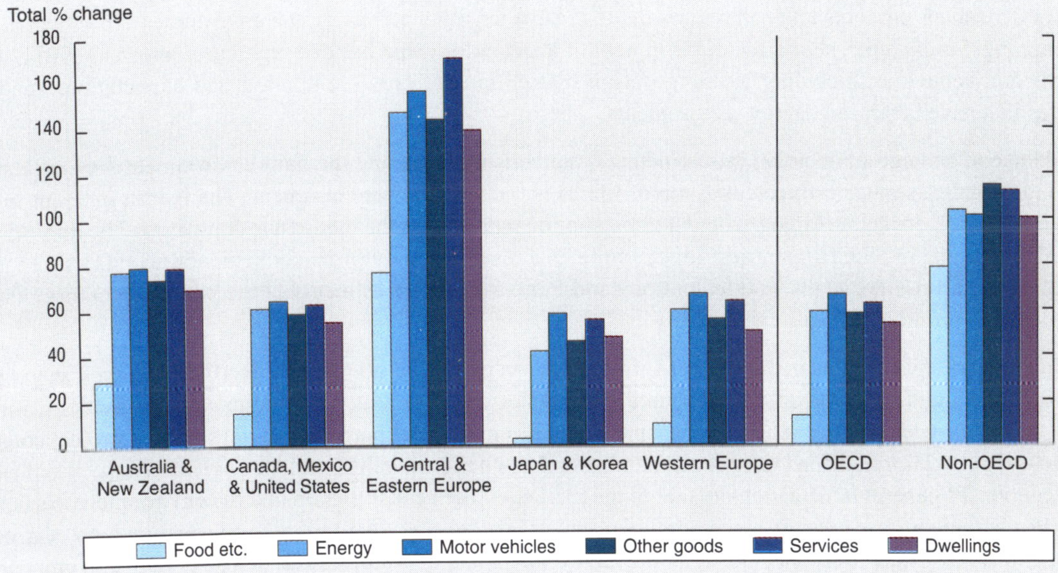

Source: Reference Scenario.

Lifestyles

Changing social and cultural contexts over the last thirty years have influenced the scale and structure of consumption patterns in OECD countries. Important recent trends include a rising number of women in the labour force, expanded school and extracurricular activities for children, longer and better financed retirement, and an increasing "individualisation" of activities and time schedules within households. The number of single-person

1. Final private expenditure here includes final consumption of households plus final consumption of private non-profit institutions.

households in OECD countries is also growing. In the European Union, for example, the share of single-person households is projected to increase from its current level of 30% to approximately 36 % in 2015, and in Japan from 23% to 28% in 2010 (OECD, 1998).

Each of these trends has contributed in turn to changes in patterns of space and time organisation that create both consumption constraints and opportunities. Common trends across OECD countries include the demand for more and bigger housing and related land and infrastructure, convenience "time-saving" appliances and services (household electrical appliances, processed and pre-prepared food), individualised product design and packaging, and greater short-distance daily and weekend travel.

The ageing of the population in OECD countries may significantly change the lifestyles to 2020 (see also Chapter 2). Potential effects of ageing are not well known, but it is expected that it will lead to an increase in the number of single person households and increased time available for recreation for a large part of the population.

Technology, infrastructure, information and institutional arrangements

Technology, infrastructure, information and institutional arrangements are a third important set of factors that influence consumption patterns. They create the prevailing conditions faced by consumers in their everyday life, and can either expand or constrain consumption options.

A number of technological and product innovations have produced new consumer products and services that reduce the environmental impact from consumption (see also Chapter 16). But new technology also creates a network of dependant products and behaviours that can have negative effects on the environment. The history of air conditioning, for example, shows the development of a technology that has had repercussions on a set of corollary systems and behaviour, including building design that eliminates cross-ventilation and expectations for indoor climates that have increased energy requirements.

Transport and communication infrastructures, the housing stock and the built environment generally are, for most individuals, as much a fixed condition of life as is the "natural" environment. The inertia inherent in infrastructure can lock consumers into particular consumption patterns for the medium to long term. The importance of personal car travel in OECD countries, for example, is a result not only of its convenience and status, but also of the policies, land-use decisions, and technology and infrastructure investments of the past half century that have facilitated and promoted private car mobility.

In the marketplace, consumers receive information via prices, labelling and advertising, as well as via retailers, distributors and other "third party" sources (consumer organisations, environmental organisations, industry associations). Providing information to consumers serves a number of purposes, in particular to protect consumer interests, support informed consumer choice, and stimulate change in behaviour. Consumers also seek and receive a vast amount of general information outside of the market on both consumption and the environment, particularly from the mass media – television, screen, radio, print, and increasingly the Internet. Information helps shape individual and social perceptions and values on the environment and consumption. It can be effective in empowering consumers to make environmentally sustainable choices. Currently, however, much environmentally relevant information is confusing or lost in the midst of media messages that largely encourage indiscriminate consumption (OECD, 2001).

Consumption decisions are also influenced by an often complex set of government policy and institutional arrangements. These include environmental and other policy measures, such as regulatory measures that constrain different consumption options or set minimum product standards which, together with economic instruments, incorporate some of the external environmental impacts in the prices faced by consumers. They also include macroeconomic fiscal and monetary policies intended to influence savings or stimulate consumption, particularly of consumer durables or real estate. The link between macroeconomic policy and the scale and structure of environmentally significant consumption patterns has not been explored sufficiently to date.

Public sector consumption

The public sector in OECD countries accounts for a significant share of consumption expenditure, ranging from 8% of GDP in Mexico to 25% in Denmark (OECD North America average: 14%; OECD Europe average: 17%) (OECD, 1999*b*). Although the size of OECD governments vary, most have large building and vehicle stocks, manage vast land areas and are major consumers of goods and services. In all OECD countries, the government is the single largest consumer. Government final consumption is extremely diverse: it involves both standard items – from pencils to vehicles – and major capital equipment and infrastructure – from electricity generators and power plants to asphalt for roads and defence equipment. Up to 75% of government expenses are for consumable goods and services (OECD, 1997). The volume and nature of government spending makes the public sector a critical actor in reducing the environmental impacts from consumption patterns.

5.3. Environmental implications of consumption patterns

OECD countries have made important progress in a number of areas to reduce the environmental impacts related to consumption patterns. These include increased efficiency in the energy and materials used for consumer products, a wider diffusion of pollution prevention and cleaner production strategies, and waste minimisation. New markets have emerged for environmentally preferable products. Governments have reduced the environmental impacts from their day-to-day operations through better environmental management and green purchasing. Also, a larger number of local government and community initiatives now focus on changing consumer behaviour than a decade ago.

However, environmental pressures from consumption continue to grow due to scale pressures from the persistent increase in the volume of goods and services consumed and discarded, and the structure of consumer demand in key areas (*e.g.* the preference for heavier, more powerful vehicles that are less fuel-efficient). Added to the natural resource pressures to meet the demand for more and larger housing, more roads, shops and tourism destinations, is the pollution associated with increasing energy consumption for space conditioning, water heaters, food storage and travel, and the waste generated from the disposal of both consumer durables (automobiles, household appliances, office and home computers) and disposable products and packaging (see Chapter 16).

Growing environmental pressures from consumption include increased use of natural resources, pollution associated with increased energy and transport use, and waste generation.

Since many environmentally significant consumption patterns are closely tied to rising per capita incomes and accompanying lifestyle changes, there is reason to believe that environmental pressures from consumption will increase to 2020. Two key developments will determine future environmental pressures from consumption. The first will be whether the demand for certain goods and services will stabilise (saturate), and at what levels. For example, some analysts expect a slowing (but still increasing) demand for household energy consumption as many OECD countries reach 100% ownership rates for major household electrical appliances. However, there remains significant demand potential for a myriad of small electrical appliances, air conditioners, personal computers and Internet accessories. Also, earlier estimates of car ownership per household have been overtaken by the observation that households, in at least some OECD countries, now purchase and use a diversified set of vehicles for different mobility needs (commuting to work or school, carpooling, shopping, recreation). Current trends and projections to 2020 suggest little saturation in the next twenty years for personal mobility, either for local use or long-distance transport for tourism.

The second key development will be the structure of technology and product options in the consumer market. The characteristics of future products and technologies can either increase or reduce environmental pressures. Although more attention is being given to environmental implications in the design and use of many products, historical trends have demonstrated that product and technological innovations to date have not been sufficient to redress environmental impacts related to consumption patterns in many areas.

5.4. Policy issues

> **The environmental effects of current consumption patterns remain a peripheral issue in most OECD countries, treated in an ad-hoc fashion.**

Early environmental policy in OECD countries focused on pollution prevention and control in the productive sector (see Chapter 1 and 24). More recent policies seek to promote eco-efficiency, closed-loop production and lifecycle environmental management. In view of persistent environmental and natural resource pressures, however, OECD governments increasingly see the need to address the impacts from public sector and private household consumption patterns. To date, economic and regulatory measures, and information and education schemes, have brought about limited changes in behaviour. Several OECD governments have also taken important steps to reduce the environmental impacts from their consumption patterns, including in day-to-day government operations (energy and water consumption, reduction of toxic chemicals, transport policy, green purchasing). Some governments are also mandating or facilitating the establishment of environmental management systems in government agencies and facilities. For the majority, however, such initiatives are very recent and not co-ordinated across government agencies. Similarly, the environmental effects of current consumption patterns in general remain a peripheral issue in OECD countries, treated in an *ad-hoc* fashion. The share of consumers that are willing and equipped to reduce the environmental impacts of their consumption patterns is much lower than necessary to make headway on key environmental problems such as climate change, habitat loss, waste generation and air and water pollution.

Given new and better understanding of the factors that influence consumption patterns, OECD governments will need to embrace a broader strategy to promote less resource- and pollution-intensive lifestyles that fulfil consumer needs at lowest environmental costs. Such a strategy would include drawing clear links between environmental quality objectives and consumption patterns, providing strong signals where the market fails to do so (*e.g.* via the price mechanism to more fully incorporate external environmental costs), technology and product innovation and infrastructure investment that reflect the environmental costs of products and services, and stimulating public awareness and action.

Governments will face a number of challenges in each of these areas. Demonstrating more clearly which consumption patterns are significant for environmental quality will be central to more effective policy development and implementation. An important step in this direction is improving environmental impact data and indicators related to consumption, not only to provide a static view of past and current consumption patterns but also to highlight trends and dynamic changes in consumption patterns over time. Indicators of both relative and absolute trends in consumption (*i.e.* consumption per unit of GDP or per capita and absolute consumption) will be needed. The link between the two categories of consumption is not linear, as rebound effects (*e.g.* increased use of appliances when running costs are reduced by increased energy efficiency) can reduce progress made on a per unit basis.

A second important challenge will be to promote innovation in technology and product design related to consumer products and behaviour. The continuing shift in OECD economies towards greater reliance on services offers significant potential for creating less environmentally intensive consumption options through new consumer goods and services (personal "mobility packages"; leased carpets; re-used water plumbing). Nevertheless, some market analysts have noted that most product innovation to date has gone little beyond marginal tinkering on relatively static product conceptions. They point to a tremendous potential for applying principles of design for the environment and for finding new ways to satisfy consumer needs at lower environmental costs. Not all consumer products can or need to give priority to environmental considerations. Resources should be targeted to those areas where lifecycle analysis identifies material flows or pollution from consumption, use or disposal patterns which have significant implications for the environment.

The third and perhaps most difficult challenge in the next twenty years will be further stimulating public awareness and action. In essence this means helping consumers to go beyond current efforts to recycle their waste or make select environmentally preferable purchases and to recognise the broader connections between their lifestyles and associated pressures on the environment and natural resources. Nearly every government, NGO, or private sector strategy for addressing current environmental problems calls for a better informed and active public. However, providing information and stimulating action may be hampered by a number of opposing factors. Public interest in the environment has been cyclical in the past, and appears low as we enter the 21st century, despite an apparent rise in general concern for the state of the environment across OECD countries. This decreasing interest is undoubtedly linked to consumer reactions to the growing volume and complexity of information on the environment, both in the market (eco-labels, green marketing) and in the media, and their scepticism *vis-à-vis* the credibility of most information sources. In 1999, for instance, Europeans reported having little confidence in the majority of sources of information on the environment and reported levels of confidence lower than in 1995 for nearly all sources of environmental information. They showed little inclination to become better informed (EC, 1999). Governments will need to resolve this information dilemma. In doing so, they must address the consumer's view of both the personal relevance and urgency of environmental problems, and the importance of individual contributions in addressing those problems. Innovative ways to call on the consumer as a citizen in decisions regarding product development in the private sector, and technology research and development and investment decision-making in the public sector, will be important.

Consumers in OECD countries are concerned about environmental quality, but their concern has often not been translated into behavioural changes.

REFERENCES

EC (European Commission) (1999), *Environnement ? Ce que les Européens en pensent*, Sondage Eurobaromètre, EC, Brussels.

EEA (European Environment Agency) (1999), *Environment in the EU at the Turn of the Century*, EEA, Copenhagen.

Geyer-Allely, E. and H.S. Cheong (2001), "Consumption Patterns", Background document for the OECD *Environmental Outlook*, OECD, Paris.

OECD (1997), *Greener Public Purchasing: Issues and Practical Solutions*, OECD, Paris.

OECD (1998), *The Future of Food: Long-term Prospects for the Agro-food Sector*, OECD, Paris.

OECD (1999*a*), *OECD Communications Outlook: 1999 Edition*, OECD, Paris.

OECD (1999*b*), "Towards More Sustainable Household Consumption Patterns: Indicators to Measure Progress", ENV/EPOC/SE(98)2/FINAL, OECD, Paris.

OECD (2000*a*), *National Accounts of OECD Countries Volume 1 1988/1998: Main Aggregates*, OECD, Paris.

OECD (2000*b*), *National Accounts of OECD Countries Volume 2 1988/1998: Detailed Tables*, OECD Paris.

OECD (2001), *Information and Consumer Decision-Making for Sustainable Consumption*, OECD, Paris.

UNDP (United Nations Development Programme) (1998), *Human Development Report 1998*, Oxford University Press, New York.

WRI (World Resources Institute) (1997), *Resource Flows: The Material Basis of Industrial Economies*, WRI, Washington DC.

6 Technological Change

KEY SIGNALS

- Over the past decades, scientific advances and technological development have brought about major changes in society, and these are expected to continue to be significant in meeting the needs of future generations.
- Although the environmental implications of technological change can be either positive or negative, it is generally recognised that technological change has a crucial role in de-coupling environmental degradation from economic growth.
- In order to induce an environmentally sustainable technological trajectory, government policies should provide adequate incentives for firms to invest in environmental research and development and to adopt environmentally beneficial technologies in products and processes.
- While the role of government policies in technological development is vital, radical and continuous technological change can only be achieved by the co-ordination of environmental policy and technological or innovation policy, exploiting their complementarity.

6.1. Introduction

Technological change, which involves both changes in the means by which products are produced and changes in the characteristics of the products themselves, has brought about major changes in society throughout history. The Industrial Revolution and the technological changes it brought, such as internal-combustion engines, electricity, and electronics, have changed the economic system as well as life-styles in modern society. However, recently the pace of this change seems to have become much faster. Technologies such as computers, the Internet, and modern pharmaceuticals have transformed the way we live, work, shop and communicate. Globalisation of the world economy, which is leading to more rapid technological change, has transformed national economies, both in developed and developing countries. These impacts are projected to be even greater in the 21st century. New developments in several critical areas, such as information and communication technology, biotechnology, and energy technology will bring further changes to OECD countries as well as to the global economy.

The environmental impacts of social and economic activities are greatly affected by the rate and direction of technological change. This linkage occurs because new technologies may either create or mitigate pollution, and either reduce or increase the use of scarce natural resources. In the past, technological changes were often viewed as being the source of environmental degradation, while recently they have been generally recognised as being one of the major means by which "de-coupling" of environmental degradation from economic growth could be attained. In fact, technology is a double-edged sword. On the one hand, technological development has provided crucial means to address environmental issues through the use and diffusion of cleaner technologies, and shifting the structure of the economy towards becoming less environmentally damaging. On the other hand, technological changes can aggravate environmental pressures. In some cases, "better-performing" technologies, such as supersonic aircraft and intensive farming, cause adverse environmental impacts. As discussed further in Chapter 23, increases in aggregate economic activities associated with technological change can result in a deterioration of environmental quality, even as the pollution intensity and resource use intensity of the economy falls. As technological change together with the accumulation of physical and human capital increasingly drives economic development, ensuring that new technologies contribute to environmental sustainability will be crucial in the future.

6.2. Drivers of technological change

A number of factors drive technological development, including research, development and the diffusion of new technologies. The scientific community – universities and public research institutes – generates a massive flow of scientific discoveries. At the same time, the direct exploitation of scientific research by business has increased, partly through the growth of new science-based sectors, such as biotechnology, and also through changes in the innovation process and the greater availability of research results.

In the innovation process, the private sector is an important driver, currently funding more than 60% of total spending on research and development (R&D) in most OECD countries. While government-funded R&D as a percentage of GDP fell across most of the OECD countries during the 1990s, the relative proportion of privately-funded R&D, which stagnated in the early 1990s, picked up over the 1995-1997 period (OECD, 1999*a*).

For the private sector, R&D investment decisions are governed by the cost of R&D and its expected return. However, R&D investment entails greater uncertainty and intangible outcomes than other investments. Unlike other forms of physical investments, the distribution of expected returns from R&D investment varies greatly and the value of the assets acquired from R&D investment combine very low probabilities of success with potentially very high profits for those which succeed. In addition, R&D investment differs from physical investment in that assets produced by the research process, such as new skills and knowledge about how to make and do things, can be copied and used by others quite easily. As such, there are important R&D "spillover" benefits for others in the economy (Jaffe *et al.*, 2000).

Recently, increasing competitive pressures in the private sector have encouraged firms to try to reduce uncertainty in research and development, and internalise some of the "spillovers" directly. Firms are integrating their R&D efforts more closely with their business strategies. The desire to reduce product development time and costs, and to achieve positive results more rapidly, is driving this trend. At the same time, the business sector is increasingly relying on co-operation with universities and public laboratories to keep abreast of fundamental research. Synergies may not only exist between the private and the public spheres, but also between different agents in the private sphere. While much of this is undertaken autonomously by firms, many OECD governments are promoting "technology clusters", private-private partnerships, and other mechanisms whereby different firms are encouraged to collaborate in developing new technologies.

The long gestation period required for many R&D investments, along with the accompanying high costs and uncertainty involved, plus the difficulty most firms face in generating sufficient financial returns from basic science, mean that governments will need to continue to support long-term research. However, governments can also play an important role indirectly. Patents and other forms of intellectual property rights can be used to provide incentives for firms to invest in research through market rewards. In fact, the existence of effective intellectual property rights (IPRs) regimes is one of the preconditions for a successful national policy on technological innovation. Protecting IPRs is important to researchers since it protects their new ideas and products, and acts as an incentive for them to continue doing research. Nonetheless, there are trade-offs. When determining the level of protection, governments have to balance the benefits of new invention and innovation, with the costs in terms of reduced technology diffusion during the life of the patent.

In addition to ensuring that firms are able to appropriate the benefits from R&D through IPRs, governments can influence other important variables such as the after-tax costs of R&D, the size of the market, and technology opportunities (Jaffe *et al.*, 2000). The after-tax cost of R&D can be changed via tax incentives, direct subsides, grants for research, and accelerated depreciation schedules. Fiscal incentives can also be used to further encourage the take-up of new technologies since industries generally bring new technologies into play only when existing capital equipment is replaced.

Demand for new technology is one of the most powerful drivers for technological change since it is often not a lack of research, but a lack of demand, that limits technological development. A technology pull from consumers and markets for good quality products and for the protection of the environment can drive firms to be more innovative. Growing public awareness of environmental quality has increased the demand for environmentally friendly goods and services, as well as for new environmental technologies. In OECD countries, however, R&D related to the environment continues to represent only a small share of public research portfolios: about 2% of R&D budgets for research are directly devoted to the environment, and an estimated 5% when environment-related research in other areas is included (such as energy, agriculture, and atmosphere). Public policy also has a role to play on the demand side, through environmental regulations. Furthermore, governments can affect the market for new technologies via direct government purchase, as well as through subsides for the installation of capital equipment incorporating particular technologies (Stoneman, 1987).

In addition to supporting research and development programmes, government support for the diffusion of technology and dissemination of information on technologies is essential for promoting the enhanced use of new technologies by a greater number of enterprises. This is particularly important for small and medium-sized firms that do not have the resources to collect and process relevant technological information. At the microeconomic level, the institutional environment of employment relations, and antitrust or other competition policies, also affect the incentives and capability of firms to develop and adopt newer technologies. At the broad macroeconomic level, governments can support R&D and technology adoption by ensuring stable macroeconomic policies and well-functioning financial, labour and goods markets.

Recent trends of economic globalisation are influencing the pace and direction of technological change. Countries that are now exposed to intensified global competition are taking advantage of their increasing technological specialisation. On the one hand, globalisation is making the innovation systems of different countries more and more interdependent, and knowledge, technology and people are moving across borders more freely. This affects scientific research as well as technology strategies in the business sector. On the other hand, certain technological developments, especially information and communication technologies and transportation technologies, are directly promoting globalisation (see Chapter 3).

6.3. Technological change: trends and outlook to 2020

In the recent past, technological change has occurred at a rapid pace, owing to the strong performance of the scientific community in OECD countries and the increasing efficiency of enterprises in turning research results into successful products. A range of indicators show that OECD economies are shifting toward a knowledge-based economy, and technologies are becoming more crucial to economic development. Investment in information and communication technologies (ICTs), one of the key sectors in the knowledge-based economy, grew twice as fast as GDP between 1987 and 1995, and represented 4% of GDP in OECD countries in 1997 (OECD, 1997 and 2000*a*). Investments in intangible assets, such as education, R&D, software, and training, are also increasing. Scientific outputs continue to rise across OECD regions and patent data show a surge of innovation in all OECD countries and across many technology fields. New technologies are being introduced and product life cycles continue to decrease in many industries. New structures and networks are emerging to produce, disseminate and manage knowledge and information. The development of the so-called "new economy" will continue to push technological change forward in OECD countries for years to come.

Consequently, the structure of the OECD economies has changed significantly. The share of the manufacturing sector in their GDP continues to decline. Although high-technology manufacturing (*e.g.* aerospace, computers, electronics, pharmaceuticals) has been able to maintain its share in the economy, medium-technology (*e.g.* chemicals, energy production) and low-technology (*e.g.* food products, textiles, paper and wood products) manufacturing have declined significantly in recent years (OECD, 1998). The service sector, which was traditionally regarded as being less dynamic with little innovation and productivity growth, has recently grown rapidly. Several services are innovative and new service jobs increasingly require skilled personnel. ICTs are enabling productivity improvements in many sectors, including transport, communications, wholesale and retail trade, and finance and business services. Between 1985 and 1997, around two-thirds of GDP growth in the business sector of OECD countries resulted from growth in the service sector (OECD, 2000*a*).

While technological change is an important driver of economic growth and environmental change, projecting future technological change is difficult. Governments and private institutions have undertaken a range of studies, including technology foresight activities. These examine the longer-term future trends of science, technology, the economy, and society, with a view to identifying the emerging technologies that are likely to yield the greatest economic and/or social benefits. They suggest that a limited number of technologies will contribute to future socio-economic development. Areas such as ICT, particularly high-density components and new types of software and health and life science technologies, including genomics and combinatorial chemistry, are expected to be important technological developments. Equally, genetics, energy physics, space research, and superconductivity are expected to continue generating a significant stream of scientific discoveries (OECD, 1998). Among technologies which are more directly environmentally-significant, the OECD projects that advanced sensors, biotechnology, clean car technology, product recycling, smart water treatment, smart waste treatment, cleaner industrial processes and micro-manufacturing, renewables and new energy technologies, and photovoltaics will experience significant technological development in the near future (OECD, 1999*b*).

> **The environmental goods and services market is projected to grow rapidly in the future, although it is currently still small.**

While these technological changes will continue to have a major impact on the structure of OECD economies, this will vary from country to country. Broadly speaking, the OECD has identified four industrial sectors that may experience particularly significant technological changes (OECD, 1998). The ICT market is projected to produce many new technological advances in the future and remain a very dynamic part of the economy. The health-care sector is also projected to continue exhibiting rapid growth over the coming years. The ageing of OECD populations and greater health awareness, stimulated by income growth, are increasing the demand for health services (see Chapter 2). Furthermore, advances in biotechnology and other health-related technologies are expected to create a range of new health-related products (see Chapter 21). The environmental goods and services market, which is still relatively small, is also projected to grow rapidly. With increasing environmental awareness in many OECD countries, demand for environmental goods and services, including those related to the energy market, are likely to increase continuously. The environmental industry now links several economic sectors, including parts of agriculture, manufacturing, energy, construction, transport, and certain services. Specialised business-sector services are a fourth market that is likely to see continuing rapid growth. This market covers sectors such as management consulting and auditing. The growth of business services and the outsourcing of services are one sign of an increasing interaction between manufacturing products that are more and more dependent upon associated services, making services the main users of advanced technologies.

6.4. Environmental effects of technological change

Technological change affects the environment in various ways. The effects include: structural adjustment within the economy (structural effects); efficiency gains in the use of resources and the reduction of pollution emissions by unit of production (efficiency effects); and increasing information and improving awareness of the complex impacts of technological change on the environment (information effects).

Technological developments have been an important driver of structural change within economies. Emerging demand for new technologies may increase the role of the sectors that produce these technologies. As mentioned before, the structure of OECD economies is continuing to shift towards services and knowledge-based industries with a declining share of manufacturing sectors. In addition, productivity continues to increase, enabling cost and price reductions in several parts of the economy (OECD, 1998). These structural changes have environmental implications. A positive trend in the structure of OECD economies is their continuing improvement in productivity by shifting towards high-quality goods and services with greater value-added and more flexible responses to changing demands. The development of many service sectors, including information processing, transport, communications, wholesale and retail trade, and finance and business services, provides the potential to continue economic development without excessive damage to the environment through better performance of the economy in general. However, adequate environmental incentives, such as the correct pricing of natural resources and the internalisation of external environmental costs, continue to be important (see Chapter 23).

Technological change can result in improvements in the environmental performance of products, processes and activities due to efficiency gains. Considerable saving and substitution in energy consumption and the use of materials (*e.g.* a reduction in energy consumption/material input per unit of output), as well as a substantial reduction in emissions and waste generation, can be achieved by optimising technological systems. Dematerialisation or a reduction in the input of scarce materials (*e.g.* metals, groundwater and wood) per unit of output can also be expected from improved industrial production systems and technological changes. Even larger efficiency improvements can be realised either by radically changing the design of contemporary technological systems or by developing new systems to take over the functions of an existing system. While the improvement of resource use efficiency driven by technological development is beneficial to the environment, efficiency savings can at the same time promote consumption, stimulate additional demand for leisure travel, and generate products with short lives such as junk mail. In fact, the aggregate consumption of natural resources in OECD regions has kept on increasing in spite of the continuous decrease in resource use intensity in recent decades. Thus, the volume effects of total economic growth have outweighed the efficiency improvements achieved (see Chapter 23).

Technological development has broadened the understanding of the relationship between pollution and its impacts on human health and the environment. New technologies help to identify emerging environmental problems and ultimately to provide solutions. They have also contributed to the formulation of environmental policies. Technological development can increase the understanding of the highly interrelated, systemic character of the environment and has become a driving force for an institutional transition to integrated permitting approaches. Advanced control equipment, sensors, satellites and other sophisticated ICT systems could contribute to reducing resource inputs. Furthermore, the growing awareness of environmental quality makes information effects more important. Providing environmental information on products to consumers and encouraging them to act on the basis of informed decisions is necessary for changing consumption patterns (see Chapter 5).

Development in information and communication technology is a good example of the significant environmental effects of technological change. Expected structural effects are both positive and negative. Positive effects come from the fact that ICT and related services are enabling efficient use of energy and resource inputs in many manufacturing and service sectors through improving information processes, extensive networking, and co-operation among firms (Romm, 1999). However, the potentially negative impacts include greater consumption of electricity and paper for ICT devices and increasing waste generation from the short lifecycle of computers and consumer goods in society in general. The efficiency effects are projected to be positive and will occur in the form of improved management of environment-related material inputs in firms. The information effects are also projected to be positive, since they can help organisations monitor different aspects of their environmental performance at reduced cost. Advances in information technologies can improve the way information is treated, stored, and diffused. They can increase the concerns and awareness about the environment through improving the way the public accesses and analyses environmental information. In order to assess the aggregate environmental impacts from ICT, more integrated investigation is required.

The environmental impacts of technological change can be more easily understood by examining individual technologies. Table 6.1 shows the economic and environmental impacts of current and potential future technological change in selected environmentally-sensitive sectors. Most new technologies are generally positive in terms of economic implications. However, their environmental impacts are diverse: they can be positive, negative, and sometimes ambiguous. For example, the electric arc furnace, which is used in steel production, shows generally positive environmental effects (see Chapter 17). It reduces energy and raw material consumption, as well as air emission, in production. In addition, it increases the recycling potential of steel scraps. Some of the technological developments in fishing, such as sonar radar and global positioning system (GPS), may entail negative environmental impacts (see Chapter 9). They are generally geared towards improving the efficiency and cost-effectiveness of fishing activities or fish processing. These technologies increase the share of living marine resources that are potentially harvestable, and have contributed to the overfishing of many fish stocks. The development of new and efficient drilling techniques for fossil fuels clearly enlarges the stocks of economically accessible fuels (see Chapter 12). However, these techniques encourage continued reliance on fossil fuels, with associated air pollution emissions.

Table 6.1. **Economic implications and environmental effects for selected technologies**

Sectors	Technologies	Economic implications	Environmental effects
Agriculture	Pesticides and fertilisers	Agricultural productivity improvement	Pollution of ground and surface waters
	Modern biotechnology (use of GM crops)	Improved crop productivity	Potential adverse health and ecosystem impacts; Potential for limiting adverse impacts from pesticides and other chemicals
Fresh water	Drip-irrigation	Reduced water expenditures; High equipment installation costs	Decrease in water intensity; Increase in energy intensity
Wastewater	Tertiary wastewater treatment	Increased infrastructure requirements; Reduced costs of pollution clean-up downstream	Reduced water pollution
Fisheries	Computer-aided fishing (*e.g.* GPS, sonar radar)	More efficient fish harvesting	Decreased fish stocks; Overfishing
Forestry	Better tree breeding techniques and use of biotechnology	Faster tree growth; More desirable tree characteristics	Increased forest area; Increased carbon sinks; Monoculture forests
Energy	New drilling techniques for fossil fuels	Enlarged stocks of economically accessible fuels	Reduced opportunity of being used by future generations; Continued fossil fuel reliance
Transportation	Hybrid, electric and fuel cell vehicles	Fossil fuel saving	Reduced air emissions
Steel	Electric arc furnaces	Reduced energy and raw material consumption	Increased recycling of scrap steel; Reduced air emissions
Pulp and paper	Totally chlorine free (TCF) bleaching	High capital investment; Improved paper quality	Reduced pollutant emissions (*e.g.* dioxins and BOD loading); Reduced energy consumption

Potential biotechnology advances in agriculture have raised social, economic, and political questions related to their potential health and environmental impacts.

In some cases, the environmental impacts of technological change are ambiguous. Modern biotechnology has been increasingly applied to crop plant breeding. The combination of rising population and decreasing productivity growth rates have led many observers to emphasise the long-term need for the use of modern biotechnology in agricultural production. While such biotechnological advances are promising, they have raised major social, economic, and political questions related to their potential adverse health and environmental impacts. Many of these questions remain unresolved. When these new technologies are brought to the market, provisions to ensure that they are safe for both human health and the environment need to be introduced at the same time.

Again, it is not easy to project the precise nature of the environmental effects of new technologies. However, technological changes are expected to play a crucial role in meeting the needs of current and future generations, and in de-coupling environmental degradation from economic growth. Improved efficiency in the use of resources, as well as increased knowledge of the environmental impacts of human activities, will probably bring positive impacts. When the efficiency gains of technological change are linked with structural adjustment, society can achieve even greater efficiency improvements.

6.5. Policy issues

Since policies that encourage general technological change can be positive, negative or neutral with respect to their environmental effects, it is important to design and implement policies that provide long-term incentives for the development and diffusion of technologies which are less environmentally damaging. Governments must improve framework conditions so as to provide the right incentives and price signals to firms, and influence consumer awareness and behaviour with respect to environmental concerns. Improving these signals involves making sure that existing prices better reflect the full marginal costs and benefits to society of different technical approaches. This will include removing government subsidies, changing relative tax rates and reforming other inappropriate public policy interventions. Thus, government subsidies (*e.g.* to energy) may stimulate overuse of inputs or over-production of certain fuels, locking-in prevailing and often inefficient technologies. In addition to getting the prices right, government policies in the areas of environment and innovation could be better integrated to stimulate the development of clean technologies and more integrated approaches to pollution prevention (OECD, 2000*b*).

Traditionally, environmental policies in most OECD countries have relied on rather direct forms of regulation, such as emission standards, technology specifications and product bans. These regulations have been effective in many cases. Assuming that the specifications remain constant through time, they provide firms with "insurance" against the risk of undertaking inappropriate investments. Furthermore, they create direct pressure on firms to meet the required environmental regulations and technology specifications. However, concerns have grown over the innovation-stifling effect of technology specifications and the economic inefficiency of rigid application of environmental regulations. In recent decades, environmental regulation has evolved in three distinct directions, which have all played a role in encouraging more innovation. First, the use of "adaptive" technology specifications, such as best available technology (BAT) and BAT not entailing excessive costs (BATNEEC), has helped to encourage innovation and diffusion of more efficient technologies by giving firms greater incentives to develop and invest in environmentally-beneficial technologies (OECD, 1999*c*). Second, the shift toward flexible performance-based standards has strengthened these incentives since they are neutral with respect to the type of technology that is adopted by firms. And finally, economic instruments such as pollution charges and tradable permits have further stimulated environmentally-beneficial technological change since innovators and adopters are given strong and continuous incentives to develop and invest in the most efficient and effective technologies. Indeed, the experience with economic instruments in the past decades illustrates in particular the importance of appropriate pricing of negative environmental impacts in enhancing environmental sustainability.

In addition to choosing appropriate instruments, general principles that guide environmental policy design will be required to improve the long term effects of technological change. For instance, the degree of predictability of a particular policy measure may be more important than the nature of the instrument itself. In sectors which are capital-intensive and in which technologies are long-lived, policy unpredictability can be a significant barrier to technological change. Thus, it is important that regulatory authorities give clear and unambiguous signals. The scheduling of policy changes can also be important for technological change. On the one hand, if forewarning is provided, firms can schedule investments in cleaner technologies in line with capital turnover, reducing costs. On the other hand, this may also result in increased environmental gains since firms will be better able to change their production processes fundamentally rather than through more limited measures, such as add-on equipment.

Policy measures also need to be complementary. The practice of targeting discrete environmental problems in a sequential manner has often led to excessive costs and may even misdirect the path of technological development. Once a particular trajectory has been adopted to address one particular environmental problem, opportunities for synergies may be lost, and in some cases the costs of meeting subsequent environmental objectives may even rise. For instance, in the electricity supply industry, the sequential and separate introduction of policies to address sulphur dioxide emissions and carbon dioxide emissions has led to very costly investments, which can even have perverse secondary environmental consequences.

In this context, it is also important that environmental policy and technological/innovation policy be co-ordinated, particularly since responsibility for the two sets of policy measures often rest in different ministries. The current trends of incorporating environmental objectives in technology policies, such as public and private partnership and the use of foresight studies, show the potential in this field. Finding efficient and effective ways of realising synergies between the need to internalise negative externalities associated with the environment, and the need to internalise positive externalities associated with research, development and diffusion, is one of the biggest tasks facing OECD governments today.

The global nature of environmental pressures requires further international technology co-operation. Environmental performance at a global level will depend heavily upon the application of clean technologies in non-OECD as well as OECD countries. That is partly because pollution emissions in non-OECD regions is likely to increase more rapidly than in OECD regions, and partly because the capacity of these countries to take advantage of cleaner technology options is limited. As indicated in the Reference Scenario (see Chapters 2 and 3), most of the global population growth and much of economic growth over the next two decades are expected to take place in non-OECD regions. As a result, their share of global pollution emissions, such as carbon dioxide and municipal waste generation, are likely to increase continuously. However, international science and technology co-operation to encourage the development and diffusion of environmental technologies faces many barriers. Restrictions remain on foreign participation in government-funded R&D and technology programmes, and differences in IPR regimes also continue to limit technology diffusion and co-operation among countries. As a result, it has become crucial to assess current international science and technology co-operation programmes from the perspective of developing countries' needs. While developing countries must take an active role, developed countries can assist in vital areas like capacity building and the formulation of policy frameworks conducive to increasing demand for cleaner technologies. In this context, integrated approaches such as promoting public/private partnerships, need to be further explored (OECD, forthcoming 2001).

REFERENCES

Jaffe, A., R. Newell, and R. Stavins (2000), "Technological Change and the Environment", NBER Working Paper W7970, National Bureau of Economic Research, (NBER), Cambridge, October, *http://papers.nber.org/papers/W7970.pdf.*

Johnstone, N. (2001), "Technological Change, Public Policy and the Environment", Background document for the OECD *Environmental Outlook*, OECD, Paris.

OECD (1997), *Information Technology Outlook*: 1997, OECD, Paris.

OECD (1998), *Science, Technology and Industry Outlook*: 1998 *Edition*, OECD, Paris.

OECD (1999*a*), "Fostering Scientific and Technological Progress", OECD *Observer*, June.

OECD (1999*b*), "Technology Foresight and Sustainable Development: Proceedings of the Budapest Workshop, 11 December 1998", DSTI/STP/TIP(99)8/FINAL, OECD, Paris.

OECD (1999*c*), *Environmental Requirements for Industrial Permitting* (Vol. 1), OECD, Paris.

OECD (2000*a*), *Science, Technology and Industry Outlook*: 2000 *Edition*, OECD, Paris.

OECD (2000*b*), *Special Issue on Sustainable Development*, STI Review No. 25, OECD, Paris.

OECD (forthcoming 2001), OECD *Analytical Report on Sustainable Development*; OECD, Paris.

Romm, J. (1999), "The Internet Economy and Global Warming: A Scenario of the Impact of E-Commerce on Energy and the Environment", The Centre for Energy and Climate Solutions, *www.cool-companies.org/ecom/index.cfm.*

Stoneman, P. (1987), *The Economic Analysis of Technology Policy*, Oxford University Press, Oxford.

Section III

PRIMARY SECTORS AND NATURAL RESOURCES

This section examines selected primary sectors and renewable natural resources. The sectoral chapters (agriculture, fisheries and forestry) present the trends and projections for the overall developments in the sector, with an analysis of how these developments might impact on the environment. The natural resource issues chapters (freshwater and biodiversity) examine the pressures on the resources and present an analysis of recent trends and future projections in the state of the resource. All the chapters conclude with an indication of the policy options available for addressing the problems identified.

7 Agriculture

KEY SIGNALS

- Agriculture represents a declining share of GDP in OECD economies although agricultural production is continuing to expand. Increased demand for food is expected both worldwide and in OECD countries to 2020.
- Future increases in agricultural yields and production efficiency are likely to be slower than in the past, although greater use of biotechnologies could contribute to yield increases.
- Agriculture can have both positive and negative effects on the environment, depending on how and where it is practised. Potential positive effects include land conservation, flood prevention, landscape amenities, and carbon sequestration. Potential negative effects include degradation of soil, biodiversity habitats, and water and air quality, as well as the release of greenhouse gases.
- Consumer concerns about health and environmental effects of agricultural production are likely to play an important role in determining production activities, including the spread of organic agriculture and the future use of biotechnologies (*e.g.* genetically modified organisms).
- The most promising policy options for reducing the negative environmental effects of agriculture include reforming environmentally harmful subsidies to agriculture; increasing use of regulations and charges for chemical and nutrient use; and disseminating information about the use of environmentally friendly technologies and farm management practices.

7.1. Introduction

Food is essential for human life. Food production, however, has complex links with the environment, and can either reinforce or degrade the ecological conditions upon which it is based. Thus, depending on how and where agriculture is practised, it can help prevent or contribute to flooding, provide attractive or unattractive landscape, filter or buffer rainwater or add contaminants to it as it passes through or across the soil, provide wildlife habitats or destroy them, sequester carbon or release it and other greenhouse gases to the air. The environmental impacts of different food production systems are site specific, and their importance varies between and within countries.

As the nature of the food production industry changes, and demand for certain agricultural products increases, the environmental impacts are also undergoing major changes. For example, the intensification of agricultural production has been driving greater energy intensity, greater risk in the use of agrochemicals, and accelerated land conversion in many areas. Persisting high levels of agricultural support and government trade policies continue to distort the relative prices of agricultural inputs and outputs. On the other hand, more recent regulations and pricing incentives in some OECD countries have led to less use of water, fertilisers and pesticides per unit of production, as well as encouraged the adoption of agricultural measures to protect soil, habitat and landscape values.

7.2. Developments in the agriculture sector

The scale and structure of agricultural activity are driven by a number of often inter-related factors. While all humans require a subsistence level of food, the main factors influencing the types of food demanded and any additional demand above subsistence level are linked to demographic factors, socio-cultural influences, consumer incomes, and product prices. On the food production side, the main factors driving changes are technological developments, government policies in agriculture, environment and trade, agro-food industry dynamics, consumer preferences and trade liberalisation.

Demand for agricultural products

Agricultural products are primarily used for direct human consumption, but as much as 36% of world cereal demand (670 million tonnes) is used for animal feed. The amount of food available worldwide for direct human consumption (after accounting for non-food uses and losses) has risen dramatically over the last 20 years – by over 10% from 1980 to 1998 – and it is expected to increase by a further 10% to 2020, reaching 3 000 kcal/person/day on average by 2020 under the Reference Scenario (Table 7.1). Already, average per capita availability is well above human requirements, but undernourishment continues to be a problem, affecting as much as 18% of the population of developing countries, or about 800 million people (FAO, 1999). As populations continue to increase (though more slowly) worldwide over the next two decades, and efforts are directed towards further reducing undernourishment, overall demand for agriculture will continue to grow globally. As populations in OECD regions are stabilising, the expected increases in agricultural demand over the next two decades will be slower there than in non-OECD regions; nevertheless, some continued increases in per capita food consumption are expected even within OECD regions.

Table 7.1. **Key agricultural sector statistics and projections**

		1980	1998 *(or latest available year)*	2020 projected	Total change 1995-2020
Gross production (1995 US$ billion)	OECD	..	1 022	1 429	40%
	World	..	1 902	3 689	94%
Share of value added in the economy (%)	OECD	..	2.6%	2.3%	–13%
	World	..	4.3%	4.9%	13%
Direct and indirect employment (million persons)	OECD	58	47	..	..
	World	1 069	1 302	..	..
Per capita food consumption (Kcal/capita/day)	OECD	3 138	3 397	3 500	5%
	World	2 525	2 781	3 000	10%

Sources: Alexandratos (2001), GTAP database, and Reference Scenario.

Recent increases in per capita food consumption have been accompanied by significant changes in diet. Cereals continue to be the dominant source of calories, providing on average 52% of world calorie intake. In recent years, however, there have been significant increases in the shares of total calories coming from livestock products (meat, milk, eggs) and vegetable oils, and reductions in those from roots, tubers and other products (see Chapter 16). This trend is expected to continue over the next two decades, with changes in diet most pronounced in developing countries.

Food consumption patterns in OECD countries have also changed as consumer awareness of food quality and safety issues has grown (see Box 7.1). Largely in response to health and environmental concerns, organic agriculture is growing in importance in many OECD countries. Although organic agriculture has been experiencing growth rates in excess of 20% per annum in a number of countries in recent years, it still occupies less than 2% of total agricultural area in most OECD countries (OECD, 2000).

Although environmentally friendly agricultural systems still represent a small share of total agricultural production systems, they are increasingly used in OECD countries.

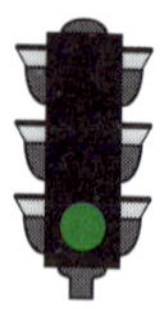

Box 7.1. Agriculture and food safety concerns

Consumer awareness of food quality and safety issues has broadened recently in some countries, and now includes concerns about agricultural production processes, animal welfare and the environmental effects of agricultural production. Food safety crises, such as those involving salmonella in eggs and dioxins in poultry in several countries, adulterated cooking oil in Spain, E. coli in beef in the US, and listeria outbreaks in soft cheeses, as well as the highly profiled case of BSE (Bovine Spongiform Encephalopathy) or "mad cow disease", have sharpened consumer fears. The current debate in many countries over genetically modified crops (soya, maize, etc.) is raising questions about the potential environmental and health effects of these crops. Common to all these issues is concern about the appropriate paths for providing consumer information about agricultural production processes, and the environmental and health effects of agricultural production.

Agricultural sector production

Worldwide, agricultural production is expected to grow by 94% between 1995 and 2020 in monetary terms, while in OECD regions it may increase by almost 40% (Table 7.1). The increase in the value of agricultural production in OECD regions will follow in part from changes in the composition of production, with particularly large increases in production for the highest-value products. Thus, the production of meats (especially poultry meat), some cereals (*e.g.* wheat and coarse grains) and vegetable oils and seeds, is particularly expected to increase in OECD countries in the coming decades (Alexandratos, 2001). Production volumes are expected to increase more rapidly in non-OECD countries, with particularly significant increases in meat production expected to 2020 (Figure 7.1).

Figure 7.1. **Agricultural production, 1995-2020**

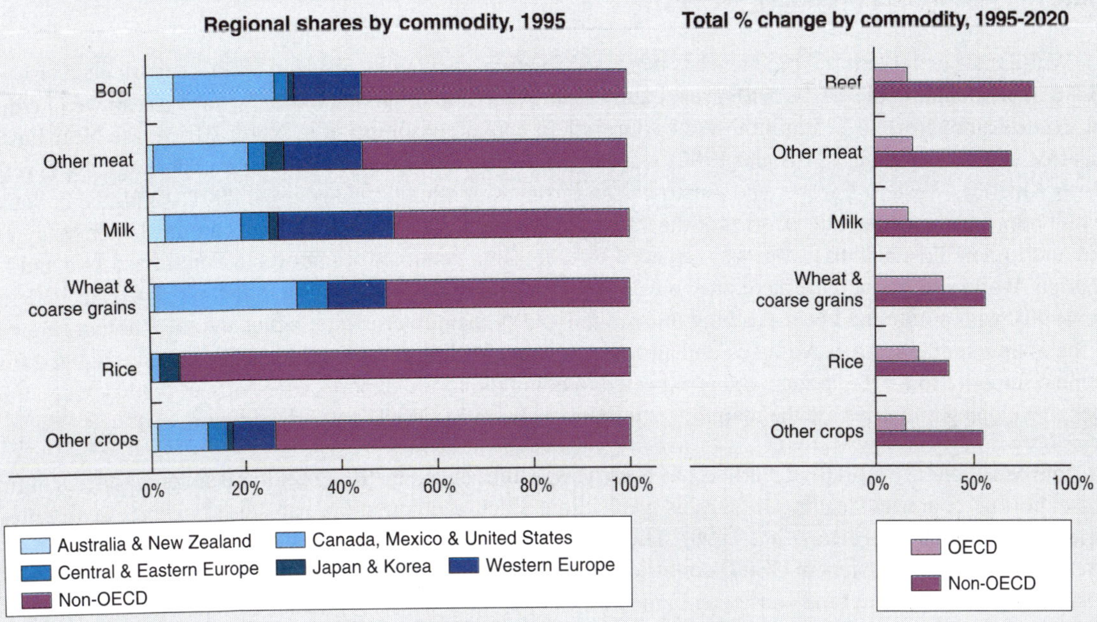

Source: Reference Scenario.

Over the last twenty years, significant increases in agricultural production in OECD countries have been achieved. These have largely been realised through the intensification of agricultural production methods that rely on both technological improvements and efficiency gains in the use of purchased inputs (*e.g.* animal feeds, land, water, seeds, pesticides, fertilisers, energy, services) and on greater labour productivity. Recent developments in technology and farm management systems have resulted in better timing and targeting of input applications (*e.g.* through drip irrigation systems), more efficient feeding of livestock, advances in animal husbandry and in plant breeding, and genetic improvements. Upstream and downstream industries of the agri-food sector have also experienced significant technological developments, often directed towards maintaining market shares through product differentiation, or lowering production or delivery costs (see Box 7.2) (OECD, 1995).

Box 7.2. **Structural changes in the agri-food industry**

There have been a number of key structural changes in the agri-food industry in recent years. Food producers are increasingly dependent on upstream and downstream linkages, with a global revitalisation, for instance, of contract farming, where producers work directly for agri-business corporations (McMichael, 1994). Integration is also occurring between distributing, wholesaling and retailing in some countries. The sector has seen a concentration of market power at all levels of the agri-food chain (OECD, 1995).

Improved refrigeration, new technologies for processing and packaging foodstuffs, and speedier and better organised delivery systems mean that products now reach supermarket shelves more quickly, and can generally be kept longer, enabling wider distribution, longer shelf life and usually less risk of contamination. Distribution and retailing have also undergone significant changes. Increased concentration and competition, and the relative inelasticity of food demand in OECD countries, have led to an emphasis on non-price factors, such as advertising, value-added product creation and differentiation, and near-market distribution and retailing, as a way for food manufacturers to boost profits. Food systems have become increasingly specialised, and niche markets for fresh and processed food have expanded globally.

Trade in agricultural products

World trade in agricultural products has increased fivefold since 1970. Trade in cereals experienced very rapid growth, in particular in the 1970s, with gross exports more than doubling. Since then, growth has slowed, with current exports amounting to 225 million tonnes. Growth in cereal consumption in North Africa, the Near East and transition economies in the 1970s and 1980s was largely fed by imports from the major exporting OECD regions (North America, Western Europe, and Australia) and from Argentina (Alexandratos, 2001). With the collapse of the oil boom and the systematic reforms of the transition economies, demand for imports diminished in many European and Middle East countries, but was replaced by increasing demand for imports in South East Asia and Latin America. World exports of meat have grown from 7% of world meat consumption in the mid-1980s to 10% in the mid-1990s, with poultry and beef growing more significantly than other meats. Japan and the Russian Federation are the main importers, while New Zealand and Australia are leading meat exporters. The US has changed from a net meat importer to a net exporter, and is expected to dominate future exports, especially of poultry meat. For oil crops, developing countries are the main net exporters, while most OECD countries import.

The behaviour of world food markets has been heavily influenced by the agricultural support policies of major food-exporting countries, leading to surplus production, stock accumulation, subsidised exports and depressed world market prices (de Haen *et al.*, 1998). The 1980s saw a rapid expansion of both agricultural subsidies and agricultural non-tariff barriers in OECD countries. The multilateral trading rules established by the 1994 Uruguay Agreement on Agriculture (and subsequent amendments), and numerous regional and bilateral trade agreements concluded in recent years, have been instrumental in starting to re-orient the agriculture sector towards market conditions (see Box 7.3).

Box 7.3. Agricultural subsidies in OECD countries

In 1998, total support to agriculture in OECD countries was estimated at US$362 billion, or 1.4% of GDP (OECD, 1999*a*). Most of it was in the form of support to producers through artificially high prices for their products (accompanied by trade barriers), or through subsidies to the use of inputs. These types of support – linked to output levels or the use of particular inputs – are most likely to result in a large proportion of the support accruing to non-target groups (*e.g.* primarily large farmers or the input suppliers) and tend to encourage excess production or use of often environmentally damaging chemical inputs. As much as three-quarters of market price support may leak away from the intended recipient (the farmer) largely through extra expenditures on inputs (OECD, 1995). Although such measures continue to dominate OECD agricultural support, the share of support which is provided by measures that are de-coupled from current production or inputs is gradually increasing. In addition to "de-coupling" support from environmentally damaging activities, many countries now tie support directly to compliance with specific environmental goals (agri-environmental measures).

7.3. Effects of agricultural production on the environment

Agricultural intensification has led to a growing use of energy, water and agrochemicals, increased soil erosion, and reduced habitat provision in OECD countries.

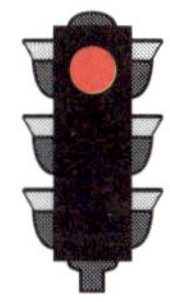

Agricultural practices can have both positive and negative effects on the environment. While all farming systems, from intensive conventional farming to organic farming, have the potential to be locally sustainable, whether in fact they depend on farmers adopting appropriate technology and management practices (OECD, 2001). Although existing evidence of the effects on environmental quality is partial and incomplete, recent trends in agricultural development in OECD countries – particularly the more intensive use and increased risks of agrochemicals, irrigation, and the use of larger and more powerful farm machinery – appear to have had negative effects on the environment. They have contributed to the higher energy intensity of agricultural production, pollution of ground and surface water by pesticides and fertilisers, soil erosion in some regions, and to increased resistance of pests to chemical control.

Air quality and climate change

Air pollution from the agricultural sector is primarily related to intensive livestock production and the resulting emissions of ammonia from manure. Agriculture is not the main source of acid rain, but air-borne ammonia can acidify soils many kilometres downwind (see Chapter 15).[1] Total nitrogen loading to the environment (air, soil and water) from livestock production in OECD regions is expected to increase by about 30% between 1995 and 2020, with particularly large increases in Central & Eastern Europe, and levels in Western Europe actually declining under the Reference Scenario (see Figure 7.2). Another source of air pollution from agriculture is topsoil blown off erodable lands. Deposited topsoil can pollute rivers and lakes, damage buildings and machinery, cause respiratory problems and increase cleaning costs.

Agriculture also affects global atmospheric conditions – negatively through the emission of ozone-depleting substances and greenhouse gases, and positively through the absorption of carbon in agricultural soils (see Box 7.4). Agricultural production contributes about 39% of methane and 60% of nitrous oxide emissions released in OECD countries (OECD, 2000). Methane emissions from agriculture are mainly produced by ruminant animals and the handling of manure, while the main source of nitrous oxide emissions is nitrogen fertilisers. To 2020, methane emissions from livestock in OECD countries are expected to increase by over 18%, while

1. In fact, acid rain pollution from other sectors has a negative impact on agricultural lands in many regions.

emissions from crops increase by roughly 10% under the Reference Scenario. Overall, agriculture-related methane emissions in OECD regions are expected to increase by almost 9% from current levels, and worldwide by over 22% to 2020 (Figure 7.2). Excessive fertiliser application and production of livestock manure can lead to the generation of the greenhouse gas nitrous oxide (N_2O), as well as nitrate pollution of watercourses, contributing to the eutrophication of water bodies (see Chapter 8). As indicated above, nitrogen loading from livestock production in OECD regions is expected to increase to 2020. However, agriculture is estimated to contribute only 1% of total CO_2 emissions in OECD countries, and with further improvements in farm equipment fuel efficiency, OECD agricultural CO_2 emissions are expected to decline by almost 15% to 2020.

Figure 7.2. **Changes in environmental effects of the agriculture sector, 1995-2020**

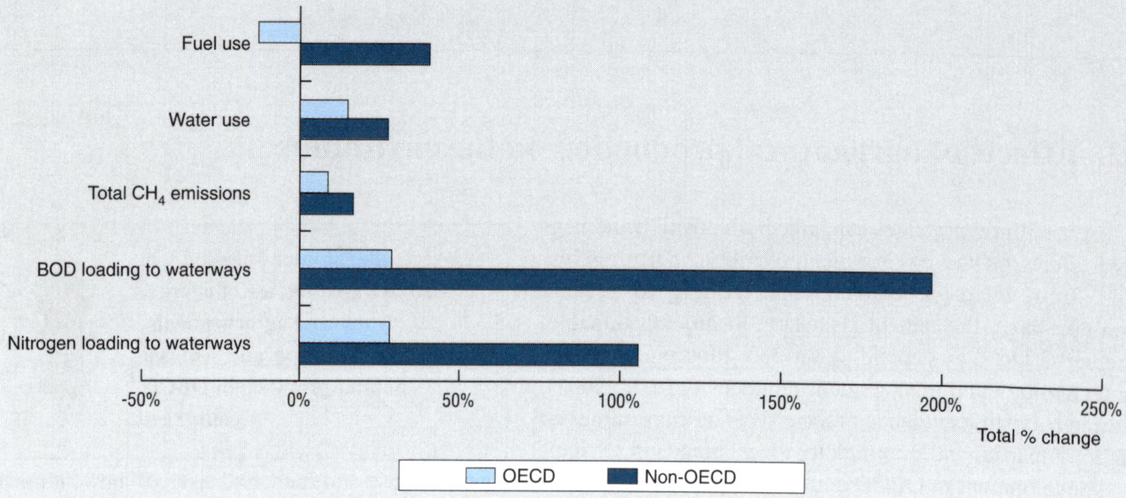

Source: Reference Scenario.

Box 7.4. Agricultural soils as carbon sinks

While agriculture contributes to the emission of some greenhouse gases, mainly nitrous oxide and methane, it can also contribute to the capture and storage of carbon. Agricultural practices that improve soil quality can also increase the amount of carbon-rich organic matter in the soil, thus providing a global depository for CO_2 drawn from the atmosphere by growing plants. As of yet there are no systematic and agreed measurements of agriculture's role as a sink for greenhouse gases, but it has been estimated that it may be possible to sequester 40-80 Pg of carbon in cropland soils over the next 50-100 years (CAST, 2000). It is clear that soil carbon sequestration may be significant and can be influenced through the use of particular farm practices. This role may increase in the future in OECD countries, particularly as most land use changes are from arable land to permanent grassland, forests, and wetland, all of which imply increased carbon sequestration potential.

Freshwater use and pollution

Agriculture is by far the largest water user, accounting for 69% of global freshwater extractions. In recent years, significant improvements in the efficiency of irrigation water use have been realised in some countries, driven by price incentives, infrastructure refurbishment and improved agricultural management techniques. Thus, while total irrigated lands have increased by 16% in OECD countries since 1980, the amount of water used for irrigation has been stabilising, with only a 4% increase since 1980 (compared with worldwide increases of 60%). As yet, however, the adoption of sustainable irrigation techniques has been limited: drip-irrigation, for example, is employed in less than 1% of the world's irrigated areas. As a result, almost half of the water withdrawn for agricultural purposes worldwide is lost through inefficient pipes and irrigation systems before it can be used. Over the next 20 years, total agricultural water use in OECD regions is expected to increase by a further 15% (Figure 7.2).

Agrochemicals run-off is one of the main contributors to groundwater pollution in OECD countries, and is expected to worsen to 2020.

Agricultural production can contribute to the pollution of water bodies, primarily through the leaching of nutrients, pesticides and animal wastes, and soil run-off and sedimentation. Currently, contamination by nitrates from fertiliser use and livestock waste is perhaps the most serious threat to groundwater sources in OECD countries (see Chapter 8). Elevated levels of nitrates and pesticides are recorded in most OECD agricultural areas, with quality standards being exceeded regularly where farming is particularly intensive (OECD, 1998*a*). In OECD countries, agriculture contributes on average 40% of nitrogen and 20% of phosphorous emissions to surface waters (OECD, 2000). To 2020, agricultural contribution to water pollution through nitrogen and biochemical oxygen demand (BOD) loading of waterways is projected to increase by over 25%. Although the apparent use of pesticides has declined in OECD countries, water bodies continue to be contaminated in intensively farmed land, areas vulnerable to run-off, and in humid regions. A more significant problem, however, is pesticide contamination of groundwater: several of the most widely-used pesticides (*e.g.* atrazine, simazine, aldicarb) have the potential to leach through soils under normal agricultural use. Integrated pest management (IPM) strategies, which are increasingly being adopted by OECD agricultural producers, offer the potential to reduce both the need for pesticides and the risk of pesticide contamination of water supplies.

Soil degradation

The loss of soil organic matter and fertility remains a serious problem in cereal and intensive livestock production regions in OECD countries. Soil degradation occurs principally through wind and water erosion, compaction, salinisation and acidification, and biological erosion (*e.g.* decline in organic matter). These degradation processes are linked to changes in farm management practices, climate and technology. For example, wind and water erosion have been accelerated by several conventional agricultural practices, such as monoculture production, fewer rotations with forages, shorter rotations, deep or intensive tillage, the cultivation of marginal lands, and overgrazing from intensive livestock production. Soil compaction can occur through the continued use of heavy farm equipment when soils are passed over or tilled when wet. Compaction contributes to lost soil fertility by reducing water and nutrient retention and movement. Salinisation and acidification contribute to the degradation of soil chemical processes and ultimately reduce yields. Saline soils are common to most arid or semi-arid areas of the world, but irrigation has contributed to the build-up of salts on irrigated cropland.

While the area of agricultural land at high or severe risk to water and wind erosion is not extensive in OECD regions overall, for some OECD countries it represents more than 10% of agricultural lands (OECD, 2000). Over the past decade, water-related soil erosion appears to have declined in a number of OECD countries, largely because farmers are adopting conservation or no-tillage practices, crop production is becoming less intensive in some areas, and marginal lands are being removed from production. Similarly, evidence suggests that other soil problems (*e.g.* acidification, salinisation, soil compaction and toxic contamination) may be beginning to improve in some OECD countries (OECD, 2000).

Natural resource use, biodiversity and habitats

Overall, there has been a marginal decrease (1%) in OECD countries of the total land area devoted to arable and permanent crops in recent decades (OECD, 1999*b*). Only the Australia & New Zealand region has experienced a real increase in lands devoted to cropping. In the future, OECD regions overall are expected to see a very small (0.1%) increase in crop lands from 1995-2020 under the Reference Scenario, particularly in the regions of Australia & New Zealand, and Canada, Mexico & the United States. The region of Western Europe, by contrast, is expected to experience decreases in lands devoted to cropping. Worldwide, croplands are expected to increase by just over 1% to 2020, while pasture lands will increase by 6%.

The increased agricultural production that has occurred recently, despite marginal decreases in cropland, indicates a wider trend in agricultural productivity gains, largely realised through agricultural intensification. While intensification has the potential to remove pressure from natural lands for agricultural expansion, it also has negative effects on wildlife habitats and biodiversity. Large-scale monoculture in cereals and grains production, the modification of grasslands, wetlands, and other specialised biotopes, land consolidation measures, farm mechanisation, intensive livestock production, and increases in the use of fertilisers and pesticides, have resulted in lower population levels of many species and a reduction in plant and animal biodiversity and habitats.

7.4. Policy options and their potential effects

As population growth slows and there is less scope for per capita increases in food consumption than in the past, it is expected that the demand for food and agricultural products will grow at slower rates in the next 20 years than over the last 20, particularly in OECD countries. However, the absolute increments in output will still be fairly large. The pressures on the agricultural resource base will continue to increase through further intensification. From an environmental perspective, the main challenge facing agricultural development in OECD countries will be to reduce the negative environmental impacts of continued agricultural intensification. Priorities will include reducing run-off of agrochemicals used for fertilisers and pesticides, reducing pesticide toxicity, improving the management of livestock manure, addressing concerns about genetically modified crops, improving soil management techniques, and developing practices that are more compatible with habitat protection.

Technological development and diffusion

While there is still scope for developing technologies and management practices to increase agricultural productivity, the rates of increased efficiency and yields have been decreasing in recent years.[2] The development of such new technologies is slowing, and there are indications that uptake of the existing environmentally sustainable technologies has been shallow and slow. Stronger incentives are needed to encourage a more wide-spread application of existing agricultural technologies. Appropriate measures include policies that discourage excess use of inputs, for example through ensuring full-cost recovery for irrigation water use or by levying charges on the use of polluting pesticides, surplus nutrient loading to soils, etc. Policies can also be used that directly encourage the adoption of the desired technologies and management practices, such as the use of GIS/GPS systems as tools for precision farming. The benefits of a more wide-spread adoption of available technologies could be significant: a recent study of the International Water Management Institute found that about 50% of the increase in world water demand to the year 2025 could be met by increasing irrigation effectiveness (Seckler *et al.*, 1998).

Although considerable progress in input efficiency and productivity can be achieved through the wider adoption of existing technologies and farm management practices, more significant increases are likely to occur only through the development of new biotechnologies (*e.g.* GMOs), particularly those which require less chemical inputs or water, increase crop growth, and are resistant to adverse environmental conditions. As agreed in the Biosafety Protocol, the development and application of such technologies by Parties to the Protocol should take into account precautionary approaches. The extent of their future use will depend on the analysis of their potential health and environmental effects, and consumer acceptance of their use.

2. While average yields of the main arable crops continue to augment, increases were significantly lower for maize, rice and wheat in the last decade than they had been in the previous two (Alexandratos, 2001).

Regulatory instruments

The agro-food sector is already subject to various environmental regulations, constraints and standards, but more will need to be developed or applied to address emerging concerns. In particular, these will need to concentrate on the effects of agrochemicals (pesticides, fertilisers) on the environment and human health, food safety, the control of particular site-specific activities, and water quality standards. Several OECD countries are already defining and adjusting regulations to control the size and intensity of agricultural operations. Animal waste storage, handling and spreading, for example, is controlled in several countries through regulations that set limits on herd size, storage capacity and spreading times. Most OECD countries have also designed more restrictive pesticide registration and use regulations, while at the same time trying to speed the spread of environmentally superior technologies. Further analysis of the effects of farm chemicals on the environment and human health is needed, and should be accompanied by agreements on the use and handling of farm chemicals that are damaging to the environment.

Economic instruments

Subsidies to agriculture in OECD countries are very large, and the majority are tied to potentially environmentally damaging production or inputs.

Although agricultural support remains high in OECD countries, a number of steps have been taken to move the sector closer to market conditions, or to reform agricultural policies in ways that will reduce some of the environmental damage caused by agriculture. While support reforms have been implemented in many OECD countries, their effects have been limited and most of the support continues to be tied to production levels. Irrigation water use also remains heavily subsidised in OECD countries (see Chapter 8). Further moves towards support that is de-coupled from potentially damaging activities could significantly reduce negative environmental impacts of agriculture and in some cases improve economic efficiency. This might be done by increasing the proportion of agricultural support that is based on land area (rather than on crop production or headage) or, even better, that is provided through direct income payments. Furthermore, increasing the share of support that is provided through agri-environmental measures – which provide payments or rewards to farmers for undertaking specified environmental activities, such as land set-aside schemes, tree break planting, soil protection or habitat conservation schemes – would also reduce the negative environmental effects of agricultural subsidies. Although agri-environmental measures make up a small percentage of OECD agricultural subsidies, they have already led to the removal of the environmentally most sensitive or ecologically most valuable lands from production in some countries (OECD, 1999*a*). A modelling simulation comparing the effects of removing agricultural subsidies in OECD regions to 2020 found significant environmental benefits from such a policy compared with the Reference Scenario (see Table 7.2). The reductions in irrigation water use and nitrogen loading from agriculture are expected to be particularly significant, especially in the regions of Western Europe, Japan & Korea and Central & Eastern Europe.

Table 7.2. **Environmental effects of agricultural subsidy removal and a tax on agrochemical use in OECD regions**

(% change from Reference Scenario in 2020 for OECD regions)

	Irrigation water use	Agricultural methane emissions	Agricultural nitrogen loading
Agricultural subsidies removal[1]	–10%	–1%	–7%
Tax on agrochemicals use[2]	–2%	–1%	–21%

1. This policy simulation explores the impacts of eliminating all subsidies on agricultural products used as inputs of final demand, all agricultural producer and export subsidies, all subsidies on inputs to agriculture, and market price support to agriculture in OECD regions. A proxy was used to simulate the removal of OECD market price support, combining a tax on agricultural production equal to market price support estimates with an equivalent subsidy to household consumption of agricultural products.
2. This policy simulation explores the impacts of applying an *ad valorem* tax increasing 2 percentage points each year (*i.e.* the tax rate reaches 50% by 2020) on the use of all chemical inputs to OECD agricultural production.

Sources: Reference Scenario and Policy Simulations.

While efforts have been made in a number of OECD countries to more broadly apply the Polluter Pays Principle (PPP) to agriculture and internalise the external costs of agricultural production, in most countries charges for agricultural pollution emissions are not applied. Partly this reflects difficulties in monitoring the emissions from individual farms; the individual source of agricultural pollution is rarely identifiable because it is often diffuse pollution from agricultural run-off of pesticide and fertiliser residues.[3] Such charges are more easily applied, however, for animal waste (*i.e.* manure) or when they are based on the agrochemicals used as a proxy for the resulting pollution. Thus, taxes and charges on fertilisers and pesticides, and caps on animal waste storage, have been applied recently in some OECD countries, although their application is limited by fears of a loss of competitiveness of domestic agricultural products on the international market and low responsiveness of farmers to such charges (*i.e.* low price elasticities). A policy simulation which examined the application of an *ad valorem* tax on chemical input use to OECD agricultural production found significant reductions in nitrogen loading from agriculture compared with the Reference Scenario (Table 7.2), and similar reductions in run-off of pesticide residues could be expected.

Voluntary agreements

Farmers, together with other landholders, have begun to form community-based, voluntary associations (*e.g.* landcare groups, water user associations, conservation clubs, etc.) to improve the environment in several OECD countries (OECD, 1998*b*). Governments are increasingly finding that for many environmental problems, especially those for which a small group of people can be held collectively responsible (*e.g.* salinisation or excessive pesticide use in a given area), implementing policies through groups can be more effective and less costly than those aimed at individuals. In most OECD countries, governments have so far been observers or facilitators in the processes of developing these associations and their programmes, rather than participants. It is only in a limited number of cases that they have taken a more active role in developing their agricultural policy actions to complement the voluntary programmes. To ensure that the best use is made of such programmes, governments need to develop policies that encourage and complement voluntary programmes, and work with the associations to develop targets and other regulatory and economic policies that can act as a framework within which the voluntary programmes can function.

Information and other instruments

Eco-labels and agricultural certification schemes are increasingly used in OECD countries to inform consumer food choices.

Certification and eco-labelling schemes are increasingly used in OECD countries for indicating agricultural products that have been developed following set criteria above and beyond those required by national regulations, for example environmental and health criteria (*e.g.* organic produce) or animal welfare criteria (*e.g.* eggs from free ranging chickens). Such schemes in general aim to improve labelling of the nutritional value and origin of food products. They tend to be developed by agricultural producer associations in response to growing concerns by consumers, but require some independent body to verify that the criteria are met and guarantee the certifications. Essentially, the schemes create a "market" for the additional environmental and health qualities associated with certain production processes, allowing concerned consumers to pay above market prices for these products. Consumers are increasingly also demanding the labelling of agricultural products made from genetically modified crops. With proper monitoring and independent certification, such schemes can let consumers directly influence the extent of sustainable agricultural activities in place, but care must be taken that they do not develop into trade barriers (*e.g.* by using an eco-label as a requirement for importation). For these schemes to be successful, co-operation between agricultural trading partners is needed, with regional or even internationally agreed certification schemes likely to be most successful.

3. It is often difficult to link any one farmer's fertilisation techniques, for example, to the nitrate contamination of local ground or surface waters. As a result, allocating the costs of such pollution to individual farmers in a way that efficiently and fairly internalises the environmental cost is often impossible.

As indicated above, large inefficiencies in agricultural production systems continue to exist, even in OECD countries. In many cases there are alternatives to current practices that could minimise environmentally damaging impacts without significantly increasing management costs to the farmer. Many OECD countries employ active farmer education and extension services to help disseminate the relevant information. Such programmes include, for example, individual consultations to establish better-targeted chemical and water inputs to farm processes, workshops or seminars on practices to reduce methane emissions from manure management systems, and longer-term educational and research programmes.

REFERENCES

Alexandratos, N. (2001), "Agriculture and the Environment", background document for the OECD *Environmental Outlook*, OECD, Paris.

CAST (Council for Agricultural Science and Technology) (2000), "Storing Carbon in Agricultural Soils to Help Mitigate Global Warming", Issue Paper No. 14, April, CAST, Ames, Iowa.

FAO (Food and Agriculture Organisation of the United Nations) (1999), *The State of Food Insecurity in the World* – 1999, FAO, Rome.

de Haen, H., N. Alexandratos, and J. Bruinsma (1998) "Prospects for the World Food Situation on the Threshold of the 21st Century", in OECD, *Food for the Future*, OECD, Paris.

McMichael, P. (editor) (1994), *The Global Restructuring of Agro-Food Systems*, Cornell University Press, New York.

OECD (1995), *Technological Change and Structural Adjustment in OECD Agriculture*, OECD, Paris.

OECD (1998*a*), *Water Management: Performance and Challenges in OECD Countries*, OECD, Paris.

OECD (1998*b*), *Co-operative Approaches to Sustainable Agriculture*, OECD, Paris.

OECD (1999*a*), *Agricultural Policies in OECD Countries: Monitoring and Evaluation*, OECD, Paris.

OECD (1999*b*), *OECD Environmental Data: Compendium* 1999, OECD, Paris.

OECD (2000), *Environmental Indicators for Agriculture: Methods and Results Executive Summary*, OECD, Paris.

OECD (2001), *Adoption of Technologies for Sustainable Farming Systems, The Wageningen Workshop*, OECD, Paris.

Seckler, D., U. Amarasinghe, D. Molden, R. de Silva, and R. Barker (1998), *World Water Demand and Supply*, 1990 *to* 2025: *Scenarios and Issues*, Research Report 19. Colombo, Sri Lanka: International Water Management Institute.

8
Freshwater

KEY SIGNALS

- There is abundant water worldwide, but significant scarcities exist on a local level. Some 25 developing countries (almost one-quarter of a billion people) are expected to suffer high water stress by the year 2020.
- Increased local water scarcities will lead to competition for resources within countries (between sectors or regions) and between countries, which could potentially lead to security threats. The increasing pollution of some water bodies, particularly groundwater reserves, will further restrict available supplies.
- Many OECD countries have reduced per capita water consumption in recent years, but the scale effects of population growth have led to increases in total water use. While these trends are expected to continue, slowing population growth will begin to stabilise total OECD water abstractions by 2020.
- Many of the worst polluted surface water bodies have been cleaned up in OECD countries, and point-source pollution from industrial and municipal systems has been reduced. However, many OECD water bodies still do not meet basic water quality standards.
- Policies are particularly needed to ensure full cost recovery for water services to agriculture and to address diffuse pollution and the contamination of groundwater aquifers. Many OECD countries need to upgrade their water supply and treatment infrastructure, which may require substantial financing.

8.1. Introduction

Freshwater resources are essential for human health, economic productivity, and social development. Freshwater is a recyclable but finite resource: with careful use and treatment, it can be managed in a sustainable manner. It is currently abundant on a global scale, but scarce in a number of countries or regions. While the water resources of the Earth are constantly recycled by the hydrological cycle, available freshwater resources for human or environmental use are declining as many water bodies become contaminated with pollutants. As a result, local and regional incidences of water scarcity are likely to increase over the coming decades.

Agriculture is responsible for the majority of global water withdrawals, although industrial water use in OECD countries now surpasses agricultural use. The wider application of existing water saving technologies and practices could significantly increase the efficiency of water use, but low water prices in most countries provide poor incentives for adopting these technologies and techniques. Despite increases in the connection of OECD households to sewerage treatment systems and greater regulation of industrial emissions, ambient water quality standards are still not met for many water bodies. Non-point source pollution, such as that from agricultural run-off of farm chemicals, is increasing in OECD countries and has particularly severe impacts on groundwater reserves.

8.2. Pressures on freshwater resources

Total freshwater abstractions

Global water withdrawals have increased significantly over the last few decades, in part because of increasing population pressures, but also because of significant increases in per capita water use worldwide. Over the last

50 years there has been a four-fold increase in global water withdrawals, while the total world population roughly doubled over the same period. According to the Reference Scenario, future world water withdrawals are projected to increase by 31% between 1995-2020.

Currently, there are significant regional differences in the use of freshwater. East Asia, Latin America, Africa & the Rest of the World utilise approximately one-third the amount of water per person that is used on average in OECD countries, and almost one-fifth of what is used in North America. There are also significant variations in per capita water use between the OECD regions, with countries such as Denmark, Luxembourg and the UK using as little as 180 m^3/capita per annum or less, while the US and Canada consume almost ten times this amount (OECD, 1999*a*). Central & Eastern Europe is the OECD region that is projected to realise the largest growth in water withdrawals to 2020, largely due to expected increases in per capita water use combined with continued population growth (Figure 8.1).

Overall, per capita water consumption has decreased in OECD regions by almost 11% since 1980, with just over half of all OECD countries realising a net decrease in per capita water use (OECD, 1998*a*). This indicates an encouraging trend in de-coupling water consumption from economic growth. However, while per capita water abstractions have been declining in many countries, the effect of net growth in population levels has resulted in increasing total water abstractions overall in most OECD countries. Thus, only nine OECD countries – primarily in Europe – reduced total water abstractions between 1980 and 1997 (OECD, 1999*a*).

Water abstractions in OECD countries may increase by 10% to 2020, while worldwide they may rise by over 30%.

Under the Reference Scenario, per capita water withdrawals are projected to remain relatively stable overall in the OECD area from 1995 to 2020, with some decreases experienced in North America and relatively large increases taking place in Central & Eastern Europe. The regional variations in the growth of per capita water use

Figure 8.1. **Water withdrawals per capita, 1980-2020**

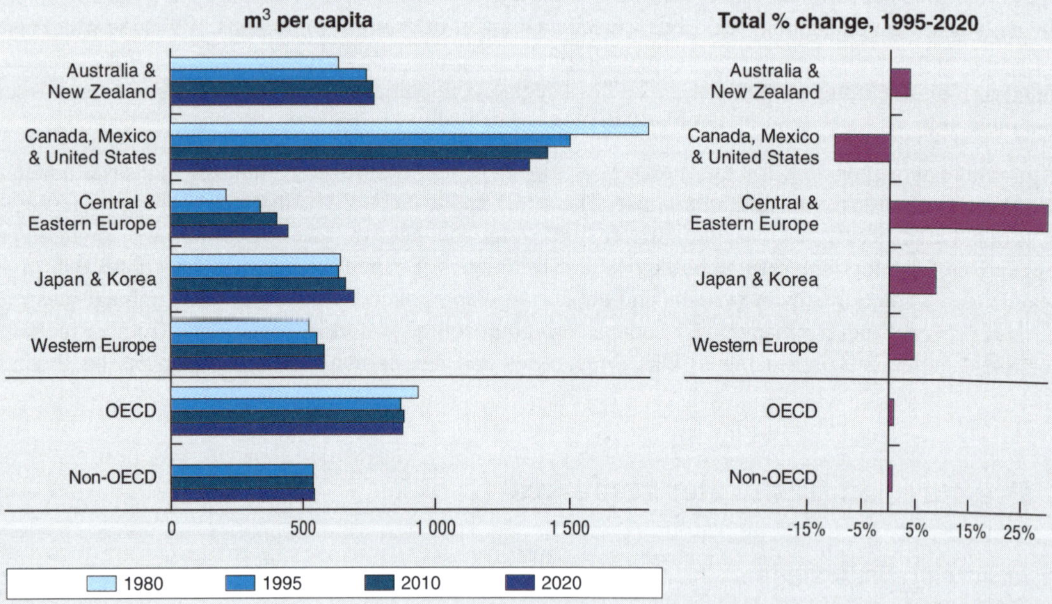

Sources: OECD (1999*a*) and Reference Scenario.

will to some extent balance the current disparities in use between the different OECD regions. Water use per person in non-OECD regions is also expected to stay relatively stable, with per capita use remaining far below the average for OECD regions.

Sectoral water demands

As total water abstractions increase, competition has been growing in most countries for agricultural, municipal, industrial, and environmental uses of limited freshwater resources. Worldwide, agricultural use is responsible for the largest proportion of freshwater withdrawals (69%), followed by industry (23%), and domestic use (8%). In low-income countries, agricultural water use makes up a much higher proportion of total use, while in OECD countries it is now surpassed by industrial use. Over the next 20 years, agriculture and industry are expected to lead to the largest absolute increases in world water abstractions, although industrial water use will grow quickest (see Figure 8.2).

Figure 8.2. **Water withdrawals by sector, 1995-2020**

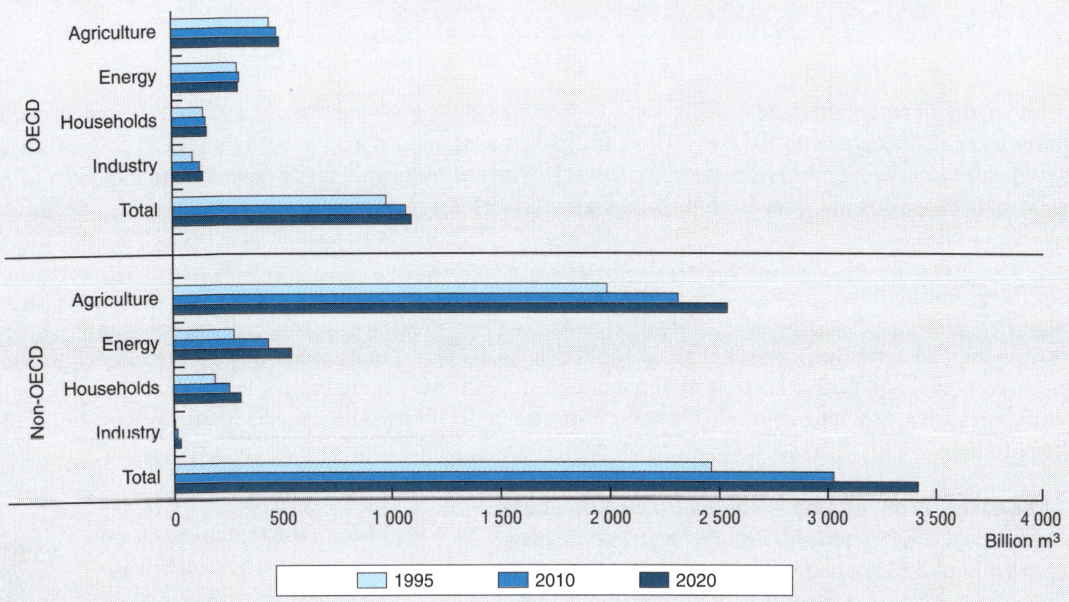

Source: Reference Scenario.

Freshwater abstractions for industry and energy use have declined by approximately 12% in OECD countries overall since 1980, while irrigation water and public supply abstractions have increased marginally (OECD, 1999*a*). To a large extent, these trends reflect the significant water efficiency improvements that have been realised in the industrial sector in recent years in response to increased water prices (see Box 8.1), as well as the increasing pressures on municipal systems from continuing population growth and domestic demand for luxury water uses. A number of OECD countries have experienced a decline in water use per household in recent decades, most likely reflecting the wider adoption of volume-based water charges that provide incentives for households to minimise water use (OECD, 1999*b*). Worldwide municipal water use is likely to continue to grow, however, particularly in regions experiencing significant urban expansion.

The relative stabilisation of irrigation water abstractions that has been occurring in OECD countries, and is expected to continue, indicates that irrigation expansion may have peaked in many industrial countries. Worldwide agricultural demand for water is projected to increase substantially over the next few decades, as much of the additional food that will be needed to feed the world population is expected to come from irrigated land (see Chapter 7). Agriculture is likely to remain the primary abstractor of freshwater in the near future, but the Reference Scenario indicates that industry will be the fastest growing water users driven largely by rapid industrialisation in many non-OECD countries. While the declines in industrial abstractions experienced in OECD countries in recent years are primarily the result of increased water use efficiency by industries, they may also reflect a shift of some of the water-intensive industries to non-OECD countries.

Box 8.1. **Increased water use efficiency in OECD industries**

OECD industries have significantly increased the efficiency with which they use water, reducing total industry-related water use by 9% since 1980, and increasing water recycling and reuse. To a large extent, these developments have been in response to greater industrial water charges across OECD countries and stricter ambient water quality standards (OECD, 1999*b*). For instance, the pulp and paper industry has traditionally required large amounts of energy and water (see Chapter 18). Through a wider use of closed-loop water systems and recycling, some mills have reduced water consumption from 80 m^3 in 1970 to 10-15 m^3 today (CEPI, 1995). Since energy needed for drying purposes is highly correlated to water consumption, this has also led to substantial reductions in energy demand.

An important emerging trend in many OECD countries is the growing use of freshwater resources for cooling purposes in electricity production. Over 50% of freshwater abstractions in at least eight OECD countries are attributable to electrical cooling purposes (OECD, 1999*a*). Water withdrawals for energy use are expected to increase by almost 100% worldwide from 1995 to 2020 under the Reference Scenario.

Freshwater pollution

Many human activities have a negative impact on freshwater quality, endangering human health and freshwater ecosystems, and further restricting the amount of freshwater available for human use. Largely as a result of public pressures, stricter controls have been imposed in most OECD countries on the main sources of freshwater pollution.

Most OECD households are connected to basic sewage treatment systems, and more advanced wastewater treatment processes are increasingly used.

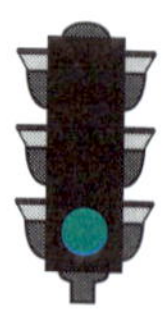

The main factor driving municipal water pollution abatement in OECD countries has been the increasing number of households connected to basic sewage treatment facilities and better treatment of wastewater (*i.e.* the use of biological or other advanced treatment technologies). The total share of the population connected to public wastewater treatment plants in OECD countries rose from 51% in 1980 to almost 60% in the mid-1990s, although connection rates vary among countries from a low of under 10% to a high of almost 100% (OECD, 1997*a*). Those households that are not connected to public systems increasingly utilise local systems to treat their sewage waste. The situation is less encouraging in developing countries. Only an estimated 29% of the populations in low-income countries are connected to sanitation systems (World Bank, 1999). With continuing expansion of secondary and tertiary wastewater treatment systems in OECD regions, OECD household contributions to biochemical oxygen demand (BOD) in freshwater systems are expected to increase by only 15% to 2020 (see Figure 8.3). BOD loading from households in non-OECD countries, however, is expected to increase by over twice as much.

Agriculture is an important contributor to water quality problems, particularly nitrate pollution in surface and groundwater, phosphorous levels in surface water, contamination with pesticides, and the harmful effects of soil sediment and mineral salts (see Chapter 7). In a number of OECD countries, agriculture accounts for more than

40% of all sources of nitrogen emissions and over 30% of phosphorous emissions into surface waters (OECD, 2000). Although the trend in nutrient surplus from agriculture through fertiliser use and livestock waste is declining in most OECD countries, the contribution of agriculture to the overall level of nutrient contamination of water is growing. Projections indicate that while agriculture will remain by far the dominant source of BOD and nitrogen loading to waterways both in OECD regions and worldwide, emissions from the sector will increase at a slower rate in OECD regions to 2020 than in the past (Figure 8.3). While the use of pesticides has begun to decrease in many OECD countries in the last decade, the long time lag between their use and their detection in groundwater means that the situation could deteriorate before it starts to improve (see Chapter 7). Recent trends in soil erosion losses in OECD countries indicate that soil sediment deposition in waterways continues to be a serious problem in many countries, but in general is probably declining.

Figure 8.3. **Emissions of water pollutants by sector, 1995-2020**

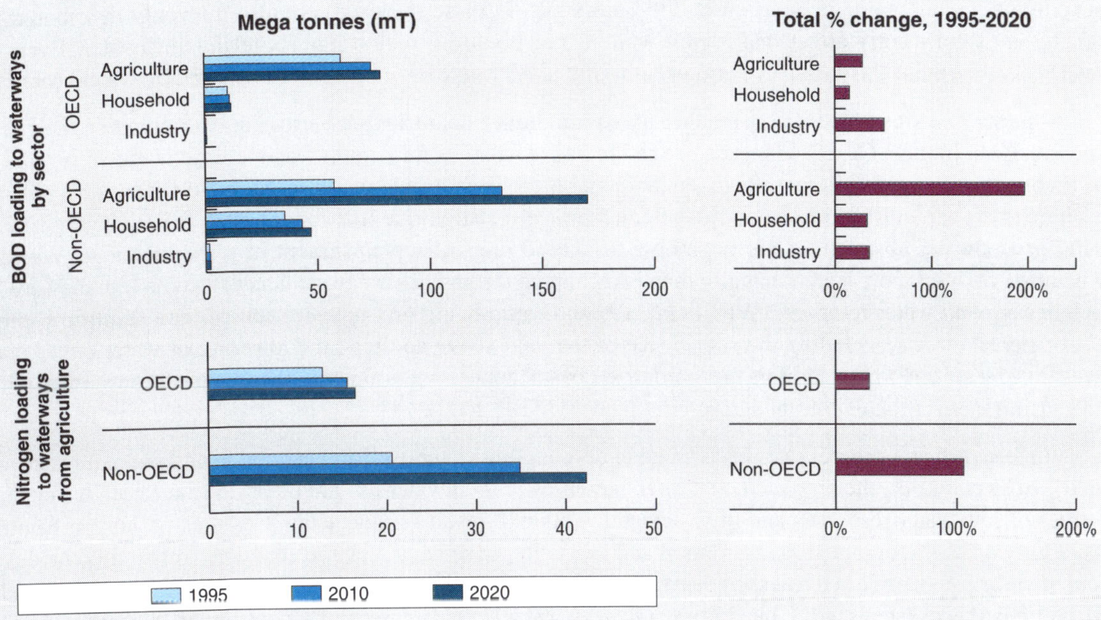

Source: Reference Scenario.

Although a relatively small contributor to BOD loading to waterways, industry remains one of the largest polluters of water resources in OECD countries through other pollutants, and worldwide it is expected that there may be a four-fold increase in industrial pollution to water courses by 2025 (OECD, 1997*b;* SEI, 1997). Industrial impacts on the aquatic environment vary considerably, depending on the types of substances released and their quantities. Industrial sites can be particularly important point sources of pollution to watercourses because of their often large operational scales, even if the individual pollutants are not dangerous in smaller quantities. Some industrial processes (*e.g.* power generation) can also result in environmental degradation through thermal pollution, which reduces dissolved oxygen while accelerating oxygen-demanding biochemical processes. In response to increasing regulation of industrial emissions, OECD industries have had to develop better and safer systems of wastewater disposal, often utilising their own in-house purification systems before releasing wastewater back to the environment (if it is of sufficient quality) or to a wastewater treatment plant. The use of technologies to reduce air emissions has also helped to reduce industrial water pollution, such as the use of sulphur scrubbers in coal burning plants which reduce the acidification of lakes and other surface water bodies.

8.3. The state of freshwater resources

Freshwater scarcity

By 2020, over one-quarter of a billion people worldwide are expected to be living under high water stress.

Water scarcity can have direct negative effects on human health, the economy, and the environment. Over the last 50 years there have been dramatic reductions in the global per capita availability of freshwater, from 17 000 m^3 per capita per annum in 1950 to 7 300 m^3 in 1995. This is largely a result of increasing population pressures, but has also been influenced by the decline in availability of uncontaminated freshwater. Current trends indicate that the level of per capita available water resources is likely to decline even further in the near future, as most countries continue to expand their populations and pollute available water sources. As a result, by 2020 over one-quarter of a billion people are expected to be living under high water stress, representing an increase of 75% from 1995 levels.[1] Because freshwater resources are distributed very unevenly within and among countries, and the pressures on these resources are also unevenly distributed, water scarcity can significantly affect one region while a neighbouring region has abundant freshwater. Even within OECD countries there exist extensive arid or semi-arid areas where development is restricted by water scarcity.

As human use of surface waters reaches maximum limits, countries are increasingly drawing on their groundwater aquifers. In most OECD countries, irrigation has reached its maximum feasible limit as far as surface water use is concerned, and abstractions are taken more and more from groundwater sources. However, most groundwater aquifers are replenished only slowly, with an average recharge rate that ranges from 0.1-0.3% per year. As a result, groundwater abstractions are beginning to exceed rates of replenishment in some locations. Worldwide, 17 countries extract more water annually than is recharged through their natural water systems (*i.e.* they are "mining" their ground water reserves) (WRI *et al.*, 1999). Overdrawing groundwater sources can result in significant environmental effects, including the subsidence of the land above aquifers, the lowering of water tables, and the intrusion of seawater where aquifers are sited near coastal zones, contaminating the freshwater resources with salt and causing salinisation of coastal agricultural lands (UNEP, 2000).

While significant progress has been made in de-coupling water use from economic and population growth in some OECD countries, the projected continued increases in total water use are likely to exacerbate water scarcity in areas of low water resources and high demand. This will result in increasing pressures on human health (the availability of water for human consumption and sanitation in some areas), the economy (restricting the development of various economic activities, *e.g.* agriculture), and the environment (increased soil salinisation, degradation of water-based ecosystems, etc.). In addition, growing water scarcity has already contributed to conflict within and between countries for scarce resources, and these conflicts may intensify in the future (see Box 8.2).

Freshwater quality

In addition to further limiting the water available for human use, the pollution of water bodies also degrades ecosystems and impairs their ability to provide valuable ecosystem services. The discharge of inadequately treated sewage in large cities causes severe deoxygenation and possibly ammonia toxicity, while nitrate pollution can stimulate rapid algal growth in waterways, leading to eutrophication in both inland waters and the sea (UNEP, 1997; UNEP, 2000). Many of the algae produce toxins that, once ingested by molluscs and fish, either kill or accumulate in them, endangering their predators in turn (WRI *et al.*, 1994). Industrial wastes can lead to contamination with heavy metals (lead, mercury, arsenic and cadmium) and persistent organic compounds. Air borne pollution can also be detrimental to water quality. For example, hundreds of lakes in Scandinavia are still suffering from acidification that was largely created by sulphur and nitrogen emissions from fossil fuel combus-

1. Water stress is considered high when the ratio of withdrawals (minus wastewater returns) to renewable resources exceeds 0.4.

tion. The situation is even more serious in developing countries. Trends there indicate that there is accelerating contamination of available water supplies, especially in areas undergoing rapid urbanisation.

Box 8.2. Water conflicts and environmental security

Many rivers, lakes and underground water aquifers cross national boundaries, with the potential for (geo-political) conflicts to arise where there is transboundary water pollution or scarcity of water resources. Over the last few decades, OECD countries have made considerable progress in resolving transboundary water issues: bilateral, regional or multilateral agreements have now been reached for most such waters. Since 1974, OECD countries have adopted OECD general principles related to transfrontier water pollution and integrated water management. OECD countries have agreed on and are implementing numerous bilateral or multilateral plans and strategies (*e.g.* for the Great Lakes, Lake Geneva, the Rhine). Several large basins (*e.g.* Rhine, Danube) are now managed in the framework of international agreements, with a basin committee and permanent secretariat. However, the issue of equitable apportionment of waters between riparian states of a given river is still not settled, though progress in this area has been achieved for selected rivers. Outside the OECD regions, disputes over the management of transfrontier water may well increase in the future as water pollution and regional scarcity increase unless management plans can be agreed.

Source: OECD (1998*b*).

The pollution of OECD groundwater resources – particularly from agro-chemical run-off – is expected to worsen to 2020.

In OECD countries water pollution continues to be a problem. Despite major efforts to clean up many of the worst polluted water bodies over the last few decades, few OECD countries satisfactorily meet the baseline quality standard for inland waters (*i.e.* suitable for fishing and swimming). Most OECD countries have been experiencing particular difficulties with the protection of groundwater quality, especially from non-point source pollution such as agricultural run-off or arsenic poisoning from mining. Thus, nitrate concentrations (most commonly linked to fertiliser use) in excess of WHO drinking water guidelines are now widespread in European and some North American aquifers. Available evidence suggests that there is a trend towards a worsening of aquifer water quality in OECD regions. Once groundwater sources are contaminated, they can be very difficult to clean up because the rate of flow is usually very slow and purification measures are often costly. While many of the regulations and policies currently in place to address surface water pollution are

Box 8.3. Institutional frameworks for water management

In many countries, responsibility for water supply and sewage treatment is fragmented among numerous federal, state, and local government bodies, all of which apply different methods and standards. The lack of an integrated regulatory approach threatens water supply systems in many areas. A number of countries have been developing new water management frameworks to address these concerns, including better integration of surface and groundwater management, addressing both supply and demand-side pressures, and the adoption of river-basin management approaches.

There has also been a general trend in OECD countries away from government as the "provider" of water services and toward government being the "regulator" of these services. In many cases, this has meant increasing autonomy of water utilities and, in some countries, this has been accompanied by privatisation of water service systems. The majority of countries have opted for the "concession" model instead, whereby the private sector participates in managing some services, but the public sector retains ownership of the system (OECD, 1999*b*).

likely to improve surface water quality in OECD regions, the lack of comprehensive plans to manage groundwater resources and address non-point source pollution will probably result in continuing contamination of groundwater aquifers over the next few decades. As groundwater aquifers are a major source of drinking water and water for agricultural and industrial purposes, it can be expected that this will further contribute to water scarcity problems and lead to higher economic costs for water purification in many OECD countries.

8.4. Policy options and their potential effects

OECD countries are increasingly managing their water resources on a river basin or catchment basis.

International concern about the potential for a world water crisis during the next century is high, as was confirmed at the Second World Water Forum held in the Hague in March 2000. For OECD countries, however, water scarcity is likely to be a problem only in selected sub-national regions over the next 20 years, with most OECD countries continuing to have generally low water stress. In order to ensure that water use stays at sustainable levels, policies are needed to ensure appropriate cost recovery and management of water demands. In the regions where water scarcity is an issue, co-operative approaches for implementing the right incentives to increase water use efficiency and reduce demand will be necessary. In terms of water pollution, OECD countries will need to focus more on diffuse sources of pollution (*e.g.* agricultural run-off and transport-related depositions) and, in particular, take urgent measures to stop the degradation of groundwater resources. Conflicts over water use may arise in areas where transboundary water resources are scarce, or where water pollution crosses national boundaries. The management of water resources on a regional level, particularly reflecting the natural catchment area, (*i.e.* river-basin management approaches) can help to facilitate the management process and ensure ecological health. Such systems are already widely used within a number of OECD countries, and should also be applied where possible to transboundary river basin catchments.

Technological development and diffusion

Most of the recent reductions in water use in OECD countries can be attributed to efficiency improvements in the distribution and use of water resources. Industrial users have already made significant progress in the development and adoption of technologies and processes that are more water efficient. This has largely been driven by their own cost concerns where water is a major input and where water charges have been increasing. While industrial water users have been quick to develop and adopt new water efficient technologies, significant improvements in irrigation use efficiency can still be made in OECD countries. Recent agricultural developments, such as drip-irrigation systems, provide the potential to increase irrigation efficiency considerably, but their diffusion has been limited (see Chapter 7). Drip irrigation is not suitable for all crops or regions, but to some extent the wider adoption of such technologies has been hampered by their high relative costs compared with the rather low, subsidised price of irrigation water in most countries. As a result, the use of such technologies brings only relatively small benefits to farmers. To better encourage the uptake of more efficient systems, governments should both increase research and development efforts into more cost-effective systems, and take measures to recover the full costs of water provision (see economic instruments below), thus increasing incentives for agricultural water users to reduce their abstractions.

New technologies for wastewater treatment and drinking water purification have also been developed, and are increasingly used in OECD countries. A large majority of households in OECD countries are now connected to public or independent sewerage treatment systems, although for most of these countries under 50% of domestic sewerage receives tertiary treatment (advanced chemical treatment processes). In the future, greater use of tertiary wastewater treatment will be needed to reduce pollution from domestic wastewater. Further research into the effects of some industrial and agricultural chemicals (*e.g.* persistent organic pollutants and new pathogens) on the environment and human health is needed, as well as the development of cost effective technologies for preventing their emission to, or allowing their removal from water bodies.

Regulatory instruments

OECD countries have designed and implemented regulatory frameworks that are fairly successful in addressing some of the main sources of water pollution. However, in order to raise ambient surface water quality in many areas and meet basic water quality standards, further efforts will be required, particularly in terms of the enforcement of existing regulations and the greater integration of permitting processes. Many OECD countries have begun to realise that monitoring the sources of pollution entering groundwater resources is not sufficient, particularly where many non-point sources are involved. In the future, more emphasis will be needed on integrated pollution management and ambient water quality plans. Those described in the new European Union Water Framework Directive will require EU countries to both monitor and address sources of pollution to groundwater reserves and also to directly monitor the quality of groundwater in aquifers.

In the coming decades, significant efforts and capital expenditures will continue to be required in OECD countries to reduce water pollution. In particular, these will include decreasing the amounts of untreated sewage, agrochemical and other multi-source pollutants that are released into waterways. Regulatory instruments will generally be most effective in addressing industrial point-source pollution and municipal wastewater, while diffuse agricultural run-off and non-point source pollution may require other policy approaches. For the more toxic pollutants, such as certain agricultural chemicals, restrictions on their use can nevertheless be a very direct and effective measure to achieve the desired reductions in their release to the environment.

The use of regulations and legal requirements for addressing water scarcity problems is likely to be more limited. Some countries have started placing upper limits on the total amount of water withdrawals in regions suffering from water scarcity, and this practice is likely to expand as water scarcity increases in some of these regions. Once limits are placed, quotas can be allocated among different types of uses, and in some cases these might be tradable between individual users.

Economic instruments

In recent years, most OECD countries have been striving to raise the price of water and decrease subsidies to water use in order to reflect the full costs of water service provision (see Box 8.4). This has largely been the result of pressures to finance essential water infrastructure repairs, extensions, and operating costs, but also reflects a growing desire to manage water "demand" in the face of potential scarcity and to employ the Polluter Pays Principle in wastewater charging systems. Despite increased water prices, few OECD countries achieve full cost recovery of the operating and maintenance costs of water services provision, let alone any additional externalities.

Box 8.4. **Developments in OECD water prices**

Over the last decade, OECD countries have seen a general increase in domestic water supply and sewage treatment prices, with significant price increases occurring in some countries. Of 19 OECD countries with data available on trends in domestic water pricing, all but one experienced real per annum increases in water supply prices over the last decade, and five experienced average rates of real price increase of 6% or more per annum (OECD, 1999*c*).

Indications suggest that industrial water prices in OECD countries may also be increasing, either through water tariff charges where consumers draw water from public networks, or through direct abstraction charges. Water for irrigation use is associated with the lowest charges and the highest subsidies in most OECD countries. Currently, most agricultural water tariff structures are based on the surface area irrigated, and are charged either as a flat rate or according to crop type, and thus do not reflect the amount of water actually used. Some OECD countries apply no charges at all to irrigation water abstractions (OECD, 1999*d*).

Many countries have resisted applying full cost pricing to domestic water services because of social equity concerns. Analyses indicate that the wider use of household water meters and the use of innovative water tariff systems in some countries have allowed the simultaneous achievement of equity and full cost recovery objectives

(OECD, 1999*b*). Changes in the structure of water tariffs, particularly replacing flat fees for water services with consumption-based pricing, can also provide strong incentives to minimise water use. Evidence indicates that raising domestic water prices in a number of OECD countries has reduced water consumption levels, especially when price increases were coupled with moves towards volumetric or marginal cost pricing (OECD, 1999*c*). A key element in implementing these charging systems is the widespread introduction of domestic water meters to measure – and implement charges based on – actual water use.

Although water charges in OECD countries have been increasing, in most cases the full costs of providing water services are still not covered by water prices.

In most OECD countries, water for irrigation use is currently subject to the lowest charges and receives the highest subsidies. Although the share of water abstractions for agricultural use is declining in OECD countries, significant reform of water pricing policies will be required to ensure local sustainability of irrigation use in regions with water scarcity. Industry is probably the sector that is most responsive to higher water prices, largely because industrial users have a wider range of technological options that are cost-effective for them because of economies of scale. As municipal charges continue to increase and industrial users choose to directly abstract their own water "off-line" to avoid paying these fees, direct abstraction charges and limitations on total abstractions will be increasingly required.

Wastewater charges are calculated as a fixed percentage of water consumption (which is generally an accurate proxy for household, but not industrial, wastewater) in most OECD countries, but some are moving towards the direct measurement of wastewater produced and, for industrial users, graduating charges based on toxicity. A few OECD countries have also begun to implement charges for agricultural water pollution, based on a proxy of fertiliser use or nitrogen application. While such charging systems are only just developing, they may hold the greatest potential for tackling sources of diffuse pollution to water bodies. In many cases, however, reducing subsidies to agricultural or industrial water use (particularly for energy intensive industries, such as aluminium smelters), and applying pesticide or fertiliser taxes will require multilateral co-operation to address fears over a possible loss of competitiveness.

As mentioned above, limitations on total water use and the design of tradable permit schemes for water quotas are underway in a few OECD countries. While these are generally allocated between individual users in a single sector (*e.g.* agricultural users), there is also potential for using them to allocate resources among competing sectors (*e.g.* industry, agriculture and municipal demands). Such schemes have been implemented in some countries to manage an orderly shift from rural to urban water use through a reallocation of water use rights, water trading and even buying out farmers' water rights and redistributing them. Tradable permit schemes are also sometimes used for allocating emissions for allowed levels of specific industrial pollutants.

Information and other instruments

Providing information and educational support to assist water users in increasing water use efficiency and reducing pollution can be a powerful policy tool. Programmes already exist in some OECD countries to inform farmers of more efficient irrigation techniques, industrial users of process improvements, and households of measures they can take to reduce their water use. The latter include the development of water-efficiency labelling schemes for household appliances, and information on efficient lawn watering and gardening practices. In general, however, these instruments are most effective when coupled with water pricing tariffs, thus also providing an economic incentive for the targeted actors to reduce water use or pollution.

Voluntary agreements also exist in a number of OECD countries between the government and industrial or agricultural water users, and can be used to encourage water users and polluters to take pro-active steps towards reducing water abstractions and emissions to waterways. For industry, voluntary agreements are primarily used to regulate pollution to waterways, often based on the achievement of a given water quality target. If this target is not met, however, stronger policy actions (*e.g.* regulatory measures or fines) will need to be imposed.

Investment in infrastructure

In recent years, major efforts in many countries have also reduced water losses in municipal networks, which are now as low as 10-12% in some OECD countries, although they remain high in others (OECD, 1998*c*). Significant investments will nonetheless be required to maintain these low levels of loss (and achieve them in those countries with high losses), particularly where the pipe infrastructure is now reaching maturity. In quite a few OECD countries, the economic limit in terms of sewerage connections has now been reached, and in the future alternatives will have to be found for serving small, isolated communities whose connection to the main sewerage system is not economically feasible.

The implementation of full cost recovery charging systems for irrigation water use could help to generate funds for essential pipe repairs and renewals, as current pricing structures generally do not even cover operating costs, let alone capital costs, in most OECD countries. In order to maintain or improve the current levels of sewage treatment in OECD regions, many countries will need to invest heavily in expanding and renewing pipe infrastructure.

REFERENCES

CEPI (Confederation of European Paper Industries) (1995), *The European Paper Industry: Competitive and Sustainable*, CEPI, Brussels.

Mountford, H. (2001), "Freshwater", background document for the OECD *Environmental Outlook*, OECD, Paris.

OECD (1997*a*), OECD *Environmental Indicators* 1997, OECD, Paris.

OECD (1997*b*), "Water Subsidies and the Environment", OCDE/GD(97)220, OECD, Paris.

OECD (1998*a*), *Towards Sustainable Development: Environmental Indicators*, OECD, Paris.

OECD (1998*b*), *Water Management: Performance and Challenges in* OECD *Countries*, OECD, Paris.

OECD (1998*c*), *Water Consumption and Sustainable Water Resources Management*, OECD, Paris.

OECD (1999*a*), OECD *Environmental Data: Compendium* 1999, OECD, Paris.

OECD (1999*b*), *The Price of Water: Trends in* OECD *Countries*, OECD, Paris.

OECD (1999*c*), "Household Water Pricing in OECD Countries", ENV/EPOC/GEEI(98)12/FINAL, OECD, Paris.

OECD (1999*d*), "Agricultural Water Pricing in OECD Countries", ENV/EPOC/GEEI(98)11/FINAL, OECD, Paris.

OECD (2000), *Environmental Indicators for Agriculture: Methods and Results*, Executive Summary, OECD, Paris.

SEI (Stockholm Environment Institute) (1997), *Comprehensive Assessment of the Freshwater Resources of the World*, SEI, Stockholm.

UNEP (United Nations Environment Programme) (1997), *Global Environmental Outlook*, UNEP, Nairobi.

UNEP (2000), GEO 2000 *Overview*, UNEP, Nairobi.

WRI, UNEP, UNDP, and the World Bank (World Resources Institute, United Nations Environment Programme, United Nations Development Programme) (1994), *World Resources* 1994-95, Oxford University Press, New York.

WRI, UNEP, UNDP, and the World Bank (1999), *World Resources* 1998-99, Oxford University Press, New York.

9
Fisheries

KEY SIGNALS

- Global production from the fishery sector increased by just over 70% between 1980 and 1998, reaching 127 million tonnes in 1998. Future growth is expected to be slower, if not declining. At most, production may reach 160 million tonnes by 2020, with almost all of the increase coming from aquaculture production.
- Limiting overfishing in capture fisheries is a major challenge: 50% of fish stocks are fully exploited, 15% are over-fished, 7% are depleted, and 2% are recovering. Marine ecosystems are also under threat from pollution (primarily from land-based sources), excessive by-catch, and destructive fishing methods.
- The main national policy options to encourage sustainable fishing activities include the setting (and enforcement) of limits on total catch levels, types, and fishing seasons; rationalising fishing capacity (including reforming environmentally damaging subsidies to the fishing sector); and reducing degradation of marine environments.
- International co-operation will be essential for implementing these policies, particularly in dealing with migratory and straddling fish, regulating fishing in the high seas, and agreeing on multilateral reductions in subsidies to fishing that encourage overfishing or environmentally damaging activities.

9.1. Introduction

Fish represent a major component of the food supply for the world's population, accounting for almost one-fifth of all animal protein in the human diet (WRI, 1998). Fish are a renewable resource and, as such, can be harvested in a sustainable manner over the long term. However, current fishing levels and methods place significant pressures on world fisheries and marine ecosystems. The harvesting of fish stocks has moved successively from one fishery to another as yields from each reach their peak and then decline. These practices have resulted in an estimated 50% of the world fish stocks being fully exploited, 15% that are overfished, 7% which are depleted, and 2% that are recovering from over-fishing (FAO, 2000*a*). The remaining 25% of the main world fish stocks are still capable of producing higher yields.

In addition to affecting the stocks of commercially valuable fish, fishing activities often contribute to the depletion of non-target fish (through by-catch), marine pollution, and habitat destruction (*e.g.* through bottom trawling and dynamite fishing). Marine degradation is further exacerbated, and fish stocks negatively affected, by pollution and other ecosystem changes originating from non-fishery sector sources, such as agriculture run-off, water pollution, the deposition of chemicals from air pollution, coastal zone development, and global warming. The introduction of invasive species to fishery (*e.g.* through the release of ship ballast water primarily from commercial shipping activities) can also lead to serious repercussions for fishery ecosystems.

9.2. Developments in the fishery sector

Several factors determine developments in the fishery sector. The biological limitations on marine fish stocks are the ultimate constraint on fisheries sector development, although they can be counterbalanced to some extent by increases in fishing capacity, technological developments in harvesting, and the development of aquaculture production. While supply in the sector is limited by biological constraints, demand for fish and fish products by consumers continues to rise. This demand is influenced by human population-levels, eating habits, available disposable income, and fish prices.

Demand for fish products

Worldwide consumption of fish and aquatic products has been increasing in recent years as populations grow and as per capita consumption of fish increases both in OECD and developing countries. The main products include fish for food consumption and fish for fishmeal and fish oil (for use as livestock feed, for example). FAO projections indicate that by 2010 total demand for food fish will be about 110-120 million tonnes a year, compared with the current demand of 92.5 million tonnes, while demand for fishmeal and fish oil will remain stable at about 30 million tonnes of fish per annum (FAO 1998*a*).

Table 9.1. **Key fishery sector statistics and projections**

		1980	1998 *(or latest available year)*	2020 projected	Total change 1995-2020
Gross production (1995 US$ billion)	OECD	..	100	127	27%
	World	..	153	215	41%
Share of value added (%)	OECD	..	0.3%	0.2%	–22%
	World	..	0.3%	0.3%	–21%
Fish production (million tonnes)	OECD	31	32	35	–15%
	World	72	127	167	35%
Per capita fish consumption (kilogrammes)	OECD	23	27	21	–18%
	World	..	14	13	–12%

Sources: FAO (1997; 1998*a*; 2000*b*) and Reference Scenario.

Worldwide, per capita consumption of food fish increased from 11.4 kg per annum in 1980 to 15.3 kg in 1997 (FAO 1998*b*). In OECD countries, per capita consumption of fish for food increased by 18% between 1980 and 1997, to an average of 27 kg per capita, with North America and Australia & New Zealand showing particularly strong growth (OECD, 1999). In the OECD regions examined, most seem to have reached a peak in per capita fish consumption around 1990, with stabilising or marginally decreasing consumption levels since then. Per capita consumption of fish products in OECD regions is projected to decline to 2020 under the Reference Scenario, particularly in the regions of Japan & Korea and Western Europe, in part reflecting limitations on the availability of fish supply (see Table 9.1).

Fishery sector production

Total fish catch from world fisheries amounted to approximately 127 million tonnes in 1998, representing more than a 70% increase compared with 1980 levels (FAO, 2000*b*).[1] The majority (72%) of fish production currently comes from capture fisheries – the capture of fish in their natural environment – and the remainder from aquaculture farms. Most of the increase in production over the last 15 years has come from aquaculture, with the share of aquaculture in total world fisheries production increasing from 8% in 1984 to 27% in 1998 (FAO, 2000*b*). Aquaculture is expected to account for most of future growth in fish production, as world capture fishery production remains stable or even declines to 2020 (see Figure 9.1).

The FAO projects that total fish supplies will probably remain around the level of 125 million tonnes for the next 3-4 years, and then increase towards a total of 130-150 million tonnes by 2010 (FAO, 1998*a*). The share of

1. These figures include finfish, shellfish and marine mammals, but exclude aquatic plants (*e.g.* seaweed). Aquatic plants constitute a small component of aquaculture production.

value added in the economy from the fisheries sector is expected to decrease significantly to 2020, by over 20% in both OECD regions overall and worldwide, reflecting a decreasing role of the fishery sector in the economy (see Table 9.1). On the other hand, gross worldwide production from the fishery sector, in monetary terms, is projected to grow by 40% from 1995 to 2020 in the Reference Scenario. This growth in value compared with the relatively stable production levels may reflect increased focus on the production of high value aquaculture products and efforts in capture fisheries to focus on high-value species, such as tuna and squid.

Figure 9.1. **World fish production by region and production type, 1950-2020**

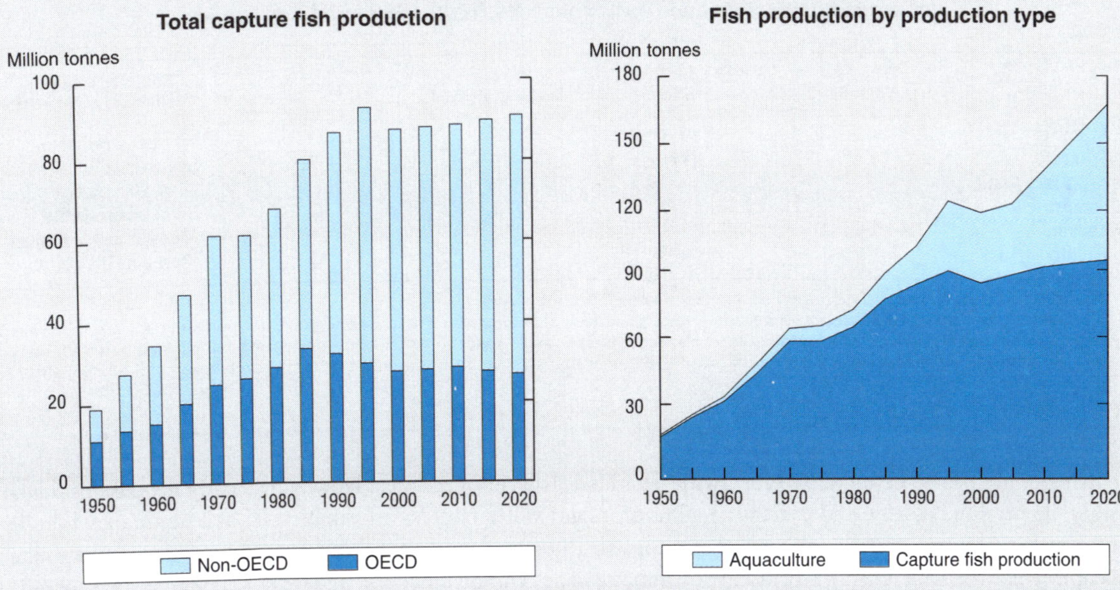

Sources: FAO (1997, 1998*a*, 2000*a*) and Reference Scenario.

Production from capture fisheries

While global production from capture fisheries grew significantly between 1950-1990, it now appears to be levelling off. Since the early 1990s, the average annual rate of growth in capture fisheries has been essentially zero, with a significant drop experienced between 1997-1998.[2] As a proportion of total capture fisheries production, the contribution of OECD countries has been declining steadily in recent decades, from about 42% in 1980 to 31% in 1998 (FAO, 2000*b*). To 2020, world capture fishery production is expected to remain fairly stable under the Reference Scenario.

Fish catch from capture fisheries is likely to remain stable, or possibly decrease, to 2020 because of poor management of fish stocks.

Among OECD countries, two-thirds of fish catch is harvested by Japan, the US and the EU. The declining share of fish catches by OECD countries in the world total can in part be explained by the status of fish stocks in the fisheries closest to OECD countries. Of the fishery sub-regions of primary importance for OECD countries listed in Table 9.2, the fish resources of four are estimated by the FAO to be overfished, one is fully fished, and the

2. The recent contraction in fish landings can be partly attributed to the El Niño phenomenon, which has led to a decline in stocks of small pelagic species off the west coats of South America (FAO, 1998*a*).

last is believed to be rapidly heading towards being fully fished or may already have reached this point in 1998. According to these fairly rough estimates, there is only limited scope for expanding harvests from these fisheries – with a potential increase of 6 million tonnes per annum of landings from the current 43 million tonnes. Furthermore, most of this increase would need to come from better management of currently over-exploited resources. Few of the marine fisheries of primary interest to OECD countries have significant potential for additional exploitation without improvements in the current fishery management regimes. Without appropriate policies to ensure these improvements, harvesting from capture fisheries by OECD countries is expected to decrease by almost 10% from 1995 to 2020 under the Reference Scenario.

Table 9.2. **Comparison between estimated potentials and average landings for marine capture fisheries of primary importance to OECD countries, 1990-1994 in million tonnes**

	Estimated potential	Year potential reached	Average landings 1990-94	Status*	Subjective degree of reliability of regression
NE Atlantic	12	1983	10	O	Less reliable
NW Atlantic	4	1971	3	O	Reasonably reliable
Med. & Black Sea	2	?	2	F	Unreliable
NE Pacific	4	1990	3	O	Less reliable
NW Pacific	26	1998	24	I	Reasonably reliable
SW Pacific	1	1991	1	O	Reasonably reliable

* Overfished, Increasing, Fully fished (based on date when rate of increase = zero).
The status of the fisheries indicated here reflects potential and actual landings for fishery sub-regions, not particular fish stocks. Some fish stocks may be currently harvested at sustainable levels, but fall within a sub-region for which other stocks have been fished past their sustainable yield, thus earning the sub-region a status of "overfished". The status of the fishery sub-regions is only a very rough indicator of fishery sustainability.
Source: FAO (1997).

Greater use is also being made worldwide of inland fisheries resources. However, mostly because of the large number, dispersion, variety, and dynamic nature of inland water bodies, information is lacking on exact landings and the state of these resources. Recorded landings in 1996 amounted to 7.6 million tonnes, with indications that this capture has been increasing worldwide by about 2% per annum since the mid-1980s (FAO, 1998*a*). Asia is by far the most important continent for inland capture fisheries (accounting for 65% of catch), followed by Africa. Both of these regions are increasing their inland capture harvests, as is Western Europe. Harvests appear to be declining in Central & Eastern Europe and in North America, with the decline in the former primarily due to over-exploitation and loss of habitat, but also to recent political and economic changes.

Aquaculture production

World aquaculture production amounted to 34.8 million tonnes per annum by 1998, at a value of US$47.5 billion (FAO, 2000*b*). Aquaculture production has increased dramatically in recent years, realising almost a 300% increase in tonnage from 1984 to1998. Most of this increase has come from outside OECD regions, with OECD area aquaculture production growing by an annual average of almost 3% over 1990-1997, while the rest of the world realised growth rates of almost 13%. China is the most significant aquaculture producer, and currently accounts for about 44% of world production in value terms (FAO, 2000*b*). Japan is the world's second largest aquaculture producing nation, and three other OECD countries (Korea, Norway, and the US) are in the top ten (FAO, 1998*a*).

The aquaculture sector is not constrained in the same way as the capture fisheries sector and, in particular, does not face the same biological production limitation that natural fish stocks have. In addition, aquaculture is an enterprise under human control, whilst in capture fisheries man is dependent on a number of external variables (biology, weather, hydrology, past fishing patterns, etc.) for production. Furthermore, aquaculture production does not face the resource management problems confronting capture fisheries in the form of ill-defined property rights characteristic of "open access" resources. As a result, it is expected that most, if not all, future growth in the fisheries sector over the next 20 years will come from aquaculture. Projections in the Reference Scenario indicate that global aquaculture production has the potential to increase from the 39 million tonnes per year harvested in 1998 to almost 70 million tonnes in 2020. The majority of this growth is expected to take place in non-OECD regions, but OECD area aquaculture production is expected to increase as well. Aquaculture development may be indirectly limited, however, by the production constraints affecting capture fisheries, as many aquaculture activities are dependent on nutrients derived from capture fisheries.[3]

Box 9.1. Excess fishing capacity and subsidies to the fishing industry

Excess fishing capacity has contributed to unsustainable harvests in many world fisheries and poor economic performance. This has led to a problem of "too many boats chasing too few fish", as fleet capacity increases and catch rate falls (Figure 9.2). By the late 1980s, the world's large scale fishing fleet had exceeded the maximum sustainable yield (MSY) of global commercial fish stocks by 30% (McGinn, 1998). Since then, the situation has worsened: between 1992 and 1997, the world fleet increased by 3% in tonnage and 22% in technical capability (McGinn, 1998). Partly this has been the result of using more efficient harvesting technologies (*e.g.* radar, sonar, and aircraft for fish spotting), and satellite technologies for navigation; weather and sea-surface monitoring; as well as new communications and electronic data systems. Subsidies to shipbuilding and fleet enhancement have also contributed to this problem where fishery management was inadequate. A very rough estimate suggests world subsides to the fishery sector may be as much as US$14-20 billion per annum (Milazzo, 1998). Government financial transfers to fish harvesting in OECD countries in 1997 amounted to at least US$6.3 billion, representing 17% of the landed value of the catch that year (OECD, 2000). However, a good portion of these transfers is now allocated to research and the better management of fisheries.

Figure 9.2. **Global fleet capacity and catch rate, 1970-1989**

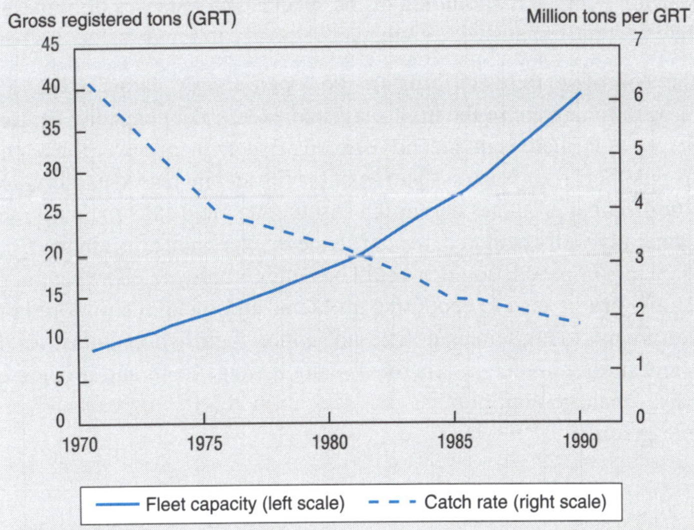

Source: McGinn (1998).

3. For example, aquaculture-produced Atlantic salmon rely upon a diet that is 45% fish meal and 25% fish oil, both derived from capture fisheries (Naylor *et al.*, 1998).

Trade in fishery products

In line with the upward trend in world fisheries production, trade in fish and fishery products has increased over time. Currently, between 35% and 40% of fisheries production is traded internationally, with export volumes having almost tripled in the last 25 years. The value of fish exports increased from US$11 billion in 1970 to US$52.5 billion in 1997 (McGinn, 1998; FAO, 1998*a*, 1999*a*). Developing economies accounted for more than 50% of the fish traded internationally, with exports largely going to OECD countries (FAO, 1998*a*). Developed countries import a large share of species such as tuna, shrimp, squid and salmon, as well as fish used for fishmeal and fish oil (McGinn, 1998). Barriers to trade in processed fish remain common (*e.g.* high tariffs on imports of already processed fish), as policies aim to keep fish processing (with its high value added) within the country where fish is landed.

9.3. Pressures on fish resources

Fish stocks

As mentioned above, many of the world's fisheries are reaching, or have already surpassed, maximum sustainable harvest limits. A good portion of world marine capture fish resources (70% according to one estimate) can now be categorised as mature or senescent, particularly the high value species (Garcia and Newton, 1997).[4] Some of the common commercial species, and in particular grundfish, such as Atlantic cod and haddock, are now threatened with commercial extinction. Some species, like tuna, are a global problem due to their migratory nature. In general, the evolution of capture fisheries can be seen as a classic example of "resource mining": starting with the exploitation of the species and fisheries of highest value or lowest harvesting costs and then, as they become exhausted, moving successively to lower value/ higher harvesting cost ones.

A number of common fish species are threatened with commercial extinction.

The main factors that contribute to overfishing are the "open access" nature of the resources (*i.e.* the rights of access to the fishery are open to anyone in the high seas) and excess fleet capacity (in terms of fish numbers and gear). Under open access regimes, there is no incentive among fishers to conserve the resource for future use, since no fisher has exclusive rights. This results in dissipation of profits and increased exploitation rates, pushing the size of the fish stock below the size that produces maximum sustainable yield (MSY). While access to fisheries adjoining OECD countries is generally well-regulated, the existence of excessive effort in the fishing industry also means that capital and labour – factors of production that could be better employed elsewhere in the economy – are occupied in a "race-to-fish". Low profits lead to economic problems and social disruption. In many cases, these problems have prompted governments to implement policies to support their fishing industries. However, many of these policies distort markets and further hinder sustainable fishing management, and it is increasingly recognised that the introduction of fisheries management policies that seek to match fishing capacity to available resources are more appropriate in the long run.

Ecosystem quality

In most capture fisheries, the target species intermingle with other species. The nature of the fishing equipment (gear) employed often has the effect that non-target species are also harvested (by-catch) along with the targeted species. The US National Research Council (1999) estimates that marine by-catch may amount to 27 million tonnes per year, the equivalent of almost one-third of the total reported harvests from marine capture fisheries. A number of policies have been implemented to reduce by-catch in some fisheries, including specifications on net type and mesh size; periodic closure of fishing grounds; limitations on by-catch or non-target species mortality

4. Mature fishery resources are those which are being harvested at the maximum average long-term yield; senescent fishery resources have been exploited beyond that limit and are, from a biological perspective, overexploited.

rates;[5] or requirements to use certain fishing gear (*e.g.* turtle and dolphin excluder devices). Certain fishing methods, including both dynamite fishing and bottom trawling, directly contribute to the destruction of marine species' habitat, including coral reefs. Global environmental changes may also impact marine ecosystems, such as rising sea levels and permanent changes in ocean temperature resulting from increased greenhouse gas concentrations in the atmosphere (see Chapter 13), and increased UV-B exposure from depletion of the ozone layer.

Most of marine pollution (about 70%) arises from land-based sources (UN, 1992). This is particularly true for coastal zone pollution from urban settlements, construction of coastal infrastructure and discharges (*e.g.* sewage treatment plants, stormwater run-off from roads), agro-chemicals run-off, tourism and industrial development. Algal blooms and red tides are also common, with the principal cause appearing to be increased nutrient loading of coastal waters (from the run-off of agricultural fertilisers) and high temperatures (see Chapter 8). Many of the polluting substances that originate from land-based sources exhibit characteristics of toxicity, persistence, and bio-accumulation in the food chain. Maritime transport and marine dumping activities (such as disposal of dredgings) represent a further significant contribution to marine pollution (UN, 1992). Roughly 600 000 tonnes of oil enter the oceans annually from routine shipping operations, accidents, and illegal discharges (UN, 1992).

While aquaculture can help alleviate stress on capture fisheries from increased demand for fish, it can also have negative impacts on local ecosystems.

The development of aquaculture also poses some threats to local environments. Where it is undertaken on a large scale at a single site, it can contribute to eutrophication,[6] reduced landscape amenity, and interference with other uses of coastal water space. The most important pollution effects from cage rearing of fish are related to organic enrichment around the sea cages, leading to reduced species diversity (GESAMP, 1992). Large-scale shrimp culture has led to a degradation of wetlands, salination of water supplies and land subsidence because of groundwater abstraction (Barg *et al.*, 1997). Often large doses of antibiotics are applied to aquaculture to maintain healthy fish populations. However, the accumulation of excess antibiotics can also affect other species and the ecosystem balance. The implications of escape of farmed fish – for example those that have been genetically modified or that are non-native species – is also a concern. Potential impacts include hybridisation, predation and competition, transmission of disease and changes in habitat (Barg *et al.*, 1997.). The introduction of invasive species which can squeeze out native populations, particularly in enclosed water bodies, can also have severe effects, such as the introduction of the Atlantic comb jellyfish in the Black Sea which caused the collapse of local fisheries.

9.4. Policy options and their potential effects

Despite the significant expected growth in aquaculture production, total world fisheries sector production may increase by only a small amount over the next few decades, if at all. FAO projects that under current management practices total world capture fishery production may even decrease to 80 million tonnes per annum by 2010 from the recent levels of 87-95 million tonnes per annum (FAO, 1998*a*). However, FAO also suggests that with better fishery management, capture fish harvests could actually increase by an additional 18 million tonnes (of which 6 tonnes would be in the fisheries of primary importance to OECD countries). To realise such a result, significant policy actions – at the local, national and international level – will be needed.

International co-operation

Global governance of fisheries is focused on recommendations by international bodies, such as the UN General Assembly, regional fisheries management organisations, and the FAO's Committee on Fisheries (FAO, 1999*b*).

5. For example, the Inter-American Tropical Tuna Commission defined a dolphin mortality limit of 7 500 for purse seine tuna fishing in the eastern Pacific Ocean during 1997.
6. The overnourishment of aquatic ecosystems with plant nutrients (mainly nitrates and phosphates) can lead to excess algae growth, hindering the development of other species.

These bodies have adopted policy and legal instruments that provide a framework for fisheries governance at various levels. A very important piece of international treaty law, pertaining to the international governance of marine fisheries, is the UN Convention on the Law of the Sea, arising from the UN Third Conference on the Law of the Sea, 1973-1982 (UN, 1982). Prior to this, coastal state jurisdiction over marine fishery resources typically extended to no more than six miles from shore. The Convention codified the general practice of states to extend their jurisdiction over marine fishery resources to 200 nautical miles from their coastlines (under Exclusive Economic Zones) from 1977 onwards. It has been estimated that 90% of world marine capture fishery harvests are based upon fishery resources encompassed by Exclusive Economic Zones (EEZs). The economic rationale for the EEZ regime lay in the hope that its implementation would lead to the mitigation of the "common pool" problem afflicting marine capture fisheries (Eckert, 1979). The results to date have been decidedly mixed, with the "common pool" problem re-emerging within many EEZs.

A particularly troublesome resource management problem emerged after 1982 in the form of transboundary fishery resources found both within the EEZ and in the adjacent high seas – straddling fish stocks and highly migratory fish stocks. In retrospect, it is clear that the relevant articles in the Convention (116-120) did not deal adequately with these resources. In response to the growing worldwide resource management crisis for this type of fisheries, the UN convened the UN Conference on Straddling Fish Stocks and Highly Migratory Fish Stocks, 1993-1995. That conference brought forth an Agreement (UN, 1995) which was designed not to supplant the Convention, but rather to support it. The Agreement calls for the management of straddling and highly migratory fish stocks on a regional basis, through regional fishery management organisations (RFMOs). While the Agreement is not yet in force, it had received 25 of the required 30 ratifications, and provided a framework for several attempts to manage co-operatively fishery resources. A key challenge now is to ensure that the Agreement is fully and effectively implemented, and that the companion 1993 Agreement to Promote Compliance with International Management Measures by Fishing Vessels on the High Seas is also implemented. Overall, it is essential that the 1995 FAO Code of Conduct for Responsible Fishing – a voluntary code of standards for the conservation of fishery resources and the management and development of fisheries both within EEZs and outside of them – is adopted and implemented by all FAO members.

A number of regional fisheries bodies exist which collect and disseminate data to support conservation and management, serve as a technical and policy forum to exchange information and disseminate "best practices", and promote harmonised policies and co-operation among members to promote sustainable management of fish stocks. In general, however, regional fisheries bodies have not been successful in conserving and managing resource stocks because of a failure to control fishing effort (OECD, 1997). A number of important issues will need to be addressed

Box 9.2. Trade measures and fishing activities

The FAO Code of Conduct for Responsible Fishing establishes the principle that international trade in fish and fish products should not compromise the sustainable development of fisheries and the responsible utilisation of aquatic resources (OECD, 2000). Two multilateral fisheries conservation agreements provide for the use of trade measures against non-contracting parties: the 1992 Convention for the Conservation of Antarctic Marine Living Resources, and the International Convention for the Conservation of Atlantic Tunas (OECD, 2000).

Bilateral trade measures have also been used to influence fishing activities harmful to the environment. Several high profile cases concerning the fisheries trade and the environment have been taken to the GATT/WTO for resolution, including the complaint filed by Mexico against the US in 1991 (regarding a US embargo on tuna caught using purse seine nets that endangered dolphins), and the complaint lodged by India, Malaysia, Pakistan and Thailand in 1996 against a US ban on the importation of shrimp and shrimp products from countries not using turtle excluder devices in shrimp nets (Tsamenyi and McIlgorm, 1999). In both cases, GATT and WTO review panels ruled in favour of the complainants. However, in the shrimp-turtle dispute, the WTO panel found that the US law requiring the use of turtle excluder devices was not inconsistent with obligations under the WTO, although it failed to meet the necessary requirements to be justified under Article XX of the GATT 1994.

to rectify this, including the failure by some states to implement key international instruments (see above); the unwillingness of some states to delegate adequate decision-making powers and responsibilities to regional bodies; and the lack of enforcement of management measures (FAO, 1999*b*).

Technological development and diffusion

Fishing technology developments are generally geared towards improving the efficiency and cost-effectiveness of fishing activities or fish processing. These technologies increase the share of living marine resources that are potentially harvestable, and so have contributed to the overfishing of many fish stocks. However, many of these technologies can also be adapted to support sustainable fishing practices, as is beginning to happen in some countries. For example, a fisheries computer-aided monitoring system has been developed that integrates a geographical information system (GIS), a global positioning system (GPS) and a database management system. It can be installed on board vessels and used to monitor fisheries effort and catch, providing valuable information to industry and fisheries managers. Australian fisheries already use a satellite-based vessel monitoring system to ensure compliance by vessel operators with fisheries management rules, and to collect data on catch and fishing. New technologies, such as turtle excluder devices, have also been developed. A more wide-spread use of these devices could prove a significant benefit to fisheries managers whose monitoring and compliance activities are often constrained by limited resources, although many of the devices are installed by the fishers themselves in order to increase their efficiency and catch.

New technologies have the potential to help improve fisheries management, but also to increase the rate of unsustainable fish harvests.

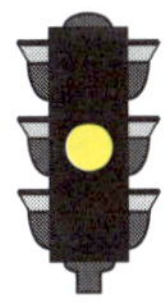

In some fisheries, policies are also being adopted that deliberately restrict the use of new technologies, either to reduce potentially damaging side-effects from their use (*e.g.* the use of purse-seine fishing nets which also catch dolphins) or to limit fishing effort (*e.g.* requirements that lobster fishers in the Chesapeake Bay in the US utilise sail-powered boats only). However, such policies have limitations. Many of the new technologies that are now being used (*e.g.* echo sounders, GPS, and automatic tracking systems) also improve safety on vessels. Furthermore, since they do not address the underlying economic incentives to increase total fishing effort in an open access resource, restrictions on the use of new technologies are unlikely to be effective in reducing fish catch in the long-run unless they are accompanied by other regulatory or economic policies.

Regulatory instruments

A number of regulatory instruments are used to limit fish catch and reduce the impacts of fishing or other activities on ecosystems and species. These include regulations to protect fish stocks, restrain fishing effort, or limit by-catch or degradation of marine ecosystems. Regulations that restrict fishing effort include limits on the fishing season, on the fishing area (particularly applicable for nursery areas), on the allowable catch (*i.e.* setting a total allowable catch – TAC – for a given fishery), or on fishing inputs (*e.g.* numbers of vessels, participants, nets, or horsepower).

In terms of the degradation and pollution of marine resources, a number of countries have adopted regulations to limit emissions or to hold polluters liable for large-scale pollution (*e.g.* from oil tankers or chemical spills). In 1995, a Global Programme of Action for the Protection of the Marine Environment from Land-Based Activities was adopted to facilitate action by states to prevent, reduce, control, and/or eliminate degradation of the marine environment, as well as to support its recovery from the impacts of land-based activities. The programme recommends approaches for dealing with, *inter alia,* sewage, persistent organic pollutants, heavy metals, nutrients, and litter. Again, current policies only address effects within national EEZs, and agreements and principles are needed to address transboundary pollution effects, or those occurring on the high seas. In some areas, regional initiatives have developed to manage fishery resources, and often extend their zone of management beyond the EEZs. However, what jurisdictional authority these regional bodies have over the open seas in their regions is not entirely clear. Further regulations are also required at the national level to deal with the potential degradation and pollution of coastal zone areas from rapidly developing aquaculture production. These regulations should be based on the Polluter Pays Principle (PPP) to provide appropriate incentives for pollution reduction.

Economic instruments

As discussed above, overcapacity in the fishing sector is a serious problem, and rationalising fishing capacity is critical to promote sustainable fisheries management. A number of subsidies to fishing activities continue to contribute to increases in capacity in the sector. Many countries provide grants and low-interest loans for the construction of new fishing vessels and to offer training and wage supplements to young people interested in working in the sector. The fishing effort is further stimulated by support that is transferred each year to fishing fleets in OECD countries through modernisation grants, fish price support, interest subsidies and fuel tax exemptions. These can encourage a build-up of capacity and an expansion of fishing activity, and some of the transfers may imbed industry expectations that may complicate future adjustment efforts (OECD, 2000). The conclusion, ratification and implementation of multilateral agreements to ensure the phase-out of subsidies to shipbuilding and fishing effort can contribute to the reduction of over-capacity in the fishery sector.[7] For these incentives to be effective, appropriate fisheries management will be necessary, and any potential spill-over effects of excess capacity into other fisheries will need to be addressed. Of course, better management of fishery resources is also in the long-term interests of the fishermen who use these resources (see Box 9.3).

A number of other economic incentives are also in use to reduce fleet capacity (*e.g.* fleet "buy-back" schemes), applying heavy vessel charges to discourage vessel introduction. OECD countries spent US$320 million in 1996 on decommissioning old or redundant fishing vessels and easing fishers into early retirement (OECD, 2000). For these incentives to be more effective, any existing counter-subsidies which encourage increased fleet capacity will need to be removed. In addition to vessel charges, many OECD countries also levy charges on permits for fish farming or commercial fishing, or on fishing quotas. A modelling simulation was used to examine the effects of removing all subsidies to fish production in OECD regions, coupled with the application of an *ad valorem* tax on fish products that increased by 2% per annum to 2020 (reaching a 50% tax in 2020). This simulation found

Box 9.3. Effects of changes in the environment on the fishery sector

The trends in over-fishing of particular fisheries, by-catch discard, and pollution of the marine environment have important economic rebound effects on the fishery sector. As seen above, production in many fisheries has progressed past the point of maximum sustainable harvests, resulting in more fishing effort required to catch fewer available fish. Furthermore, the economic consequences of overexploitation are more severe than physical levels of harvests would suggest because the current list of overexploited capture fishery resources is heavily weighted in favour of the high valued species.

In terms of marine pollution, the open sea is still relatively clean but the coastal zones are under serious threat of pollution (GESAMP, 1992). Degradation of the marine environment can have negative impacts on fish habitats and stocks, especially since many fish species spend part of their life cycle in the coastal zone. Impacts include changes in water quality, hydrological characteristics, nutrient concentrations, habitat availability and food supply. Establishing conclusively that a negative impact or change in fish stocks is the result of pollution is extremely difficult; this is further complicated by the biological factors and variation in sex, age and reproductive characteristics, as well as the mobility, of fish species.

These effects can extend beyond the harvesting sector into the post-harvest (processing and distribution) sector. Fisheries that are regulated to restrict fishing seasons tend to create incentives to race-to-fish, ensuring that the fishing season becomes progressively shorter and shorter. Processing plants find themselves confronted with short periods of glut, followed by lengthy periods of famine, leading to higher than necessary processing costs and possibly diminished quality of product (OECD, 1997).

7. The conclusion in 1994 of the OECD Shipbuilding Agreement marked an important step towards agreeing rules on government support to naval shipbuilding and establishing a dispute panel and legal instrument for dealing with signatories who contravened the Agreement, but unfortunately it has not yet been ratified, and its coverage is limited in that it does not cover the building of fishing vessels.

that gross production in the fisheries sector would decrease worldwide compared with the Reference Scenario, with decreases of over 4% for OECD regions overall and only a marginal decrease (0.1%) for non-OECD regions.

The lack of well-defined property rights in fisheries is one of the main factors leading to over-fishing. For some species, it is possible to define physical boundaries for property rights. For example, oysters stay in one place on the seabed, and so it is possible to physically define oyster beds, which can be allocated to individual fishers. Even where it is not possible to define physical boundaries, the right to catch fish can be applied through the use of individual transferable quotas (ITQs), which allocate quotas for a part of the total allowable catch (TAC) for a given fishery. The introduction of ITQs can help to increase economic efficiency of fishing fleets and vessels by reducing the fishing effort put into harvesting the allowable catch. Once fishers have the right to catch a certain amount of fish, they have a private incentive to maximise their economic value, and the setting of a total maximum of fish catch for the fishery overall can help to ensure sustainability of the resources. Further scientific capacity building is needed to better establish maximum sustainable yields from fisheries to enable the setting of total allowable catches based on a fully scientific basis, rather than on a combination of science and political bargaining. Currently, most IQ and ITQ schemes in operation are implemented at the national level in EEZs. With international co-operation, multilateral schemes could be established for high seas fisheries. The development of community-based fisheries management schemes can also be seen as a means to create quasi-property rights to limited harvests (allocated to local communities or nation states for their management), if not to the resource itself.

Voluntary agreements

Because of the open access nature of high sea fisheries, and the lack of a strong regulatory framework to back up agreements, voluntary agreements are unlikely to be a strong policy option for addressing unsustainable harvesting practices. However, such agreements may be more likely in the fish processing and purchasing industries, where firms are already adding additional buying specifications for the fish they wish to purchase.[8] These sectors may be potential candidates for voluntary agreements in the future, influencing the demand for fish from the fishing industry at the processing level. Such schemes may well emerge in conjunction with the further development of eco-labelling schemes, which help to influence consumer demand.

Information and other instruments

Eco-labelling has emerged in some OECD countries as a response to consumer demand for information specifying that fish and fishery products have been harvested and processed in an environmentally responsible manner. The Nordic Council, the European Commission and the FAO have each been holding on-going discussions on eco-labelling schemes for fish and fishery products. Some privately sponsored labelling or information programmes have also emerged, including those of the Global Aquaculture Alliance and the US National Fisheries Institute. The Marine Stewardship Council (initially set up as a partnership by Unilever and WWF in 1997 and now an independent foundation) has established a process for third-party certification of fishery or fish stock, as well as for fishing methods and practices. It is now possible for retailers and consumers to purchase fish bearing the Council's logo, indicating that the fish comes from well-managed sources. To ensure their effectiveness and credibility, eco-labelling schemes for fishery products should comply with the national regulations in the countries where the schemes are used, and they should be transparent, voluntary and non-discriminatory.

8. Some retailers now demand fish sizes that fit portion-sized packaging requirements (OECD, 1998*a* and 1999*a*). Some large post-harvest operators, such as supermarkets and processors, have established their own specifications that exceed the national sanitary and phyto-sanitary requirements ensuring food safety.

REFERENCES

Barg, U.C., D.M Bartley, A.G.J. Tacon and R.L. Welcomme (1997), "Aquaculture and its Environment: A Case for Collaboration" in D.A. Hancock, DC. Smith, A. Grant, and J.P. Beumer (eds.), *Developing and Sustaining World Fisheries Resources: the State of Science and Management. 2nd World Fisheries Congress*, CSIRO Publishing, Victoria: 462-470.

Eckert, R.D. (1979), *The Enclosure of Ocean Resources: Economics and the Law of the Sea*, Hoover Institution Press, Stanford.

FAO (Food and Agriculture Organization of the United Nations) (1997), "Review of the State of the World Fishery Resources: Marine Fisheries", FAO *Fisheries* 920, FAO, Rome.

FAO (1998*a*), *The State of the World Fisheries and Aquaculture* 1998, FAO, Rome.

FAO (1998*b*), "Fish and Fishery Products: World Apparent Consumption Statistics Based on Food Balance Sheets, 1961-1995", FAO *Fisheries* 821, Rev. 4, FAO, Rome.

FAO (1999*a*), FISHSTAT, FAO, Rome.

FAO (1999*b*), "Regional Fishery Organisations or Arrangements as Vehicles for Good Fishery Governance", Document FI:RFB/99/4, Meeting of FAO and Non-FAO Regional Fishery Bodies or Arrangements, Rome, 11-12 February.

FAO (2000*a*), "Agriculture Towards 2015/30: Technical Interim Report", Global Perspectives Unit, April, FAO, Rome.

FAO (2000*b*), FISHSTATPLUS, FAO, Rome.

Garcia, S. and C. Newton (1997), "Current Situation, Trends, and Prospects in World Capture Fisheries", *Global Trends: Fisheries Management*, in E.K. Pikitch, D.D. Huppert, and M.P. Sissenwine, (eds.), *American Fisheries Society Symposium* 20, American Fisheries Society, Bethesda: 3-27.

GESAMP (1992), *The State of the Marine Environment*, Blackwell Scientific Publications, Oxford.

McGinn, A.P. (1998), "Rocking the Boat: Conserving Fisheries and Protecting Jobs", *Worldwatch Paper* 142, Worldwatch Institute, Washington DC.

Milazzo, M. (1998), "Subsidies in World Fisheries: A Re-examination", *World Bank Technical Paper* 406, Fisheries Series, World Bank, Washington DC.

Munro, G., C. Bertuzzi, and C. Chung (2001), "Fisheries Trends: A Background Report", background document for the OECD *Environmental Outlook*, OECD, Paris.

Naylor, R.R., H. Goldburg, H. Mooney, M. Beveridge, J. Clay, C. Folke, N. Kautsky, J. Lubchenco, J. Primavero, and M. Williams (1998), "Nature's Subsidies to Shrimp and Salmon Farming", *Science* 282: 883-884.

OECD (1997), *Towards Sustainable Fisheries: Economic Aspects of the Management of Living Marine Resources*, OECD, Paris.

OECD (1999), "Towards More Sustainable Household Consumption Patterns: Indicators to Measure Progress", ENV/EPOC/SE(98)2/FINAL, OECD, Paris.

OECD (2000), *Transition to Responsible Fisheries*, OECD, Paris.

Tsamenyi, M. and A. McIlgorm (1999), *International Environmental Instruments: Their Effects on the Fishing Industry*, Dominion Consulting, Hurstville.

UN (United Nations) (1992), *Agenda* 21, Report of the United Nations Conference on Environment and Development, Chapter 17: "Protection of the Oceans, All Kinds of Seas, Including Enclosed and Semi-Enclosed Seas, and Coastal Areas and the Protection, Rational Use and Development of their Living Resources", United Nations, New York.

US National Research Council (1999), *Sustaining Marine Fisheries*, National Academy Press, Washington DC.

WRI (World Resources Institute) (1998), *World Resources* 1998-99, Oxford University Press, New York.

10 Forestry

KEY SIGNALS

- Forests provide a range of services to society, including economic benefits (*e.g.* timber, pulp for paper, cork, rubber), environmental services (*e.g.* air and water purification, biodiversity, carbon sinks, erosion control, wildlife habitat), and social benefits (*e.g.* employment, recreational opportunities, cultural values).
- World demand for industrial wood (*i.e.* non-fuel wood) is expected to increase by up to 70% to 2020, with demand in OECD countries increasing more slowly. Increased production will mainly occur in new plantation forests, primarily in the southern hemisphere.
- Total forest area in OECD regions has increased marginally in the last decade, and is likely to either remain stable or decrease slightly to 2020. Deforestation is expected to continue in tropical countries, with a loss of biodiversity and global forest coverage as a result.
- Forest quality is degrading in many areas in both OECD and non-OECD regions as natural forests are replaced with mono-culture, intensively produced plantation forests, and industrial pollution reaches forest areas.
- Policies to ensure the production of environmental, social and economic services from forests are needed. Eco-certification systems for sustainable forest management and carbon sequestration schemes can complement existing regulatory frameworks and provide incentives for more sustainable forest management. Promoting the use of timber for building and forest biomass for energy can encourage further forest expansion.

10.1. Introduction

Forests provide a range of services to humans, including wood products, recreational opportunities, and ecosystems services. Approximately half of the wood products harvested for human use worldwide are used for fuelwood and the other half for industrial purposes (*e.g.* building materials, paper products). Global demand for wood products and fuel wood has increased steadily in recent decades, but now appears to be levelling off. Wood is taken from natural forests that are undisturbed by man, semi-natural forests that are managed for timber production, and full "plantation forests" that are deliberately planted and intensively cultivated for wood and fibre production. In addition to providing wood for human use, forests provide essential environmental services, including air purification functions, carbon sinks, watershed protection, erosion control, biodiversity, and wildlife habitat. While all forests contribute to these important environmental functions, the extent to which they do so depends on their size, structure, density, and management.

Deforestation (forest clearance followed by land-use changes) and forest degradation (depletion of forest growing stock, without changes in land use) are the main factors limiting the ability of forests to provide economic and environmental services. The fragmentation of natural forest areas and the development of mono-culture plantation forests can reduce the quality of forest wildlife habitat, while intensive forest farming, with the use of fertilisers and pesticides, can affect soil, water and air quality. In many cases, natural forests can be managed so that they continue to provide a range of services – for example combining limited timber harvesting with recreational opportunities and providing high quality biodiversity habitats.

10.2. Developments in the forestry sector

Table 10.1. **Key forestry sector statistics and projections**

		1980	1998 (or latest available year)	2020 projected	Total change 1995-2020
Gross production (1995 US$ billion)	OECD	..	105	169	60%
	World	..	160	299	87%
Share of value added in the economy (%)	OECD	..	0.3%	0.2%	–1.7%
	World	..	0.4%	0.4%	10.4%
Industrial roundwood production (million cubic metres)	OECD	789	955	1 500	57%
	World	1 447	1 516	2 600	70%
Estimated fuelwood consumption (million cubic metres)	OECD	165	143	170	4%
	World	1 372	1 753	1 700	0%

Sources: FAOSTAT (2000), GTAP database and Reference Scenario.

Demand for forest products

Approximately half of wood removals worldwide is for industrial roundwood[1] use (1.5 billion cubic metres per annum), while the other half is for fuelwood use (1.5-2 billion cubic metres). Roundwood is divided fairly evenly between use for pulp and paper and other uses (*i.e.* sawnwood, plywood, veneer, etc). The main underlying driving forces for industrial wood demand are population and income levels and, for sawnwood and panels, the level of construction activity. The vast majority of fuelwood consumption occurs in the developing world, with OECD countries accounting for less than 10% of estimated fuelwood consumption. Population growth in developing countries and lack of alternative energy sources (*e.g.* electricity supply) are the main drivers of fuelwood demand.

Consumption of the main forest products rose by about 50% between 1970 and 1990, both in OECD regions and the rest of the world, but has been stagnating since the mid 1980s. The consumption of wood-based panels and of paper and paperboard has risen more strongly since 1970, while sawnwood demand has been slower, reflecting more moderate growth in the construction industry. The consumption of processed products has generally followed the economic cycle. Driven by population growth and low incomes in developing countries, world fuelwood consumption has risen consistently since the 1970s.

In OECD regions, growth in demand for the main forest products is likely to weaken over the period 1995-2020, as population levels stabilise and increased waste paper recycling reduces the need for virgin pulp. For the developing world, increasing populations and income levels are expected to contribute to a continued growth in demand for industrial wood. While projections for world demand for industrial wood in 2020 vary significantly (see Figure 10.1), most projections indicate that the increases in demand will slow after 2010.

While demand for primary fibre inputs (including industrial roundwood, recycled paper and non-wood fibre) to the pulp and paper industry is projected to increase significantly worldwide to 2020 (see Figure 10.2), an increasing portion will come from recycled paper rather than roundwood. By 2020 recycled paper is expected to account for almost half of all fibre inputs to the pulp and paper industry (see Chapter 18). The future demand for fuelwood is more ambiguous, partly because of difficulties in measuring total fuelwood use in many countries. Although the increasing incomes expected in developing countries are typically associated with shifts away from fuelwood to other energy sources, continuing population growth is likely to counter this trend such that world fuelwood use remains stable to 2020 (see Figure 10.2). In OECD countries, fuelwood use currently amounts to less than 0.2 billion cubic metres per annum, and this has been declining steadily since the mid-1980s (FAOSTAT, 2000). However, under the Reference Scenario, a small (4%) increase in fuelwood demand is expected in OECD countries.

1. Industrial roundwood includes all wood extracted from forests for purposes other than fuel use, *e.g.* sawnwood for building materials and furniture (lumber, wood panels, veneer), and fibre products (paper and paperboard).

Figure 10.1. **World industrial wood production, 1970-2020**

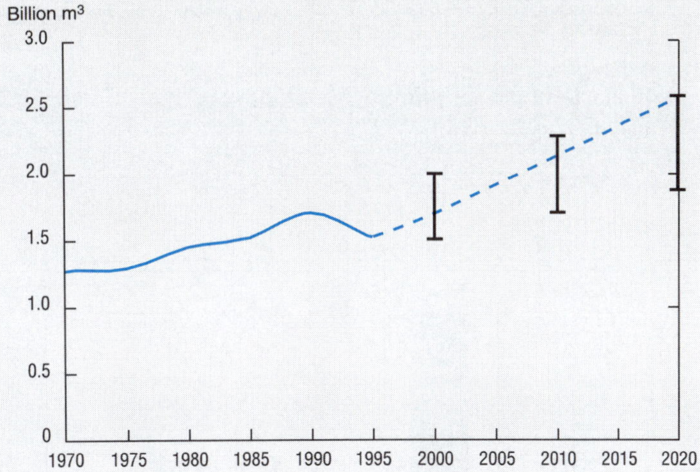

Sources: FAOSTAT (2000) for 1970-1995 numbers; Reference Scenario (dotted line) and range of other projections cited in Bonnis (2001) and Edjo (2001) for 1995-2020 numbers.

Wood production

> **Timber is increasingly being harvested from second growth and planted forests, reducing pressures on natural old growth forests.**

Approximately 22% of current world timber harvests come from natural forests, 44% from managed forests, and 34% from plantation forests.[2] Harvests in OECD countries generally reflect the global proportions, with much of Europe, the US and Japan harvesting from managed and plantation forests, and Canada harvesting primarily from natural and managed forests. Only a few countries (primarily in the OECD Pacific region, and some southern European countries) have significant production from exotic industrial plantation forests. The share of wood production from plantation forests has been rapidly increasing since the 1960s, and this trend is likely to continue, with new plantation forests developing particularly in non-OECD regions. About 25% of the world's industrial wood (15% of OECD countries' industrial wood) originates from tropical forests, particularly from the Asia-Pacific region. Increased production from second growth and planted forests has been displacing natural forest timber in recent years, with decreases seen in harvest levels from primary natural forests as a result. In many countries forestlands are publicly owned, with leases provided for private harvesting (see Box 10.1).

Significant technological developments have been realised in recent years in both forestry production and in wood processing. On the production side, there has been the use of better forest management techniques, irrigation and fertilisation systems to increase growth, more intensive plantation forests, as well as new developments in tree breeding and biotechnology to improve and standardise tree characteristics. The results are faster tree growth and more desirable tree characteristics. Plantation forests have been found to produce 5-20 times the wood growth that

2. It should be noted that there have been changes in the definitions of plantation forests over time.

can be achieved in most temperate region forests. "Fibre farms" go even further than plantation forests, growing fibre as an agricultural crop in very intensively managed short rotations. On the wood processing side, reductions in wastage in tree harvesting and the re-use of processing by-products have significantly increased efficiency.

Figure 10.2. **Demand for primary fibre inputs by type of use, 1995-2020**

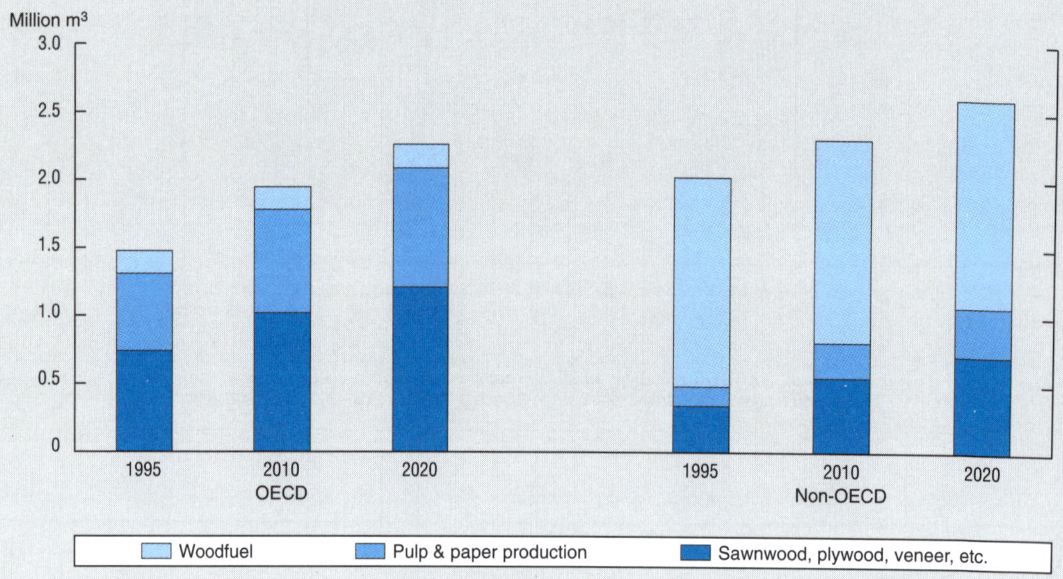

Note: These figures include the use of industrial roundwood, recycled paper and non-wood fibre.
Source: Reference Scenario.

Box 10.1. Characteristics of forest holdings in OECD regions

Almost one-third of forestland in OECD regions is privately owned, although there are large variations between countries. Thus, in North America, 7% of Canadian forests are in private hands, compared with 73% of forests in the US. In EU countries, the share varies from 23-93% of forests under private ownership. The long-term trend in many European countries has been an expansion of public forests as a result of acquisition and afforestation. That trend is now being reversed in some countries, however.

Average sizes of holdings vary considerably: 30-50 hectares in Scandinavia against 3-6 hectares in other European countries, and 10 hectares in Japan. Even in the Nordic countries, there are a large number of holdings of less than 5 hectares.

Because of the greater processing efficiency that has resulted from these technological developments, wood removals from forests have been increasing at a slower rate than wood consumption. Roundwood removals increased by a third between 1970 and 1990, while wood consumption rose by a half.

It is unlikely that wood will be in scarce supply in OECD regions in the coming decades, as only about 55-60% of annual forest growth in OECD regions is currently harvested, with some room for increased harvests if needed. Numerous studies on the economic availability of industrial wood support this projection, with no indication of future shortages.

Trade in forest products

Globally, most trade in industrial wood occurs in the northern hemisphere between industrial countries: OECD countries account for roughly 80% of total exports and 90% of total imports. Already a net importer, the OECD area is expected to increase net imports of wood products by almost 200% over the next 20 years. The three dominant industrial wood importing markets – Western Europe, Japan and North America – are all expected to more than double their net imports of wood products. While most of the industrial wood consumed in Western Europe is produced within Europe (with large volumes being exported by the Nordic countries to other European countries), by 2020 Western Europe is expected to more than double its currently modest net imports of wood. The US and Canada are two of the world's major producing regions and export (as well as import) huge volumes of industrial wood, both to and from each other and globally. While net imports there remain relatively small, they are expected to increase significantly. In Asia, Japan is the dominant consuming country and is the world's largest net importer of industrial wood. Slow growth in domestic production coupled with larger demand increases are expected to further increase net imports in Japan.

Central & Eastern Europe is a net exporter of wood products, and is expected to significantly increase this trade surplus over the next two decades as domestic production increases by over 150% and domestic demand grows more slowly. Russia has been an important exporter of industrial wood to a number of regions in the past, but considerable recent reductions in production have lowered Russian exports.[3] There are growing exports from plantation forests of the southern hemisphere (including New Zealand) to most of the world's major industrial wood markets. Most tropical wood exports flow to OECD regions (particularly to Japan, but also to North America and Europe), most often in the form of plywood and sawnwood.

In general, OECD countries tend to have low tariffs on forest products, with the situation being more variable outside the OECD area. Other trade measures apply to roundwood exports (many countries prohibit log exports) and sawnwood exports (some countries apply export taxes). Eight of the largest forest product producing and trading countries agreed at the Uruguay Round of the General Agreement on Tariffs and Trade (GATT) to a tariff reduction of 33% on wood products and the removal of tariffs on paper products by 2004. A proposal by the Asia Pacific Economic Co-operation Forum (APEC) to the World Trade Organisation (WTO) calls for a more rapid reduction in tariffs on building materials and removal of tariffs on fibre products.

10.3. Environmental effects of forestry sector activities

Forest area and deforestation

Estimates indicate that in 1995 the world's total forest area amounted to about 3.4 billion hectares, 20% of which was situated in OECD countries (FAO, 1999). Globally, forest cover represents a third of total land area. The largest remaining forest areas that have not yet been developed for large-scale commercial logging are found in the Amazon, parts of eastern Russia, including Siberia, and northern Canada. Large portions of these forests are economically inaccessible, and are likely to remain so for the near future.

Tropical deforestation is expected to continue at alarming rates over the next few decades.

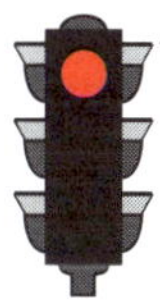

Commercial logging *per se* is rarely the major cause of deforestation. This is true for both temperate and tropical regions. Rather, permanent forest loss is typically driven by the desire to convert forestland to other uses, usually agriculture. Commercial logging in the tropics does, however, extend road infrastructure into previously inaccessible forestlands, enabling access to these lands for future conversion to agriculture. FAO data indicate that as a result of continued pressures, forest loss in tropical countries has been averaging over 15 million hectares per annum since 1980, although some of this may recover (FAO, 1999).

3. While Russian harvests are likely to increase from the current depressed levels – they fell by over 75% since the mid-1980s (OECD, 1999), they are unlikely to return to previous levels when harvest and transport costs were heavily subsidised.

In certain areas, especially dry areas where forest growth is modest, fuelwood use also exerts pressures on tropical forests. Deforestation is expected to continue in the tropics under the Reference Scenario, with non-OECD countries losing an estimated 15% of current forest area by 2020 (Figure 10.3).

Figure 10.3. **Change in forested area by forest type, 1995-2020**

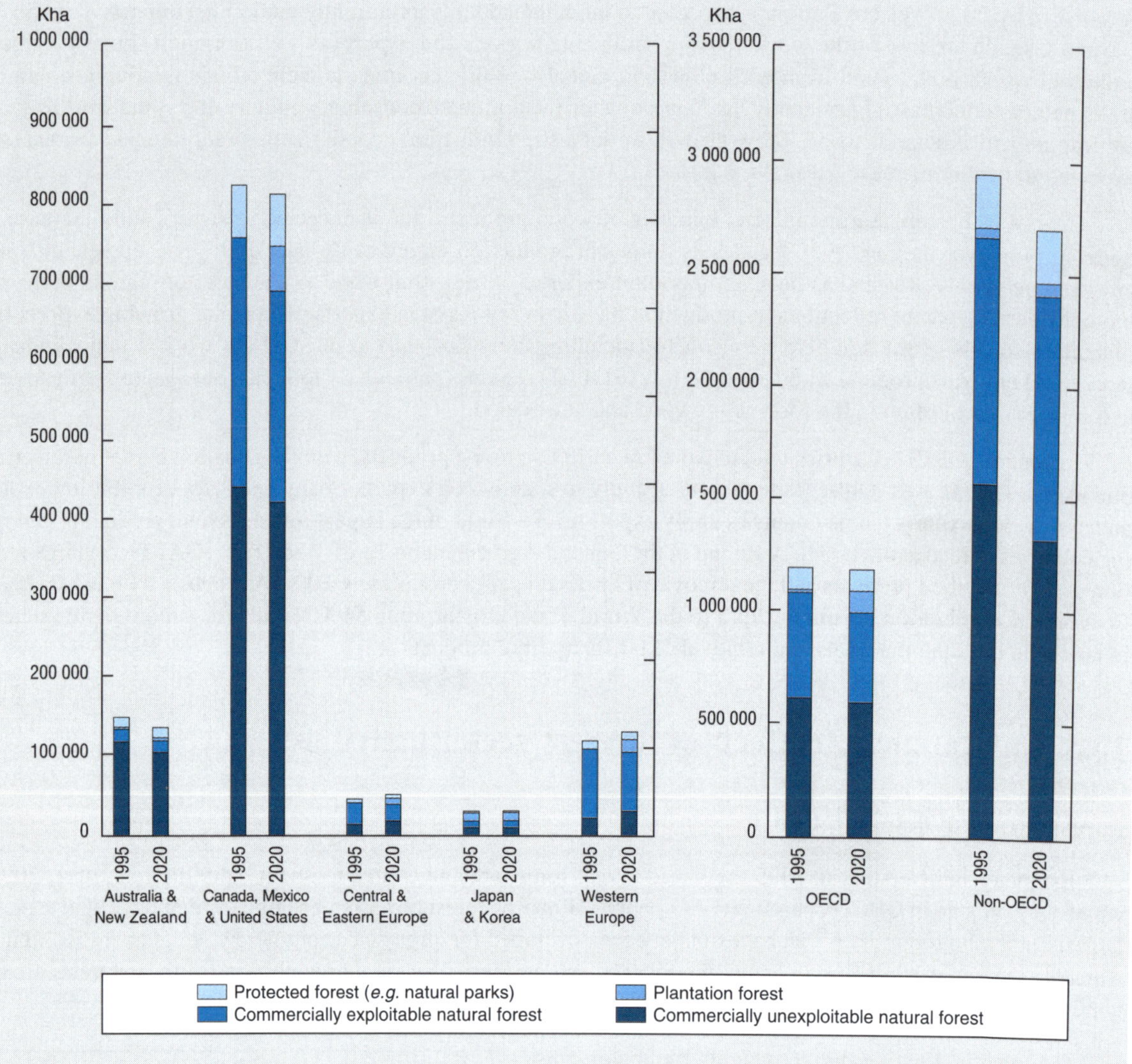

Source: Reference Scenario.

In contrast, the temperate regions – largely in OECD countries – have been marginally expanding their forest areas in recent years. The results of the recent UNECE/FAO temperate and boreal forest resource assessment and of the FAO global forest resource assessment show that the expansion of forest areas, growing stock, and growth increment is common in OECD regions (UNECE/FAO, 2000). This expansion has largely been driven by the abandonment of agricultural production on marginal lands and their natural reversion to forests, and has been further enhanced by deliberate investments in reforestation in many countries. Evidence indicates that, overall, temperate forests will continue to maintain their present size or decrease only slightly in the foreseeable future, while forest growing stock will increase.

Forest ecosystem quality

Forests provide habitat for about two-thirds of known terrestrial species, and have the highest species diversity of any ecosystem, and as such are an important reservoir of genetic material. They also have the highest number of threatened species (see Chapter 11). In addition to the rich biological diversity encountered in forests, forests also provide a wide range of ecosystem services (see Box 10.2). The environmental services they provide are dependent not only on the stability of the forest cover, but also on the conditions of that cover (type and size of forest, species mix, etc.). There are indications that forest quality or authenticity is deteriorating, although there is a clear lack of reliable information on the (current and past) state of forest biodiversity. Acid rain, caused by sulphur dioxide and nitrous oxide emissions, has been identified as an important factor in forest demise. While there now seems to be consensus that earlier worries about forest decline were exaggerated, there is still concern about the possible long-term impact on the forest soil (acidification) and consequently on the forest environment. Similarly, the effects of global warming on tree growth will vary according to the location and growing conditions of forests, but it is expected that the effects may be quite detrimental in some regions, for example around the Mediterranean. A more direct pressure comes from uncontrolled forest fires, which are often caused by man through negligence or by arson. In some countries, forest coverage is burned off deliberately to open up land for agricultural development. For some forest types, however, such as the boreal forests of Canada and Russia, and many pine and eucalypt forests (*e.g.* in Australia and the US), fire is an integral part of the natural disturbance regime.

Box 10.2. Environmental services provided by forests

Forests provide a range of essential environmental services as well as economic (*e.g.* timber, fruit and nuts) and social (*e.g.* recreational activities, cultural) services. Forests help to clean the atmosphere by acting as a pollutant filter: they trap solid matter and suspended aerosols, and can absorb some chemical compounds. Forests also provide a substantial CO_2 sink, countering the build-up of this greenhouse gas in the atmosphere. Forests regulate water flows, with tree roots reducing run-off and improving water infiltration into the soil. As a result, flood levels of streams can be reduced, and improved infiltration also helps to replenish groundwater.

Forests also improve soil fertility. Tree roots penetrate deep into the soil and add inorganic substances at the expense of parent rock. The falling leaves supply the topsoil with organic substances, which are decomposed by ground life developing in the microclimatic conditions under the trees. Forests also play a vital soil protection role. Branches and leaves intercept some of the precipitation and thus diminish the effects of rainwater erosion. In mountain areas, they protect against avalanches, while in sandy areas, shelterbelts reduce wind erosion. Finally, forests host a number of species (from hundreds/ha in tropical forests to 5 or less/ha in boreal forests).

The share of wood produced in intensive plantation forests is expected to increase in OECD regions to 2020.

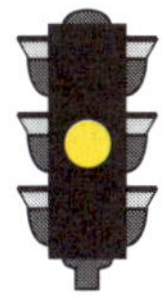

Another potential threat to forest ecosystem quality is the increasing development of intensive plantation forests. Plantation forests are made of one or a limited number of species. Compared with natural forests, their ecosystems are simplified so the wildlife is likely to be less varied. Furthermore, mono-culture plantations populated with inappropriate species may acidify some soils (*e.g.* pine forests in some regions). However, for the most part, both managed and natural forests provide many of the same environmental services. Plantation forests, for instance, provide numerous environmental services, including watershed protection and erosion and flood control. In some cases, exotic (planted) forests are superior in providing some of these services. Thus, fast-growing plantation forests may perform better in terms of carbon dioxide absorption than natural forests. More important, the increasing development of plantation forests to meet industrial wood needs may reduce pressures for harvesting from natural forests. Over the coming decades, plantation forests are expected to increase as a proportion of total forested lands in OECD countries from 1% in 1995 to almost 5% in 2020 under the Reference Scenario (see Figure 10.3).

10.4. Policy options and their potential effects

Maintaining forest coverage and meeting demand for forest products is not likely to be a problem for OECD countries in coming years, but policies will need to be established to ensure the protection of remaining old growth forests and to limit any negative environmental pressures associated with intensive production of plantation forests. To a large extent, these policies will be in the form of regulations or subsidies to encourage sustainable forest management practices, the establishment of protected forest areas with buffer zones to protect biodiversity and restrictions on access to and use of these areas. A more urgent concern is the destruction of tropical and old growth forests in non-OECD countries, with ensuing negative effects on climate change, ecosystems, and global biodiversity. OECD countries can influence these activities through internationally co-ordinated programmes or a legally binding convention, bilateral co-operation, and through informing consumer decisions (*e.g.* eco-labelling). Continuing developments in methods and technologies for timber removals, processing, and recycling of pulp and paper products can also help to reduce the energy and resource intensity of the forestry sector.

Technological development and diffusion

Significant technological developments have been realised in recent years in forestry production and in wood processing, and it is likely that further developments will occur in the future, particularly through new tree breeding techniques and biotechnologies.[4] While fibre farms are still in the fledgling stages of development, they are also likely to increase rapidly in coming years as they become a more cost-efficient source of fibrous material. These changes will to a large extent be driven by cost considerations on the part of forest producers, as they have been in the past. However, incentives and support – such as research, education campaigns, outreach, and subsidies – to encourage technological development and diffusion in the forestry sector to reduce pressures on the environment could hasten the transition to sustainably managed planted forests. Forthcoming technologies that can aid sustainable management of forest resources include software programmes to allow precision management in the application of inputs (*e.g.* chemicals, nutrients, and water) in farmed forests.

Regulatory instruments

In many OECD countries, codes of practice to protect or enhance the soil and water conservation functions of forests have long been in place, including obligations to replant after harvest and to avoid clear-cutting on hill tops and steep slopes. In the 1990s, a number of OECD countries further adapted their forest legislation to include biodiversity conservation. This was largely an outgrowth of the 1992 Rio Declaration on Environment and Development and the "Statement of Principles on Forests" that agreed that all countries should practice sustainable forest management. The further adoption and enhancement of such measures are necessary to ensure sustainable forestry management. In addition, to better incorporate biodiversity concerns into forestry codes of practice and ensure the continued provision of environmental services from managed forests, regulations will increasingly need to be developed to govern the management of intensive plantation forests – in particular, to regulate the inputs and processes used in intensive plantation forestry (*e.g.* chemical fertilisers, water, genetically-manipulated saplings, etc.) to minimise environmental damage.

OECD countries have been expanding the area of natural forest that is protected in national parks or reserves.

OECD countries have been setting aside a growing proportion of natural forest lands as protected or restricted access reserves in recent years, and some expansion of these zones can be expected. However, as managed forests and plantations are becoming more economically-viable timber sources than the remaining OECD area old growth forests, there will be only a limited need for expanding protected forest zones there. Nonetheless, in tropical countries such provisions will be necessary to reduce deforestation, and it is likely that their development will require funding from OECD countries, either bilaterally or through organisations such as the World Bank (see Box 10.3).

4. Public resistance to the use of biotechnology and genetically modified organisms in forestry production is likely to be less than for agriculture, as there are not likely to be associated health risks.

Box 10.3. International forestry initiatives

The 1992 Rio Earth Summit produced a "Statement of Principles on Forests" which recommended all countries should practice sustainable forest management. Since then, a number of international initiatives have been devoted to this issue, including the Helsinki Process in Europe, the Montreal Process and the Santiago Declaration for non-European temperate countries, the Amazon Co-operation Treaty, the African Timber Organisation and the South Pacific Forum. International standards for forest management are being discussed in the UN Forestry Council, and the International Standards Organisation (ISO) is providing guidance on how ISO 14001 could be used as a basis for developing a management system for sustainable forestry. The UN Commission on Sustainable Development (CSD) established the Intergovernmental Panel on Forests (IPF) in 1995. One of the main objectives of the IPF was the "implementation of forest-related decisions of the United Nations Conference on Environment and Development (UNCED) at the national and international level" (UN, 2000). The IPF was succeeded by the Intergovernmental Forum on Forests (IFF) in 1997 which, in turn, will be succeeded by the United Nations Forum on Forests that will look into the possibility of developing legally binding instruments for forest management.

A global initiative of the World Bank and World Wildlife Fund (WWF), the Alliance for Forest Conservation and Sustainable Use, aims to significantly reduce the loss and degradation of all forest types worldwide. The World Bank/WWF Alliance is working towards this goal by promoting forest conservation and internationally recognised best practice in forest management, and has set three targets for achievement by 2005: 50 million hectares of new forest protected areas; 50 million hectares of existing, but highly threatened, forest protected areas secured under effective management; and 200 million hectares of the world's production forests under independently certified sustainable management. If this project goes ahead, the commercial logging and destruction of the last of the major "frontier" forests of the world (*e.g.* in the Amazon, Russia and Canada) may already have reached their peak.

A new initiative by the World Resources Institute (WRI), called Global Forest Watch, has the potential to move forest management oversight further into the public domain. It combines the use of satellite imagery, geographic information systems, mapping software, the Internet and on-the-ground observation by community environment groups to record actual forest uses compared with stated uses in forest lease agreements. The information is posted on the Internet, allowing the general public to directly monitor management of the world's forests by timber companies. Already, IKEA, Home Depot, Lowes, and other major timber purchasing companies have announced their intentions to source wood supplies from companies operating in compliance with environmental laws, as indicated by Global Forest Watch.

Economic instruments

A range of economic instruments are already used in OECD countries to protect forest resources, including felling fees or charges, non-compliance fees to ensure replanting, levies or taxes on the diversion of forest lands to other uses, and subsidies for afforestation and sustainable forest management. Currently, forest management policies are dominated by subsidies in OECD countries, but should move in the future more towards charges and fees in order to better reflect the User Pays Principle for the use of publicly-owned natural resources. Such charges can help to increase the efficiency with which forest products are used, although policy simulations indicate that they may not have a significant effect on the demand for forest products and the sector. Thus, a simulation examining the effect of removing price subsidies to forestry in OECD countries and applying an annually increasing 2% *ad valorem* tax on forest products was found to lead to only a very small reduction in OECD area demand for forest products compared with the Reference Scenario.[5]

Support to forest activities remains significant in OECD countries, and in many cases has been the main driver behind the expansion of forest resources. While originally the underlying objective was to raise wood production, the emphasis is shifting towards payment for ensuring that forests provide social and environmental services. Direct subsidies of this type can be essential in reducing the profitability gap between sustainable and unsustainable forest techniques by forest managers. Thus, payments can be given to "compensate" forest owners for income loss resulting from biodiversity protection activities or for ensuring soil protection. These measures should be geared

5. The impacts of this simulation on demand for forest products were limited in part because of assumptions in the model regarding fixed shares of non-energy inputs in production (see Annex 2).

towards maximising the social, economic and environmental services provided by forests. At the same time, it is essential to remove any implicit and explicit subsidies which promote logging and access to natural forests (*e.g.* low stumpage fees, provision of roads through forested areas, support to agricultural expansion). Better co-ordination of subsidy policies is needed to minimise the risk of conflicting policies and strengthen the incentives for sustainable management. International co-operation on adverse subsidy removal through the World Trade Organisation can help to alleviate any potential competitiveness effects, although the results of the policy simulation results described above indicate that these would be minimal.

Forests have a strong potential to absorb carbon dioxide emissions, a major greenhouse gas. If forest carbon sequestration were taken as a significant tool for mitigating carbon build-up in the atmosphere, the global volume of forest would have to increase. A global market of carbon rights could provide a financial incentive for industry to finance extensive carbon-fixing afforestation programmes. Some initiatives are already underway in this direction on a local or regional basis. For example, the Sydney Futures Exchange announced that it would establish the world's first exchange traded market in carbon credits. Plantation of "Kyoto forests" has also started, particularly in tropical countries (for example, in Costa Rica, through financing by US firms). Although such plantations may be on the increase, impediments remain to their rapid development and the global coverage they would need for maximum efficiency and effectiveness. In particular, agreement on carbon credits for forest sinks need to be reached under the UN Framework Convention on Climate Change (UNFCCC) process, and the concept of carbon debits for conversion of forestland to other land uses introduced (see Box 10.4).

Box 10.4. **Carbon sinks and the Kyoto Protocol**

Under the Kyoto Protocol, Parties to the UNFCCC agreed to commitments to reduce their overall emissions of greenhouse gases by 2012 (see Chapter 13). In order to help in this process, Parties to the Convention adopted a draft decision in 1999 (COP5) to address the potential of adding in net changes in emissions and removals by sinks from afforestation, reforestation and deforestation (sometimes referred to as "Kyoto lands"). The potential of forests to act as carbon sinks is significant. The recent ECE/FAO temporal and boreal forest resource assessment estimates that the carbon sink function of temperate and boreal forests is equivalent to 16% of global emissions of greenhouse gases. If forest carbon sequestration were to become a significant tool for mitigating carbon build-up in the atmosphere, the global volume of forest would have to increase. Although much of this increase would probably be through plantation forests, attempts to develop forest carbon sequestration could bode well for efforts to reduce tropical deforestation as well.

However, at this point there is still no consensus about which methodology should be considered by the Protocol to measure the carbon sink function of forests and forest soils. The Intergovernmental Panel on Climate Change (IPCC) was asked to clarify the issue of carbon emissions from sources and removals by sinks from land use, land-use change and forestry. No final decision was taken on the incorporation of sinks in the Kyoto Protocol at the 6th Conference of the Parties (COP6) in November 2000, but the issue will be further discussed at the continuation of the Conference in 2001.

Voluntary agreements

An information-based voluntary instrument that has been gaining in importance in recent years in OECD countries is the certification of forests as meeting sustainable management standards. Eco-certification has many advantages: it guarantees that the certified wood comes from sustainably managed forests through a chain-of-custody procedure. Under certain conditions, the certified wood can be sold at a price premium. Eco-certification schemes have been developed by the forest industry (*e.g.* the Sustainable Forest Initiative of the American Forest and Paper Association); by environmental NGOs (*e.g.* the Forest Stewardship Council – FSC); by the EU (*e.g.* the recent Pan European Forest Certification Scheme – PEFC), and are being considered by the International Standards Organisation. While it is possible to maintain several eco-certification schemes at the same time, this has the tendency to overwhelm consumers and lead to mistrust of the competing schemes. Efforts should focus on the combination of existing schemes into a single internationally agreed certification scheme, and the development of a suitable monitoring and certifying agency (*e.g.* possibly under the FSC or ISO). In order to ensure compatibility with WTO rules, eco-certification will need to remain voluntary, and not prevent the importation of non-certified wood.

Information and other instruments

Information dissemination to forest managers regarding sustainable practices and the use of new technologies and processes, as well as to consumers about the sources of forest products (eco-labelling), are important tools for encouraging sustainable forest management. While further research is required into new management techniques and technologies, the main information needs now are gathering better data on forest quality and the environmental services forests provide, and the wider dissemination of this information to forest managers and consumers.

Box 10.5. Incentives for increasing the use of wood products

Further encouragement of the use of wood as a building material could lead to environmental benefits: it is a recyclable material with low energy consumption and low waste generation. Current building codes, especially when they are product-oriented rather than performance-oriented, are often discriminatory against wood. Undertaking comparative life-cycle analyses to compare wood with substitutes (cement, plastic, steel, etc.) could help to clarify the advantages and disadvantages of the different materials, including the positive and negative externalities associated with the production and use of the different raw materials. A recent study found that the implementation of a carbon tax in European countries would lead to greater substitution of wood for steel in building projects because of the higher CO_2-intensity of the latter (OECD, 2000).

The development of indicators of forest quality and criteria for sustainable management is an essential requirement for the design of appropriate forest management policies, and the monitoring and enforcement of the success of these policies. Current forest inventories essentially evaluate changes in forested areas or stocks of timber. A closer monitoring of changes in forest quality components, including biodiversity and carbon sink indicators, is necessary to facilitate sustainable forest management policy-making and implementation. Furthermore, there is a need for harmonisation of monitoring methodologies. Processes to develop sustainable forest management criteria and forest-related environmental indicators are already underway in various international fora, including the United Nations Food and Agriculture Organisation's work on forest assessments.

REFERENCES

Bonnis, G. (2001), "Environmental Outlook in the Forestry Sector", background document for the OECD *Environmental Outlook*, OECD, Paris.

FAO (Food and Agriculture Organisation of the United Nations) (1999), *State of the World's Forests* 1999, FAO, Rome.

FAOSTAT (2000), Food and Agriculture Organization of the United Nations (FAO), *http://apps.fao.org/default.htm*

OECD (1999), OECD *Environmental Data: Compendium* 1999, OECD, Paris.

OECD (2000), "Instruments and Technologies for Climate Change Policy: an Integrated Energy and Materials Systems Modelling Approach", ENV/EPOC/GEEI(99)15/FINAL, OECD, Paris.

OECD (2001), "Draft Summary Record, OECD Workshop on the Environmental Outlook and Strategy for the Forestry Sector", Jedlnia, Poland, 15-16 February 2000, background document for the OECD *Environmental Outlook*, OECD, Paris.

Sedjo, R.A. (2001), "Environmental Outlook: The Forestry Sector", background document for the OECD *Environmental Outlook*, OECD, Paris.

UN (United Nations) (2000), "Report of the Intergovernmental Forum on Forests on its Fourth Session, New York, 31 January – 11 February 2000", E/CN.17/2000/14, UN, New York.

UNECE/FAO (United Nations Economic Commission for Europe/Food and Agriculture Organisization) (2000), *Temperate and Boreal Forest Resources Assessment* 2000, FAO, Rome.

11 Biological Diversity

KEY SIGNALS

- Many human activities contribute to biodiversity loss, including habitat destruction or alteration, the exploitation of wild species, introduction of exotic species, species homogenisation, pollution, and global environmental change.
- As a result of these pressures, many natural ecosystems have been degraded or destroyed, seriously limiting the essential ecosystem services they provide for humankind (*e.g.* water purification, CO_2 sequestration, etc.).
- There are some positive trends, however: the number and extent of protected natural areas are increasing in OECD countries, partly offsetting decreases in natural areas outside of these protected areas, and a number of previously threatened species are now recovering.
- Over the next 20 years, biodiversity is expected to continue to decline in OECD countries due primarily to pollution, land use changes, and the intensification of agricultural and forestry activities. The situation is more serious in non-OECD countries, where major threats to biodiversity – including tropical deforestation, ecosystem degradation, and species extinction – are likely to continue.
- The use of regulations and access restrictions to protect biodiversity (*e.g.* hunting bans or the creation of nature reserves) is likely to be more limited in the future because these measures are often expensive to maintain. Instead, there will be greater focus on the use of other instruments – economic instruments, awareness raising, certification schemes, etc. – that allow for the productive use of biological resources, but in a sustainable manner.

11.1. Introduction

Biodiversity has been slowly and naturally evolving since the beginning of life, but human activities also play a significant role in shaping biodiversity. The impact of human activities on biodiversity has been characterised by mankind taking from ecosystems what it needed for sustenance, and at times over-exploiting certain species because of high demand for their products or because of a perception of them as pests. The main pressures on biodiversity in OECD countries come from land use changes as populations increase, the unsustainable use of the natural resource base (*e.g.* in fisheries, agriculture, and forestry), climate change and pollution. Although some of these pressures are expected to decrease in the near future, the total burden on biodiversity is expected to remain high to the year 2020. Biodiversity loss in developing countries is even more serious, with the main driver being habitat loss as natural areas are opened up to agriculture and natural resources exploitation.

The loss of biodiversity has socio-economic consequences that translate into diminishing amenities and decreasing options for future generations, threatening both current and future economic productivity and social stability. However, while it is clear that biodiversity loss affects human welfare, measuring this impact is difficult since appropriate and consistent data both on the physical/biological side and the socio-economic side are lacking (see Box 11.1). Furthermore, the benefits associated with biodiversity are often not clearly perceived, and are diffused across a wide range of actors, making them more difficult to identify than the benefits that accrue to a limited number of actors through exploiting the resources. Thus, because of the benefits all people gain from global biodiversity (through ecosystem services such as climate change mitigation, future option value such as for genetic material for medical uses, pure existence value, etc.), biodiversity losses in one country will also affect those living in other countries.

Box 11.1. Difficulties in defining and measuring biodiversity

A number of concepts and indicators have been developed in recent years to try to define and measure biological diversity and changes in this diversity. The Convention on Biological Diversity (CBD) defines biological diversity as "the variability among living organisms from all sources including, *inter alia*, terrestrial, marine and other aquatic ecosystems and the ecological complexes of which they are part; this includes diversity within species, between species and of ecosystems" (UNEP, 1994). Making this definition operational and translating it into numbers, however, is difficult. First, cultural and geographic variation across countries means that biodiversity has a different meaning for different people. Furthermore, unlike specific biodiversity resources such as fish for which physical measures are quantifiable, overall biodiversity lacks a common denominator that can be used to produce an indicator reflecting the state of the ecosystem. In addition, interactions among species within an ecosystem and with man and man-made events are often unclear, making it difficult to identify the specific pressures acting on biodiversity. For the sake of simplicity, most indicators focus on species richness instead of attempting to reflect ecosystem health more broadly (Brookes, 1998). However, research efforts are underway to establish more generally applicable measures of the state of biodiversity, with one example provided in Box 11.3.

Despite the data limitations, there is growing recognition that not only are a number of individual species under threat, but whole ecosystems in some cases. In response to this, biodiversity-related issues have gained in importance throughout the 1990s. A range of international conventions and agreements focusing on biodiversity are being discussed and implemented, and policy makers are becoming increasingly responsive to biodiversity-related concerns.

11.2. Pressures on biological diversity

Almost all human activities affect biodiversity. Table 11.1 indicates some of the potential pressures that traditional economic sectors exert on biodiversity. While sometimes these effects are positive, the reduced size of natural habitats and ecosystems in most regions indicates that in many cases the impacts have been negative and significant. The main direct causes of biodiversity loss can be categorised as: *i)* habitat destruction or alteration; *ii)* exploitation of wild species; *iii)* introduction of exotic species; *iv)* species homogenisation; *v)* pollution; and *vi)* global environmental change.

Modelling exercises of global biodiversity pressures for the 21st century indicate that land use changes will remain the main cause of biodiversity loss, with climate change and nitrogen deposition the next major pressures (Sala *et al.*, 2000). Of the economic sectors identified in Table 11.1, it is evident that in OECD countries the primary sectors – agriculture, forestry, and fisheries – play key roles in biodiversity loss, particularly contributing to the first four categories of proximate causes identified above. The agriculture sector occupies the largest share of total land area for most OECD countries, accounting for over half of all land area in most OECD countries, and currently covering approximately 39% of total OECD total lands (OECD, 2000*a*). This figure has remained fairly stable over the last few decades, although it is projected that agricultural cropland and pastures in OECD countries will increase marginally to 2020 as shown in Figure 11.1.

While agricultural lands can contribute positively to biodiversity conservation, for example through providing wildlife habitat, pollination and increasing soil fertility, many agricultural systems also negatively affect the environment (see Chapter 7). Intensification of agricultural production has resulted in the degradation of the natural resource base in many areas. This has been exacerbated by the growing reliance on fewer crop and animal species,[1] and there are concerns that a greater use of genetically modified species could further increase this

1. FAO and UNEP estimate of the domestic animal breeds for which precise population data exist, at least one-third (*i.e.* 1 350) are at risk of extinction, 119 are officially confirmed as extinct and another 620 are reported to be so (FAO/UNEP, 2000). The biggest threat to domestic animal diversity if the export of animals from developed to developing countries, which often leads to crossbreeding or even replacement of local breeds.

agricultural homogenisation and perhaps spill over into natural ecosystems, posing a threat to their stability. Finally, agricultural run-off of chemicals has also resulted in serious ecosystem damage and habitat loss in, for example, wetlands and water bodies. These pressures have reduced the number and variety of wildlife species using agricultural land as habitats in the last few decades, but there is evidence that this trend may be reversing in some OECD countries, especially where policies are implemented with the aim of ensuring habitat conservation (OECD, 2000*a*).

Table 11.1. **Pressures on biodiversity**

SECTOR	DIRECT		INDIRECT	
	Positive	Negative	Positive	Negative
Agriculture and plantation forests	• Creation of diverse ecosystems • Support for biological functions	• Natural ecosystem conversion to agriculture or forest • Fragmenting habitats • Introduction of non-native species	• Maintenance of ecosystem services, enrichment of biological diversity in some cases	• Pollution of ecosystems through farm chemicals run-off • Genetic homogenisation through monoculture use • Erosion, siltation, etc.
Fishery		• Destruction of habitats through damaging fishing practices • Potential over-fishing of target species or by-catch species • Introduction of non-native species		• Pollution of marine and fresh water ecosystems through effluent discharge, excessive nutrient and chemical loading (aquaculture), noise, etc.
Forestry (natural forests only)		• Habitat loss or fragmentation through forest clearing and infrastructure construction		• Pollution of forest ecosystems through effluents and noise • Erosion and associated effects • Colonisation of natural areas facilitated through infrastructure/ access provision
Oil production		• Pollution of ecosystems through spills • Destruction of ecosystems through infrastructure construction	• Decreased dependency on renewable natural resources (*e.g.* wood fuel)	• Pollution of ecosystems through extraction (*e.g.* effluents, noise, etc.)
Mining		• Pollution through leaching, etc. • Habitat destruction through infrastructure construction	• Reduced resource extraction through recycling	• Pollution of ecosystems linked to use of inputs in extraction (*e.g.* effluents, noise, etc.)
Transport and related infrastructure		• Facilitates access to fragile ecosystems, fragments habitats, pollution, etc. • Use of land for transport infrastructure	• Brings people to conservation sites, increases awareness	• Pollution associated with transport use, including greenhouse gas and air pollution emissions
Water and sanitation	• Creation of special habitats	• Water pollution and over-use destroys habitats and ecosystems	• Water conservation measures beneficial to ecosystems	
Industry		• Pollution of ecosystems • Loss of habitat through infrastructure development		

Source: Based on Biller (2001).

Forest quality and biodiversity habitat in OECD countries may decrease to 2020 because of more intensive forestry practices.

With the exception of a few countries, deforestation trends seem to be part of the past for most OECD regions (see Chapter 10). However, the quality of OECD forested areas in terms of natural habitat and biodiversity is decreasing in many cases due to a move from primary old growth forests towards secondary growth forests and monoculture plantations, often using intensive forestry production practices. Under the Reference Scenario, the proportion of total forested area in OECD countries under intensive plantation management is expected to increase from 1% in 1995 to roughly 5% in 2020, while the area of primary or old growth forest continues to decrease. Worldwide, total forest lands are expected to decrease by almost 6% over the next two decades. A large part of this reduction will take place in currently unexploited forests, particularly in South Asia and Africa.

Many of the major changes in land ecosystems are being brought about by infrastructure development. As indicated in Figure 11.1, while built environment remains a relatively small percentage of total OECD land use, it is expected to show the most significant growth rate in OECD regions to 2020. This reflects the increasing urban sprawl and the expansion of rural towns and cities occurring in most OECD regions (see Chapter 2). In OECD countries, the expansion of road networks poses a serious threat to biodiversity through the fragmentation of natural ecosystems, breaking them down into smaller, more isolated components. Between 1970-1997, road network lengths increased by almost 14% in OECD countries (OECD, 1999*a*). Roads also contribute to habitat loss indirectly through substantially lowering the costs of colonising an area, and so encouraging the spread of agricultural and rural developments. Pollution associated with road use is also increasing, and is expected to be one of the main contributors to OECD biodiversity loss in the future (McNeely, 1997). Humans have also significantly modified natural waterways in the last century, with dams having perhaps the greatest impact on freshwater ecosystems. Currently, dams impound 14% of the world's runoff. Furthermore, it is estimated that half of the world's wetlands have been lost in the 20th century, as land was converted to agriculture and urban areas or filled in to combat diseases (UNDP *et al.*, 2000).

Figure 11.1. **Changes in OECD land use, 1995-2020**

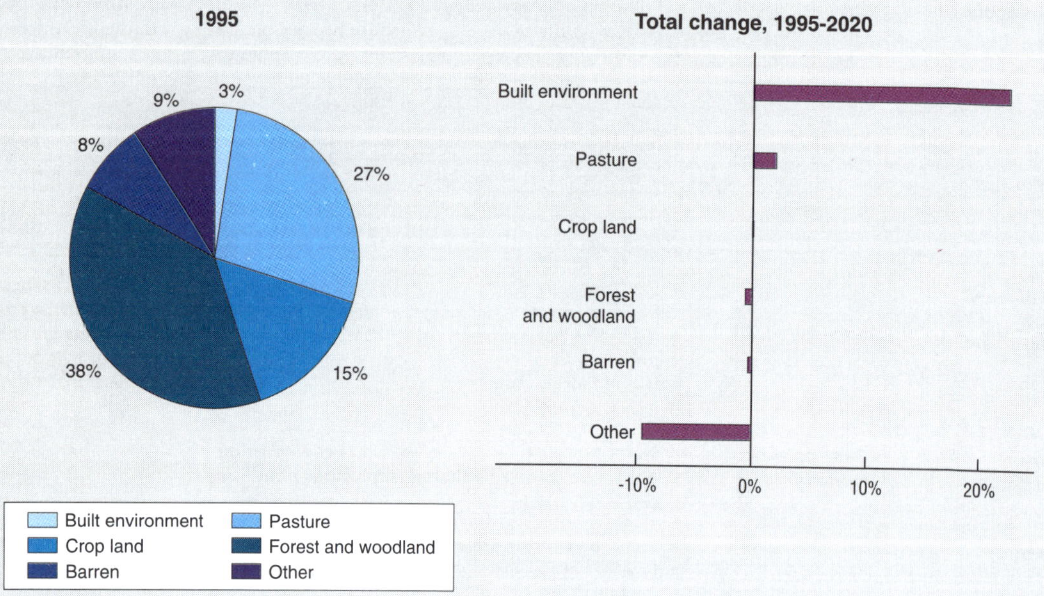

Source: Reference Scenario.

In the face of the expansion of infrastructure development and the conversion of old growth forest and natural grasslands to other land uses, protected areas have grown significantly both in size and number since 1980 in almost all OECD countries, reaching 12% of the total OECD land area in 1996 (OECD, 1998*a*). The potential for these areas to expand further in the future will depend directly on policy developments, including public and private partnerships for nature conservation. The establishment of marine parks and protected coastal zones has also been increasing in recent years, as awareness grows of the threats to these ecosystems and their species, particularly from commercial and tourism activities.

Protected natural areas are growing both in number and size in OECD countries.

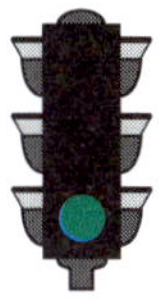

Inland waterways are under pressure in OECD countries as well, primarily from excessive siltation, industrial and agrochemical run-off, while marine ecosystems are under threat from the degradation of coastal zones, marine pollution, and over-exploitation. The main threat to reef ecosystems is over-exploitation for tourist or harvesting purposes, with land-based pollution of coastal reefs also a significant pressure. The considerable number of large-scale oil spills that have occurred in recent years have also had a strong impact on coastal ecosystems. Global climate change is likely to exacerbate pressures on species highly dependent on water temperature, such as some corals. For commercially viable fish stocks, over-fishing is the primary threat, and a number of fisheries have reached or are approaching collapse as a result of these pressures (see Chapter 9). Non-commercial fish and marine wildlife species are also threatened as by-catch in commercial fishing processes, or through the use of habitat-destroying fishing methods, or general pollution of the marine environment. In many cases adverse incentives are in place that encourage over-fishing or other environmentally damaging commercial activities (see Box 11.2).

Box 11.2. Perverse incentives and biodiversity loss

Perverse incentives, which encourage environmental damage and biodiversity loss, are present throughout OECD economies. They are often in the form of subsidies encouraging certain (environmentally harmful) activities, but also may take the form of direct payments from government budgets, tax exemptions or reductions, or the subsidised provision of private and public services (*e.g.* road infrastructure). Regarding the conservation of biodiversity, they can be particularly harmful by generating rents through the consumption of natural resource intensive goods, or are directed towards certain economic sectors (see Table 11.1).

Estimates indicate that direct subsidies to agriculture in OECD countries amounted to as much as US$361 billion in 1999, while government support for marine capture fisheries amounted to US$6.3 billion, and for coal production it was US$6.2 billion (OECD, 2000*b*; OECD, 2000*c*; IEA, 1998). While some of these subsidies – such as agricultural land set-aside or fishing vessel buy-back schemes – aim to reduce pressures on biodiversity and the environment, it is likely that the majority contribute to further destroying the natural resource base, exploitation of wild species, pollution generation and climate change.

11.3. Changes in the state of biological diversity

Measuring the state of biodiversity, let alone establishing recent trends or drawing future projections (see Box 11.1), is fraught with practical and theoretical difficulties. To a large extent, it is necessary to rely on indications of the pressures acting on biodiversity or proxy indicators of biodiversity quality (*e.g.* changes in particular species or the state or extent of specific natural resources). However, some work is underway to develop broader, operational indicators of biodiversity, such as the Natural Capital Index (NCI) framework being developed under the Convention on Biological Diversity (see Box 11.3). A provisional, pressure-based application of the NCI framework for Europe found that in that region only about 50% of the land is still natural area, of which only a small part is protected under existing biodiversity-related legislation (RIVM, forthcoming 2001). The remaining natural areas are under high pressure, particularly from pollution, fragmentation, and urban development. It is expected that biodiversity loss will continue in Europe, although at a slightly reduced rate. However, it is expected that there will be an increase in the protection of the most threatened species and in land set-aside programmes for nature protection (EEA, 1999).

Box 11.3. The Natural Capital Index (NCI) framework

The Natural Capital Index (NCI) framework is a simple index of natural and man-made areas and their quality. It was developed under the Convention on Biological Diversity as a general framework to describe and assess ecosystems in a comparable and universal manner (UNEP, 1999; Ten Brink, 2000). The framework was accepted as a starting point for further discussions at the fourth meeting of the Conference of the Parties in Bratislava in 1998, and is still evolving.

The NCI framework provides information on the state and changes in biodiversity due to human interventions. The NCI is defined as the product of the size of natural areas and their quality, using a range of 0-100%. Ecosystem quantity is calculated on the basis of what remains of an ecosystem and the percentage of total land area this represents. This calculation is relatively easy as it is made with land-use statistics. Ecosystem quality is calculated by averaging the current or baseline ratios of a representative core set of species or other ecosystem quality variables. This relies essentially on species monitoring and research. When data on ecosystem quality are not available, a pressure index can be used instead, since it is probable that a high quality ecosystem will show an inverse relation to the pressures exerted on it. The NCI is not perfect and can require significant amounts of data gathering if applied comprehensively. Nonetheless, one of the advantages of this approach is that it offers a consistent and understandable method for measuring biodiversity, which in principle is traceable and can project future trends, and which can be useful in policy-making.

Once the individual components of biodiversity have been examined more closely, particular areas of concern and the threats that ecosystems will face in the near future become more apparent. The continuing pollution of ecosystems and changes in agricultural production and land use are expected to further degrade biodiversity in OECD countries. It is estimated that about two-thirds of agricultural lands worldwide have been degraded in the past 50 years by erosion, salinisation, compaction, nutrient depletion, biological deterioration, or pollution, and that over half of this land has been strongly or very strongly degraded (UNDP *et al.,* 2000). More fish species may become commercially extinct due to over-exploitation, with species such as Atlantic cod, some tuna, and haddock already threatened. Marine habitat loss, increasing incidence of exotic or invasive species, and coral bleaching all contribute to a decline in coastal biodiversity. Biodiversity in freshwater ecosystems is also under serious threat, with an estimated 20% of the world's freshwater fish extinct, threatened, or endangered (UNDP *et al.*, 2000).

While forests have the highest species diversity of any ecosystem, many forest-dwelling large mammals, half the large primates, and nearly 9% of all known tree species are at some risk of extinction (UNDP *et al.*, 2000). Recent estimates place the worldwide loss of tropical rainforest caused by human intervention at around 12 million hectares per year. In South East Asia alone, a forest area equal to the size of Switzerland is being lost each year, while it is estimated that about 55 million hectares of the Brazilian Amazon rainforest have been cleared since the 1970s.

In many OECD regions, the introduction of non-native (exotic) species can also pose a threat to biodiversity and natural resource management, as well as to the agriculture, forestry and fisheries sectors, when these are invasive weeds or animal species. Australia and New Zealand are acutely affected by the impact on non-native species, because of their distinctive biota, but other OECD regions are also affected. In the US, a government study provides a conservative estimate of US$97 billion for the economic losses suffered from the spread of non-indigenous species over the last hundred years (Office of Technology Assessment, 1993).

11.4. Policy options

While land use changes will continue to put pressure on biodiversity in some OECD regions over the next few decades, the main pressures are likely to come from the continued pollution of ecosystems, natural and human-induced changes in the global climate, greater crop and tree homogenisation, and over-fishing. Policies for tackling biodiversity loss in OECD countries will need to focus particularly on reducing habitat degradation (*e.g.* lowering pollution levels, especially of persistent and toxic chemicals), limiting habitat fragmentation from infrastructure development, minimising agriculture and forestry intensification, and addressing the causes of over-fishing. Of more concern is biodiversity loss in non-OECD countries, which is expected to continue at alarming rates in some ecosystems and for some species.

A range of policy measures can be used to tackle these problems. Because of the uncertainties surrounding biodiversity, and the complex inter-relationships that contribute to its maintenance or loss, a "package" of policy instruments or "policy mix" will generally be required (OECD, 1999*b*). In order to ensure the stability of the whole ecosystem, biodiversity policies need to be applied within the framework of an ecosystem approach, which calls for a comprehensive strategy for the integrated management of land, water and living resources that promotes conservation and sustainable use of the ecosystem in an equitable way. The Conference of Parties to the CBD has recognised the ecosystem approach as the primary framework for action under the Convention, and OECD countries have begun responding with a policy shift that focuses less on protecting endangered species and more on developing holistic biodiversity strategy plans. The 2000-2001 edition of the *World Resources* report produced by UNDP, UNEP, Work Bank and WRI uses the ecosystem approach as the underlying basis for analysing natural resource health (UNDP *et al.*, 2000).

International co-operation

Because of the global nature of many biodiversity problems, and the benefits ecosystems bring both in terms of the direct services they provide and their associated existence and option values, there are a number of international agreements and conventions to protect biodiversity and species.[2] While many of these have been successful in ensuring the development of systems for monitoring ecosystem integrity (*e.g.* the Ramsar Convention and World Heritage Convention) or proposing appropriate policy actions for addressing threats to biodiversity (*e.g.* the CBD and Desertification Convention), few have made much headway in gaining country commitments to implement the necessary policies. Increased co-operation between the Conventions (for example, between the Framework Convention on Climate Change, the Convention to Combat Desertification and the CBD), and a greater focus on their complementarities would help in this process.

In many cases, financial or other incentives need to be employed at the international level to ensure that the necessary actions are taken. Because many of the biodiversity-rich areas are in less developed countries, and the concern for the protection of biodiversity and the funds to enable such protection tend to be available in more developed countries, increased transfers of funds through bilateral or multilateral agreements to protect biodiversity are needed. Thus, it is likely that global efforts to halt deforestation in South America and South-East Asia will only bring results once viable economic alternatives to the activities that lead to this destruction are offered. Where funds are available, they also need to be better targeted to address the most pressing threats to biological diversity, rather than simply the most popular or well-known concerns.

Regulatory instruments

The use of regulations and access restrictions to protect biodiversity or biological resources is already widespread in OECD countries and internationally. Included in this are the delineation of natural parks and conservation

2. These include the Convention on Biological Diversity (CBD, 1992), the Convention to Combat Desertification (1994), the Convention on International Trade in Endangered Species of Wild Fauna and Flora (CITES, 1973), the Convention on Migratory Species of Wild Animals (Bonn Convention, 1983), the Convention on Wetlands (Ramsar Convention, 1971), and the World Heritage Convention (1972), to name a few.

zones, and the enactment of prohibitions on the harvesting or use of wildlife species. They are conceptually easy to understand, and can contribute to the achievement of clear, pre-formulated objectives such as reducing the direct pressures on threatened or endangered species. As a result, a number of protected or natural heritage areas have already been established worldwide, and a number of threatened species have been identified. Established principles for protecting these species and areas are in many cases in place, such as the agreement by signatories to CITES (Convention on International Trade in Endangered Species) to ban trade in products of endangered species, or the World Heritage Convention list of sites whose outstanding values should be preserved for all humanity. The use of these types of measures will need to continue and perhaps even expand in the coming decades in OECD countries and elsewhere, as the integrity of natural areas outside of specifically protected zones diminishes.

While the fair and equitable sharing of the benefits arising out of the utilisation of genetic resources is one of the three objectives of the Convention on Biological Diversity, it has received little attention so far. In the future, the distributive aspects of restrictions on access to biodiversity, and the use of its components, will need to be addressed (see also Chapter 22). As protected natural areas increase, co-operative agreements which ensure the sustainability of the resources while allowing some productive use of their components will be necessary, particularly in non-OECD countries, where access or use restrictions may come into direct conflict with the economic development of local communities dependent on these resources for their livelihood. Similarly, clear intellectual property rights and agreements on the sharing of the benefits from the use of genetic resources (*e.g.* for scientific or medical purposes) is essential to establish the rights of all parties involved in bio-prospecting (OECD, 1998*b*).

Economic instruments

Because access or use restrictions tend to be costly policy instruments both to government administrations and society, they will increasingly need to be complemented with more flexible mechanisms that provide positive incentives for sustainable use. Where they are earmarked for particular environmental purposes, the charges can also help to raise revenues for maintaining the regulatory measures and access restrictions. For example, visitor fees for national parks can be introduced to raise revenues to fund park maintenance, or permits sold for the sustainable use of some components of a protected area. In most cases, however, revenues from taxes or charges should be returned to the central revenue fund and dispersed through the regular budget mechanisms. The use of agri-environmental payments is also growing in OECD countries, with payments to farmers for undertaking certain nature conservation or habitat enhancement practices (see Chapter 7). These incentives can be used to make up the profitability gap between unsustainable and sustainable use of biological resources.

Perverse incentives, including subsidies to some sectors, continue to encourage environmentally damaging activities resulting in biodiversity loss in OECD countries.

A range of economic instruments is available, but the most widely used ones in OECD countries are park fees or charges, resource extraction charges, hunting and fishing fees, subsidies for conservation efforts, taxes on unsustainable activities, and individual tradable quotas. While such instruments are widespread in OECD countries for tackling some natural resource use problems (*e.g.* wastewater treatment charges, hunting license charges, or national park fees), they are less well-developed or wide-spread for other problems (*e.g.* wetlands protection, or soil protection). Where charges or fees are levied on biodiversity-threatening activities, they are often too low to provide the appropriate incentives or to cover the full environmental costs of the activity. This may be true, for example, with timber stumpage charges in many countries, or charges for livestock grazing on public lands. Furthermore, a range of sectoral policies also have the perverse effect of encouraging environmentally-damaging activities (see Box 11.2), and may counter the effects of other instruments employed to encourage biodiversity protection. The reduction or elimination of these perverse incentives, while at times politically difficult, can serve to increase efficiency in both the economic and the environmental sense

Increasingly, transferable development rights, or rights to use certain aspects of protected lands, are being employed. These allow the conservation of the essential components of the ecosystem, while at the same time allowing the sustainable use of some components, thus generating incomes. Examples include transferable development rights in rural areas, or allowing the sustainable use of forest ecosystems for fruit and nut production, rather than for harvesting old growth timber. Again, these can be used to make up the profitability gap between sustainable and unsustainable use of the resources, while ensuring that the social, economic and environmental benefits of the resources are realised.

Information and other instruments

An important pre-requisite for all biodiversity policy-making is the gathering, processing and dissemination of information. Appropriate treatment of biodiversity information and the development of suitable indicators of biodiversity quality not only have the potential to improve enforcement through capacity building, but can also improve instrument design. In the past, the lack of physical and economic information regarding biodiversity values has hampered efforts to design policies to protect it. Increasingly, new indicators are being developed (*e.g.* the NCI discussed in Box 11.3), innovative techniques are being used for managing biodiversity monitoring or classification (*e.g.* the identification of eco-regions or biodiversity hotspots), and increasingly planning decisions require at least an attempt to establish a valuation of the environmental benefits of biodiversity and ecosystem services (see Box 11.4). Significant progress still needs to be made, however, in the monitoring of key natural resources, the understanding of the pressures that act upon them, and the design of efficient and effective policies to ensure the conservation or sustainable use of biological resources

Consumers are increasingly demanding ecologically sound products while investors worry about their reputation when searching for opportunities in which to invest. This has translated into the release of a myriad of products that claim to be less harmful to the environment than their usual counterparts. Eco-labelling and certification schemes for forest, fishery and agricultural products are gaining in importance in OECD countries to indicate the reduced environmental effects of some products or processes, and can have a significant impact on consumption patterns (see Chapter 5). These in effect assist in the creation and functioning of markets for "environmentally friendly" goods and services, allowing consumers to choose to pay higher prices for products which are developed in ways they consider more compatible with nature.[3] For similar reasons, investment funds that are aimed at ecologically sound investments have begun to appear in many OECD countries. As experience has indicated that many consumers are willing to pay more for such products, the use of these schemes is continuing to grow rapidly in OECD countries. A challenge for the future will be to ensure their legitimate use and effectiveness, and that they are clearly understandable to consumers.

Box 11.4. Revealing the economic value of biodiversity

Assessments of the economic value of biodiversity can help to raise public and political awareness of the importance of biodiversity, and to inform cost-benefit decisions regarding land use and other activities that may impact on the environment. Partly because measuring biodiversity or biodiversity loss is difficult (see Box 11.1), estimating economic values for the services ecosystems provide is particularly hard. However, since without such estimates the market value of biodiversity is often assumed to be zero or infinite, efforts are being made to establish reliable and agreed methods for calculating biodiversity value.

The main valuation methods can be divided into two categories. First, there are revealed preference techniques (*e.g.* market prices, hedonic pricing method, travel cost method, discrete choice modelling, and production function approaches), which attempt to separate out the amount people pay for given biodiversity products or services from the total bundle of services and goods they may be purchasing. Second, there are stated preference techniques (*e.g.* contingent valuation method, contingent ranking, rating or choices, and expert valuation) which rely on questionnaires to establish people's preferences. OECD is developing a handbook on the valuation of biodiversity in order to help OECD countries establish common principles and procedures for valuation (OECD, forthcoming 2001). Once economic values of biodiversity loss or the benefits that ecosystems provide have been established, these can be used to inform policy decisions or for establishing economic incentives to internalise the full costs of natural resource use.

3. OECD is currently preparing a major report on the creation and use of markets that promote biodiversity conservation and sustainable use (OECD, forthcoming 2001).

REFERENCES

Biller, D. (2001), "Biodiversity", background document for the OECD *Environmental Outlook*, OECD, Paris.

Brookes, M. (1998), "The Species Enigma", in *New Scientist*, 111: 1-4, June.

EEA (European Environment Agency) (1999), *Environment in the European Union at the Turn of the Century*, EEA, Copenhagen.

FAO/UNEP (United Nations Food and Agriculture Organisation/United Nations Environment Programme (2000), *World Watch List for Domestic Animal Diversity*, FAO, Rome.

IEA (International Energy Agency) (1998), *The Role of IEA Governments in Energy: 1999 Review*, OECD/IEA, Paris.

McNeely, J.A. (1997), *Conservation and the Future: Trends and Options toward the Year 2025*, IUCN, Gland.

OECD (1998*a*), *Towards Sustainable Development: Environmental Indicators*, OECD, Paris.

OECD (1998*b*), "Economic Issues in Benefit Sharing: Concepts and Practical Experiences", ENV/EPOC/GEEI/BIO(98)7/FINAL, OECD, Paris.

OECD (1999*a*), *OECD Environmental Data Compendium 1999*, OECD, Paris.

OECD (1999*b*), *Handbook of Incentive Measures for Biodiversity: Design and Implementation*, OECD, Paris.

OECD (2000*a*), *Environmental Indicators for Agriculture Volume 3: Methods and Results*, OECD, Paris.

OECD (2000*b*), *Agricultural Policies in OECD Countries: Monitoring and Evaluation*, OECD, Paris.

OECD (2000*c*), The *Impact on Fisheries Resource Sustainability of Government Financial Transfers*, OECD, Paris.

OECD (forthcoming 2001), *Handbook on the Applied Valuation of Biological Diversity*, OECD, Paris.

Office of Technology Assessment (1993), *Harmful Nonindigenous Species in the United States*, United States Congress, OTA-F-565, US Government Printing Office, Washington DC., Online at: *www.ota.nap.edu/pdf/1993idx.html*.

RIVM (National Institute of Public Health and the Environment), (forthcoming 2001), "Biodiversity Loss", *Economic assessment of Priorities for a European Environmental Policy Plan* , Technical Background Report 9 prepared for Directorate General XI (Environment Nuclear Safety and Civil Protection), RIVM, Bilthoven.

Sala, O., F. Chapin, J. Armesto, E. Berlow, J. Bloomfield, R. Dirzo, E. Huber-Sanwald, L. Huenneke, R. Jackson, A. Kinzig, R. Leemans, D. Lodge, H. Mooney, M. Oesterheld, N. Poff, T. Sykes, B. Walker, M. Walker, and D. Wall (2000), "Global Biodiversity Scenarios for the Year 2100", *Science* 287: 1770-1774.

Ten Brink, B.J.E. (2000), Biodiversity Indicators for the OECD Environmental Outlook and Strategy: A Feasibility Study. Report Number: 402001014, RIVM, Bilthoven.

UNDP, UNEP, WORLD BANK, and WRI (United Nations Development Programme, United Nations Environmental Programme, World Bank and World Resources Institute) (2000), *World Resources 2000-2001: People and Ecosystems, the Fraying Web of Life*, WRI, Washington DC.

UNEP (United Nations Environmental Programme) (1994), *Convention on Biological Diversity*, UNEP, Geneva.

UNEP (1999), "Development of Indicators of Biological Diversity", UNEP/CBD/SBSTTA/5/12, UNEP, Montreal.

Section IV

ENERGY, CLIMATE CHANGE, TRANSPORT AND AIR QUALITY

This section examines the closely interrelated sectors of energy and transport, and the issues of climate change and air quality. The energy and transport chapters discuss the trends and projections of the overall developments in the sectors and the environmental impacts of these changes. The climate change and air quality chapters examine past and projected future emissions of greenhouse gases and air pollutants, and analyse the possible impacts on the environment of these changes. All the chapters conclude with an indication of the policy options available for addressing the problems identified.

12
Energy

KEY SIGNALS

- Energy use in OECD countries grew by 36% from 1973 to 1998. By 2020 it is expected to grow by a further 35% and by 51% worldwide. Commercial and residential energy uses, as well as transport, represent the most rapidly growing demands for global energy use.
- The production and use of energy is the main contributor to greenhouse gas emissions, air pollution and nuclear waste generation, and also leads to noise, water pollution, and ecosystem degradation. Energy-related air pollution has significant negative effects on human health.
- In the 2020 timeframe there is a potential for significant improvements in the efficiency and environmental performance of fossil fuel combustion, and for a growing market share for renewable energy and new technologies such as fuel cells.
- However, expected total increases in energy use, combined with an increasing demand for electricity and declining use of nuclear energy, is likely to lead to increased total primary energy use and energy-related emissions.
- To achieve significant reductions in energy-related environmental pressures in OECD regions, more aggressive policies are needed to accelerate the use of natural gas, renewables and other low emission technologies, and quicken the pace of improvements in conversion efficiency for fossil fuel use in the industry and power sectors. Policies aimed at consumer demand management can improve energy efficiency in the residential, commercial, agricultural and transport sectors, as well as promote "green" power.

12.1. Introduction

Energy production and use is fundamental to modern, industrialised economies, and supports all economic activity. Energy is used to move vehicles, heat homes, and power industry. However, producing and using energy also leads to adverse environmental effects, including air pollution, greenhouse gas emissions and nuclear waste. Energy systems also result in noise, water pollution, and ecosystem degradation, as well as impact on human health. In general, the impact of energy use on the environment is a result of several factors: the mix of fuels used to produce energy; the efficiency of conversion of primary energy into useful energy (including distribution); the technology in use; and the total level of energy used. The use of fossil fuels in the fuel mix for energy with today's technologies continues to pollute the air and emit greenhouse gases, though pollution intensities are declining. Nuclear energy production entails the risk of costly nuclear accidents and the need to dispose of radioactive waste. Even renewable energy sources have some negative environmental impacts: hydropower may degrade the ecosystem by altering the natural state of river basins, with resulting effects on flora and fauna, and other renewable energy systems, such as wind, have visual and noise impacts.

Numerous new technologies are under development which could significantly transform the links between energy production and use and the environment. In addition to electricity systems based on renewable energy, these include the use of cellulose ethanol fuel in motor vehicles, and the development and use of hybrid vehicles, hydrogen fuel cells, and CO_2 capture technologies. A key challenge in the development of a sustainable future is the transition to cleaner technologies and energy sources which improve the balance between security and reliability of supply, costs of production and use of energy, and environmental externalities.

12.2. Developments in the energy sector

Demand for energy

OECD countries are achieving some de-coupling of energy use from economic growth.

It is projected under the Reference Scenario that worldwide total energy use will grow by about 51% between 1995 and 2020, with much of the increase arising in non-OECD countries. The share of OECD area energy consumption in the world total is projected to fall from 35% in 1995 to 32% in 2020. Compared with recent trends in energy demand, these figures represent a decline in the energy intensity of the economy both for OECD regions and worldwide, indicating that some de-coupling of energy use from economic activity is taking place (Table 12.1). However, per capita energy use is expected to continue to increase to 2020 in both OECD countries and worldwide.

Table 12.1. **Key energy sector statistics and projections**

		1980	1998 (*or latest available year*)	2020 projected	Total change 1995-2020
Total primary energy supply (TPES) (PJ)	OECD	170 068	213 400	275 622	30%
	World	299 817	402 569	586 193	52%
Total final energy consumption (TFC) (PJ)	OECD	123 636	145 155	197 768	35%
	World	202 096	278 244	417 460	51%
Energy intensity of economic activity (TPES/GDP) (GJ/1 000 US$ GDP)	OECD	13.0	10.9	8.0	–19%
	World	..	15.5	12.9	–13%
Energy intensity per capita (TPES/capita) (GJ/capita)	OECD	177	196	221	18%
	World	68	70	78	15%

Sources: IEA (2000*a*), IEA (2000*b*), GTAP database and Reference Scenario.

Energy consumption in OECD countries and worldwide is projected to continue to increase to 2020.

In OECD regions, energy use has been growing more slowly in North America (1.1% p.a.) and Europe (0.9% p.a.) since 1973 than in the OECD Pacific countries (2.7% p.a.) (IEA, 2000*b*). Over the next two decades, the rates of increase in energy demand will be slower, particularly in Western Europe and Japan (Figure 12.1). In non-OECD regions, the demand for energy in China and East Asia will continue to increase rapidly over the next 20 years (at roughly 3% and 2% per annum respectively), driven by strong economic and population growth. The dramatic drop in final energy use that was seen in Eastern Europe and the former Soviet Union in the 1990s[1] due to major political reforms and economic decline will be reversed, and energy demand there is expected to grow again over the next twenty years.

In OECD countries, transport accounts for one-third of current final energy consumption, but has represented almost two-thirds of the growth in consumption since the early 1970s. Industrial energy consumption constitutes almost a further third of current consumption, with the remainder split among households, services and, to a small extent, agriculture. According to the IEA *World Energy Outlook 2000*, by 2020 transport is likely to account for more than half of world oil demand, 62% of OECD country primary oil demand, and account for all oil-demand growth in OECD regions (IEA, 2000*b*).

1. From 29 100 PJ in 1990 to 16 300 PJ in 1997 for the Russian Federation (OECD, 1999*a*).

Figure 12.1. **Total final energy consumption (TFC), 1995-2020**

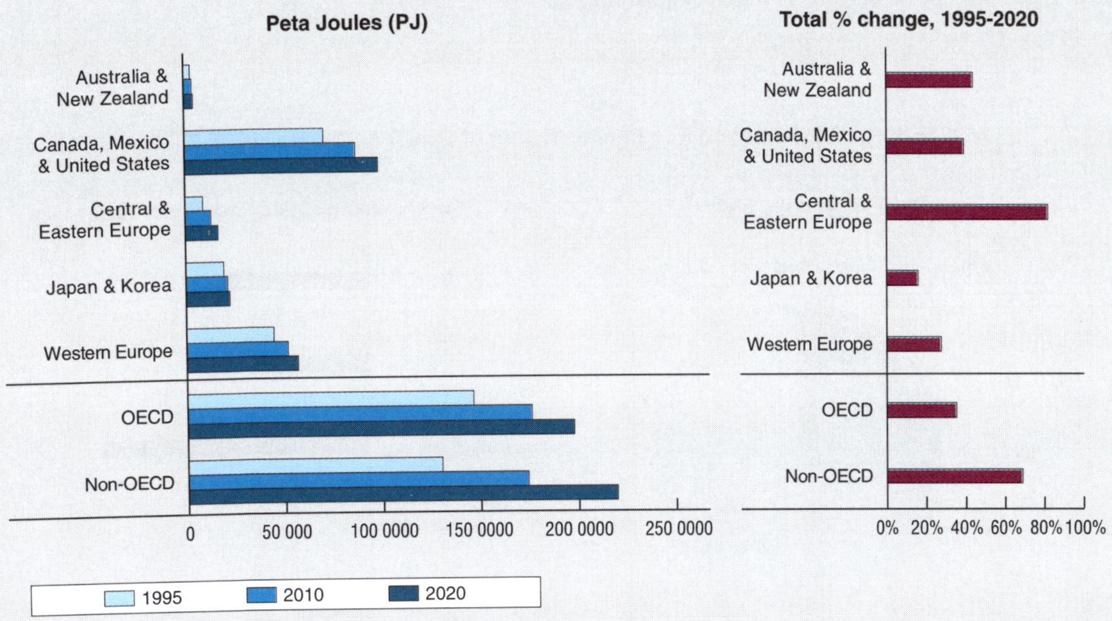

Source: Reference Scenario.

In the industrial sector, final energy consumption has been falling slightly in OECD countries, mainly due to changes in the structure of economic activities – with a shift from industrial manufacturing to services – along with increased efficiency in energy use. Structural shifts in the economies of the US, Germany, and Japan are estimated to have led to significantly lower total energy use (by at least 10%) (IEA, 1997). In parallel to this decrease in energy consumption, there has been a trend towards increasing use of fuels from less carbon intensive energy sources, for example the growing use of biomass energy in the pulp and paper sector (see Chapter 18).

The commercial and residential sectors are continuing to increase their energy use, with greater use of electrical goods (such as more computing and Internet time), more floor space per capita, and higher levels of cooling and heating comfort (see Chapters 5 and 16). Because of these trends, energy use by the household and services sectors is expected to continue to increase in OECD countries to 2020.

Energy supply

Significant changes in the types of fuels used to feed energy production have occurred over the last few decades. The current fuel supply mix for the OECD area is dominated by oil, followed by solid fuels (largely coal) and natural gas (Figure 12.2). The shares of coal, and to a lesser extent oil, in total energy supply have fallen steadily in OECD regions in recent years, reflecting increased electrification, relative reductions in the importance of energy intensive manufacturing sectors and, the substitution of natural gas for oil and coal, especially in the electricity sector. In OECD countries, virtually all the increase in demand for coal stems from power generation. Gas has increased its share in overall energy use, and a further deregulation of energy markets and technological advances in combined-cycle gas turbines is expected to continue to increase the penetration of gas in electricity production and some industrial uses over the next two decades. New renewables have been growing at a very rapid rate in recent decades, but from a relatively small base.

Figure 12.2. **Fuel shares in energy mix of OECD countries, 1995-2020**

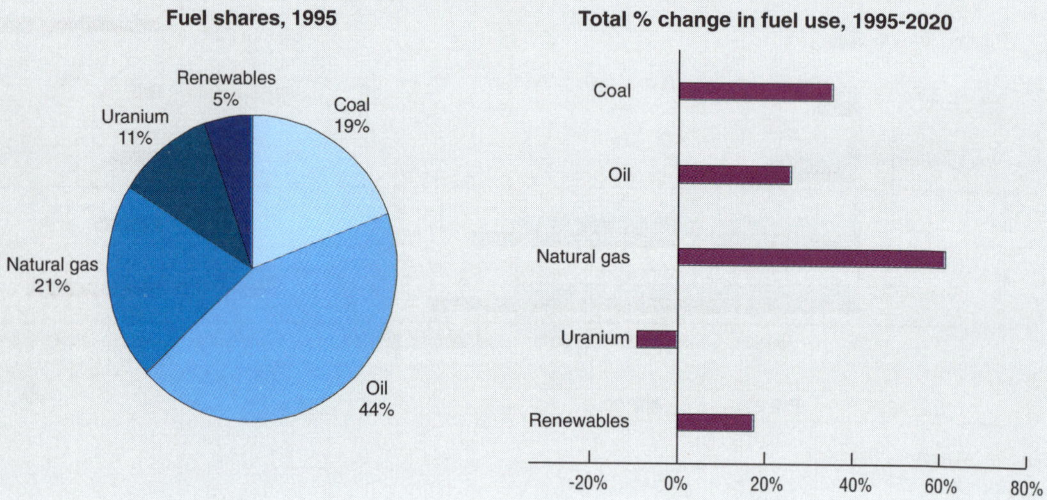

Sources: IEA (1998*c*) and Reference Scenario.

The turnover of energy generation capital has a significant influence on the changing fuel mix in different sectors. The power sector typically has a slow capital stock turnover, as fossil fuel and nuclear stations have life spans of at least 20 to 40 years. As a result, the bulk of generation facilities needed to meet demand in 2020 in OECD countries have already been built, limiting the extent to which new sources, such as renewables, can penetrate the electricity generation market.

Perhaps the most important trend in energy production over the last few decades has been the rapid increase in the share of energy from electricity, which has almost doubled its share of total energy consumption since 1975. This is partly the result of increased growth in the electricity-intense residential and commercial sectors. Many applications in these sectors, especially business equipment and residential appliances, have no realistic alternative fuel source to electricity. Furthermore, there has been a shift in all sectors to greater electricity use. This reflects the wide range of uses of electricity, its flexibility, safety and cost. In some regions in OECD countries, where the electricity supply system is too unstable to provide the continuous energy required for high-powered computer use, an increasing share of electricity is obtained from off-grid power sources.

Most of the growth in energy use in OECD countries to 2020 is expected to come from greater oil and natural gas use, with the former being driven by the demand for mobility in the oil-dependent transport sector, and the latter by increased use of gas for electricity generation. The share of renewables is expected to increase to 6% of total fuel use by 2020 in OECD countries, but this will largely depend on financial incentives from government. While nuclear fuel use has grown significantly in OECD countries over the last few decades, accounting for nearly 40% of the increase in their total energy use from 1973 to 1996, it has slowed in recent years. Political, economic and technical issues have led to a halt or reduction in nuclear power plant construction in most OECD countries, and several are phasing out their existing nuclear power facilities. From now to 2020, about 30% of existing plants are expected to be retired, although there is a possibility that some life extensions will be granted (IEA, 2000*b*). Projections indicate that total final nuclear energy use in OECD countries could decline slightly to 2020 as a result.

Energy intensity

Although total energy use in OECD regions has been rising, the energy intensity of the economy (total primary energy supply per unit of GDP) has decreased by just over 16% since 1980. This mainly reflects improved energy efficiency across the economy, and fuel and technology changes in the power sector, and has largely been driven by changes in energy prices as well as market reform policies. Energy intensity fell rapidly, especially in OECD countries, following the oil price rises of 1973/74 and 1979. Although the rate of decrease in energy intensity slowed after 1985 when real prices fell to about their pre-shock levels, there was no sharp rise in energy use following the fall in prices, indicating that some energy efficiency improvements have been "locked in" to OECD economies through infrastructure and permanent behavioural change.[2] Recent OPEC arrangements to reduce oil output were responsible for large price rises in 1999 and 2000. Future energy price trends are difficult to predict, but will depend largely on OPEC arrangements and on the pace of energy market reforms (Box 12.1).

Box 12.1. Energy sector reforms

Most OECD countries have begun a process of liberalising energy supply and distribution services. In most cases policy reform in the electricity sector is leading to lower electricity prices as industry becomes more efficient and competitive. The continuing reform of these markets can be expected to further drive down electricity prices, increasing demand. In order to counter-balance the negative environmental impacts that will result from these reforms, it will be necessary for countries to adopt policies to encourage greater energy efficiency, emission reductions, and use of cleaner fuels. Appropriate policy tools are discussed in Section 12.4.

In OECD countries, there has also been a general decline in recent years in the overall efficiency of converting primary energy to final energy. This reflects the increased share of electricity in final energy use, but is offset to some extent by efficiency improvements in the conversion from primary to final energy. These will result when available higher-efficiency technologies (*e.g.* combined cycle-gas turbines, CCGT) will be brought on stream with the turnover of power generation capital stock. Improved technologies offer significant potential for high efficiencies, and the market potential for new gas-fired cogeneration is also high. The energy efficiency of OECD economies is expected to improve to 2020 as measured by TFC/GDP (see Table 12.1).[3]

Trade in energy

As energy consumption in OECD regions rapidly outstrips energy production, the Reference Scenario indicates that more imports of fuels will be required over the next few decades, particularly of natural gas and oil. Currently, OECD regions produce roughly 74% of the amount of energy they consume. Energy self-sufficiency reached a peak of 78% in the 1980s, and has been gradually declining since. The energy self-sufficiency of different fuels varies substantially. While OECD countries are self-sufficient in the use of nuclear and renewable energy sources, and import only a small proportion (3% in 1998) of the coal they use, more than 15% of gas and more than 50% of oil used are imported from non-OECD countries.

Energy self-sufficiency is projected to drop significantly to 2020, as total energy consumption rises by over 20% and energy production in OECD countries remains stable or declines slightly. An increasing percentage of world crude oil supplies is expected to come from the Middle East OPEC countries (from 54% in 1997 to over 74% in 2020) (IEA, 2000*b*). OECD regions, particularly Western Europe, are projected to significantly expand imports of natural gas over the next 20 years, while non-OECD countries will expand exports. These changes in trade patterns, and dependence on sources of gas supply from non-OECD countries, could have implications for gas supply

2. Price elasticities tend to be asymmetric, with consumers more responsive to a rise in price than to a decline.
3. A considerable degree of uncertainty exists over the rate of this "autonomous energy efficiency improvement" (AEEI) however. For the sake of simplicity, the Reference Scenario assumes that AEEI will be 0.75% per year for all sectors and regions to 2020. See Annex 2 for more details.

security. As open regional electricity markets continue to expand, so will the inter-regional electricity trade and, depending on the fuel mix within the markets (hydro/fossil fuelled/ nuclear), could either increase or decrease local and global emissions. Domestic power pool dispatch systems that operate on an auction basis favour the power source with the lowest marginal cost which, in the case of existing plants, tends to be nuclear power.

12.3. Environmental effects of energy production and use

Energy production and use can affect the environment in many ways, with very diverse impacts from different fuel sources. The expected total increases in energy use in OECD countries will increase the environmental effects of energy use and production, while the effects of changes in the composition of fuel use will depend on the final fuel mix. Reduced reliance on coal in OECD regions will lead to lower emissions intensities for sulphur oxides (SO_x), particulate matter (PM) and carbon dioxide (CO_2), while reduced reliance on nuclear may counterbalance these effects and lead to increased nitrons oxides (NO_x) emissions depending on the fuel that is used to replace the declining share of nuclear. The decline in the share of nuclear energy in the fuel mix will, however, lower the risk of nuclear accidents and reduce the generation of radioactive waste.

Air quality and climate change

Fuel combustion remains the major source of greenhouse gas emissions and many air pollutants.

Air pollution and human induced climate change are the most pressing environmental problems arising from energy use. While situations vary significantly among individual countries, fuel combustion is the major source of air pollution across OECD regions as indicated in Table 12.2, with subsequent impacts on human health and well-being, as well as on ecosystems (see also Chapters 15 and 21). Regional and global impacts on ecosystems include acid rain through SO_x and NO_x emissions, and climate change through increased atmospheric concentrations of greenhouse gases. In 1998, the energy sector (stationary fuel combustion and transport) accounted for approximately 80% of total reported anthropogenic greenhouse gas emissions in OECD countries, excluding land use change and forestry (see also Chapter 13).

Table 12.2. **Contribution of energy use to air pollutants and greenhouse gas emissions**

% of total emissions in OECD countries deriving from energy use, mid-1990s

Air pollutant	Transport	Electricity production	Other combustion (industry and residential)	Other
SO_x	4%	23%	71%	2%
NO_x	52%	28%	16%	4%
CO	85%	2%	13%	0%
Particulates	17%	12%	26%	45%

Greenhouse gas emissions from Annex I Countries, mid-1990s (as reported to UNFCCC)

Air pollutant	Energy sector contribution	Main source within energy sector
CO_2	96%	Fuel combustion
CH_4	35%	Fugitive emissions
N_2O	26%	Fuel combustion

Sources: OECD (1999*a*), UNFCCC (2000).

Most air emissions from energy result from the use of fossil fuels.[4] Oil is the largest contributor to energy-related CO_2 emissions, accounting for 46% of energy-related CO_2 emissions in OECD regions in 1996

4. Although burning biomass can produce substantial emissions of methane (CH_4), non-methane volatile organic compounds (VOC), carbon monoxide (CO), particulate matter (PM) and soot.

(IEA, 1998*c*). More than half is from the transport sector, which is also the main source of carbon monoxide (CO) and NO_x emissions, and consequent peaks in localised ozone concentrations. The power sector is the dominant source of SO_x emissions, followed by stationary energy use in the industrial sector – which are also the sectors where coal use is highest. Emissions of particulate matter from fossil fuel combustion is also significant, but declining rapidly under strict regulations and the introduction of more effective control technology.

The Reference Scenario indicates that OECD regions will see an increase in energy-related sulphur dioxide emissions of about 26% to 2020, a significantly lower increase than that expected in non-OECD countries. CO_2 emissions are expected to increase in both the OECD area and worldwide under current policies, although more slowly than in the past, with total increases of about 33% and 61% respectively from 1995 to 2020. To some extent, the lower rates of growth in CO_2 and SO_x emissions in OECD countries in the Reference Scenario will result from an increased share of energy supplied through natural gas and the use of relatively clean gas turbine technology (with relatively low greenhouse gas, sulphur and NO_x emissions), as well as a rising share of renewables. These reduced growth rates will also reflect improvements in end-of-pipe controls for SO_x, NO_x and VOC, though not for CO_2.

Water and soil

A significant portion of the water abstracted in many OECD countries is used for cooling in electrical power generation, and the share of water used for this purpose has increased dramatically in recent years (see Chapter 8). While this water is generally returned to the source, it often has a higher temperature than when it was abstracted, and a lower oxygen level. This thermal pollution of waterways can lead to oxygen depletion in freshwater ecosystems. Thermal pollution is regulated in most OECD countries, usually through siting permits. Pollution of water and soil from energy use also occurs directly through leaking oil tanks and indirectly through acidic deposition caused by air emissions of NO_x and SO_x (predominantly produced from fuel combustion). This has led to severe effects on lakes and rivers and on forests in some regions, with damage to freshwater fish, other fauna, and habitats.

Natural resource use

Fossil fuels are essentially non-renewable resources: as they are depleted they need to be substituted by other, more abundant or renewable, resources. Over the period to 2020, however, it is unlikely that constraints on fossil fuel resources will play a significant role in energy supply. The IEA's *World Energy Outlook 2000* projects that the world oil-resource base is adequate to meet demand over the projected period to 2020, but assumes a steady price rise to US$28/barrel in 2020 to enable capital investments. Reserves of natural gas are estimated at 1.9 trillion barrels of oil equivalent, and although world demand for gas is growing faster than for oil, serious supply shortages are not expected until after 2020. World coal production is expected to match demand much beyond 2020, with any local supply deficiencies covered through the international market.

The environmental impacts of coal extraction include land degradation (and consequent changes in habitats) from open-cast coal mining and leaching of mine drainage water containing dissolved/suspended solids and acids into nearby waterways. Oil extraction and transport, particularly off-shore extraction, can result in spills and leaks, and gas and oil production results in emissions of CO_2 and CH_4. Coal mining is also a major source of CH_4. Uranium mining generates radon gas, dust, and contaminated rainwater streams, as well as mining waste which forms the bulk (in volume) of waste associated with nuclear power production. Large hydropower dams have a substantial impact on water flow, although effects on marine life are often reduced through minimum water flow regulations. Dams can also have potentially significant localised effects on bird life and other fauna or flora through the flooding of land. Recent studies indicate that dams lead to significant releases of CH_4 from accelerated degradation of organic matter in flooded areas. Visual and noise pollution, together with bird kill, are frequently cited as negative environmental impacts of wind power. However, the negative environmental impacts of small-scale renewable electricity generation are generally limited (and site-specific).

Waste

Both coal mining and the combustion of coal generate significant quantities of solid waste (slag and ashes), which may cause pollution and reduce visual amenities. The amount of waste produced per tonne of coal mined or combusted will vary from site to site, depending on the chemical characteristics of the coal seam. Coal cleaning also results in emissions of particulates to air, and in emissions of "black water" (with suspended particulates).

Nuclear waste carries unique risks in relation to its transport and disposal, as it can remain highly radioactive for hundreds of years. Furthermore, there is a threat of accidental releases of radioactive material from nuclear power generation facilities. While many OECD countries have agreed that isolation of nuclear waste in stable geological structures is the most appropriate option, implementation of this policy has been slow for political reasons.[5] In addition, there is no consensus on whether waste disposed in such a manner should be retrievable – in case another method of storage is subsequently preferred – or irretrievable, so as to minimise the risk that the storage facility could turn into an illegal source of nuclear materials.

12.4. Policy options and their potential effects

In OECD countries, there have been rapid increases in both energy production and consumption since the 1980s, although the energy intensity of the economy has been declining. This seems likely to continue, with energy use in OECD countries projected to grow by another 35% to 2020, while worldwide it increases by 51%. In order to reduce the environmental effects of energy use, particularly air pollution and greenhouse gas emissions, energy policies will need to be geared towards better demand-side management (to reduce growth in overall energy demand); ensuring a higher share of low-emission energy sources in the fuel mix; and encouraging the development and uptake of low-emission, energy-efficient technologies. The main environmental driver for new energy policies in OECD countries will be to meet agreed targets for reducing greenhouse gas emissions. Unlike emissions of "conventional" air pollutants, there are no economically viable end-of-pipe controls available as yet for emissions of CO_2. The focus will need to be on emissions abatement or reduction, which implies a reduction of energy use and/or a switch to using no- or low-emission technologies or fuel sources, such as renewable energy. On the supply side, the increasing share of electricity in energy supply makes this a particularly important focus of future policies.

Technological development and diffusion

A number of fuel sources and technologies are coming on-stream which can reduce the environmental impacts of energy use and production (see Box 12.2). For many of these, however, the still high costs of their use, compared with more traditional fuel sources or technologies, means that they are unlikely to command large shares of the

Box 12.2. Energy-related technologies

Improvements in fossil fuel combustion technologies, including advanced gas turbines and coal gasification, offer considerable emission abatement potential over older fossil fuel technologies. Falling costs of renewable energy are making wind and biomass competitive in an increasing range of applications, and improved power storage and grid management technologies will help in the market penetration of renewable energy. For transport-related energy use, hybrid electric/petrol and bio-ethanol/petrol vehicles have recently been commercialised in several OECD countries and can offer a strong potential for fuel savings and emission reductions. Hydrogen fuel cell technologies are also nearing commercialisation in the transport sector and these could radically change the sector's energy requirements and environmental impacts. Technologies are also being explored for addressing greenhouse gas emissions from the energy sector (see Chapter 13), such as CO_2 scrubbing processes which would pump CO_2 into used gas wells for storage.

5. The first installation for high-level nuclear waste storage is not expected before 2010 (IEA, 2000*c*).

market in the next 5 to 10 years. The relatively slow rate of capital turnover in the power sector hampers the up-take of more efficient technologies, although the relatively fast turnover of vehicle stocks could lead to more rapid penetration of new technologies in the transport sector.

To accelerate the process of developing and adopting sustainable energy systems, including more energy efficient end-use appliances, governments will need to provide further support to research and development in these alternatives, and to their adoption. While the share of total IEA government support to energy research and development that is devoted to energy conservation and renewables (respectively 13.6% and 8.2% of total) now surpasses the share attributed to either coal or to oil and gas, most R&D support continues to go to the development of traditional fuels such as nuclear (IEA, 2000*e*). Energy research, development and incentives to encourage the up-take of renewables and energy efficient technologies will need to be accelerated to counter-balance expected increases in the demand for energy. Ensuring price differentials that reflect environmental externalities between different fuels and energy technologies would provide significant incentives for the use of cleaner energy systems.

Regulatory instruments

In most OECD countries, the wide adoption of many end-of-pipe environmental technologies in the energy sector, in particular to reduce air emissions from fuel combustion, has primarily been driven by emission regulations. These include technical emission standards, product standards, and/or air quality standards. In turn, these regulatory policies have accelerated adoption of flue-gas desulphurisation, the use of low sulphur fuels, low NO_x burners and catalysts for NO_x reduction, as well as the phasing out of leaded petrol.

In addition to encouraging further end-of-pipe emission reductions, regulations may increasingly be used in the energy sector to set specific targets and drive changes in the fuel mix and conversion efficiency towards greater use of renewables and high-efficiency applications, including co-generation and low emission motor vehicles. A number of OECD countries are already adopting portfolio standards, often as an accompaniment to general energy sector regulatory reform, which require that a certain percentage of new power generation or vehicles be fuelled by renewable or non-fossil fuel energy sources. This could significantly affect air emissions from the energy sector: a modelling simulation of adopting a policy across all OECD regions that would require a minimum of 15% renewable energy in the energy supply in OECD regions to 2020 indicated that both CO_2 and SO_x emissions could be reduced by over 7% compared with the Reference Scenario. However, OECD research indicates that such targets may be relatively costly for reducing greenhouse gas emissions when compared with the use of economic instruments (OECD, 2001). Despite this, such targets may in many cases be the most practical option for ensuring reductions in the environmentally damaging effects of energy use, and they can provide clear and consistent signals to the public and business community of government environmental aims.

Energy efficiency standards for equipment can also play a potentially important role in reducing energy use, and consequent environmental impacts. Research indicates that energy efficiency standards can be particularly effective in overcoming market barriers to investment in energy efficient equipment (IEA, 2000*d*).

Economic instruments

Most subsidies to energy production continue to support the more environmentally-damaging fuel sources.

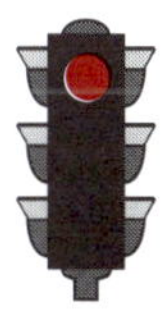

Energy is considered to be one of the most heavily subsidised sectors in the OECD area (OECD, 1998). OECD data on coal subsidies for selected countries shows that direct producer subsidy equivalents (PSE) have dropped significantly in recent years, but in 1997 the level of PSE for coal production in these countries still stood at almost US$8 billion per year. Subsidies to the entire energy sector have been estimated as being an order of magnitude higher than that, with the bulk of support going towards nuclear, coal and oil production, often for purposes of maintaining regional employment (de Moor and Calamai, 1998).[6] Subsidies to specific fuels lead to an economically inefficient energy supply level and mix, and discourage new fuel or technological developments that could reduce negative environmental effects. In OECD countries, the reform of environmentally damaging subsidies – particularly those that are tied to the use of more polluting fuels (*e.g.* fossil fuels, especially coal) or to energy production or consumption – could contribute to meeting Kyoto targets for greenhouse gas emission reductions and national environmental targets.

Following the Polluter Pays Principle, removing support to energy production and consumption should be accompanied by levying charges or taxes on fuels in order to internalise their environmental impacts. Ideally, such charges should reflect the respective environmental damage caused by different fuels, for example reflecting their carbon content as a proxy for their contributions to climate change. A number of OECD countries have discussed, or already introduced, energy levies or carbon taxes. In order to avoid any adverse effects on the competitiveness of large energy-intensive industries, however, most countries exempt these industries from such taxes, reducing the economic efficiency and environmental effectiveness of the instrument.

Table 12.3 provides the results of an energy policy simulation comprising the removal of all subsidies to energy sources in OECD countries and all subsidies to energy production, combined with adding an annually increasing ad valorem tax on fuel use in OECD countries. The ad valorem tax increases by 2 percentage points per annum for coal, 1.6 percentage points for crude oil and 1.2 percentage points for natural gas, reaching a total tax levy of 50%, 40% and 30% of pre-tax prices respectively in 2020. The yearly increases in the taxes are linked to the carbon content of each fuel, but as the starting point for the increases have no such link, this shock should not be seen as simulating a proper "carbon tax". As can be seen from Table 12.3, the policy simulation has fairly significant effects on fuel demand in OECD countries, particularly on coal use. Worldwide there would also be reductions in total coal, oil and natural gas demand compared to the Reference Scenario, although in non-OECD countries a small leakage effect would be expected, with fuel use in these regions increasing marginally. The environmental effects of the subsidy removal and energy tax implementation would be more substantial. Compared to the Reference Scenario, introducing these policies would result in 25% reductions in both SO_x and CO_2 emissions, thus significantly improving air quality and reducing climate change effects of energy use in OECD countries. The economic effects of these policies would be small, reducing expected GDP in OECD regions in 2020 by approximately 0.1% only.

Table 12.3. **Effects of energy subsidy removal and energy tax use in OECD countries**
(% change from Reference Scenario in 2020)

Effect on demand in OECD countries for:			GDP	SO_x emissions	CO_2 emissions
Coal	Oil	Gas			
–32%	–18%	–17%	–0.11%	–25%	–25%

1. This policy simulation explores the impacts of eliminating all subsidies on all energy sources and all subsidies on inputs used in production in the energy sectors in OECD regions, combined with a 2%, 1.6% and 1.2% *ad valorem* tax added each year on coal, crude oil and natural gas respectively.
2. The energy subsidies reflected in the model are based on the data available from national accounts and listed in the GTAP database (see Annex 2 for further detail), reflecting only a small proportion of actual energy subsidies. Thus, energy subsidies in OECD countries listed in the GTAP database amount to US$4.7 billion in 1995, while other estimates indicate these might be US$70-80 billion (de Moor and Calamai, 1998).

Sources: Reference Scenario and Policy Simulations.

A number of OECD countries are discussing, or have begun to establish, national emission trading schemes for greenhouse gases or other air pollutants from energy production (OECD, 1999*b*) (see also Chapters 13 and 15). When such permits are used in combination with an overall cap on emissions, and with specific allocations for indi-

6. While subsidy removal to coal production is likely to lead to negligible but positive effects on net employment in the long run (through increased economic efficiency), this does lead to unemployment in the affected sector and regions in the short term unless measures are taken to support transition to new employment or training opportunities. To allow time for adjustment, transition measures should be considered, and subsidies be removed gradually.

vidual firms or entities (cap-and-trade schemes), they can help to efficiently allocate a chosen level of emission allowances between competing firms and provide strong incentives to reduce emissions. Their wide introduction, however, requires significant investment in emission monitoring systems and regulatory oversight to provide clear incentives for compliance. In general, emission trading schemes make economic sense when there are many emitters with widely varying marginal costs of reduction, and where the environmental benefit does not vary widely with the location of the emission reduction. The latter is true for greenhouse gas emission reduction but not necessarily for other pollution problems stemming from fuel combustion, such as thermal water pollution, conventional air pollution, or noise.

Voluntary agreements

Voluntary agreements to reduce energy use in the industrial and power sectors have been in use for almost a decade. Such agreements often form part of a country's national climate change policy package (see Chapter 13). However, the effectiveness of voluntary agreements in reducing industry energy use and emissions has been questioned. Moreover, some countries have found it difficult to attract a large number of interested private-sector participants from a broad range of sectors. New voluntary, or negotiated, agreements are being set up for consumer appliances, such as clothes washers and consumer electronics (IEA, 2000*d*). The voluntary agreement on clothes washers as part of the EU's SAVE agreement succeeded in phasing-out the least efficient washers in 1998.

Information and other instruments

Energy-efficiency labels and standards for appliances and equipment– predominantly for household use – are being used in 37 countries, and the range of products to which they are being applied is continuing to expand. The use of such labels can influence consumer choice to a significant degree. For example, the sales-weighted annual average energy-efficiency index of "cold appliances" (fridges and freezers) improved by 4.5% from 1994 to 1996 during the EU's cold labelling programme (IEA, 2000*d*).

Accompanying the deregulation of energy markets, a number of OECD countries have introduced "Green Pricing" schemes whereby consumers can choose to use electricity that is produced only from renewable energy sources. While such schemes currently have very low take-up rates because the electricity is still relatively expensive, they may offer the potential for significant changes in fuel mixes driven by consumer demands in the future as the price for renewable energy falls (IEA, 1998*b*).

REFERENCES

de Moor, A. and P. Calamai (1998), *Subsidizing Unsustainable Development: Undermining the Earth with Public Funds*, Earth Council, San José, Costa Rica.

IEA (International Energy Agency) (1997), *Indicators of Energy Use and Efficiency: Understanding the Link between Energy and Human Activity*, IEA/OECD, Paris.

IEA (1998*a*), *World Energy Outlook*: 1998 *Edition*, IEA/OECD, Paris.

IEA (1998*b*), *Renewable Energy Policy in IEA Countries Volume* II: *Country Reviews*, IEA/OECD, Paris.

IEA (1998*c*), CO_2 *Emissions from Fuel Combustion*, IEA/OECD, Paris.

IEA (2000*a*), *Key World Energy Statistics from the* IEA: 2000 *Edition*, IEA/OECD, Paris.

IEA (2000*b*), *World Energy Outlook* 2000, IEA/OECD, Paris.

IEA (2000*c*), *Review of Nuclear Power*, IEA/OECD, Paris.

IEA (2000*d*), *Energy Labels and Standards*, IEA/OECD, Paris.

IEA (2000*e*), *The Role of* IEA *Governments in Energy*: 2000 *Review*, IEA/OECD, Paris.

Newman, J., N. Beg, and G. MsGlynn (2001), "Energy and Climate Change: Trends, Drivers, Outlook and Policy Options", background document for the OECD *Environmental Outlook*, OECD, Paris.

OECD (1998), *Improving the Environment through Reducing Subsidies*, OECD, Paris.

OECD (1999*a*), OECD *Environmental Data: Compendium* 1999, OECD, Paris.

OECD (1999*b*), *Implementing Domestic Tradable Permits for Environmental Protection*, OECD, Paris.

OECD (2001), "Instruments and Technologies for Climate Change Policy", ENV/EPOC/GEEI(99)15/FINAL, OECD, Paris.

UNFCCC (United Nations Framework Convention on Climate Change) (1998), "Second compilation and synthesis of second national communications, FCCC/CP/1998/11/Add.1, September 1998.

UNFCCC (2000), "UNFCCC database", *www.unfccc.de* drawn from official national inventory submissions, UNFCCC, Bonn.

13 Climate Change

KEY SIGNALS

- OECD countries contribute approximately 50% of emissions of greenhouse gases (CO_2, CH_4 and N_2O), but this share is expected to decline in the period to 2020.
- Emissions of greenhouse gases (GHG) in OECD countries increased by 4% in the 1990-1998 period. The emissions grew most rapidly in North America and in the Asia-Pacific region, while emissions in Europe have decreased by almost 5%. GHG emissions are likely to increase substantially in the period to 2020.
- Projections by the OECD show that total CO_2 emissions in OECD countries are likely to increase by about 33% from 1995 to 2020. This illustrates the challenge of achieving Kyoto Protocol targets, which could require OECD countries to reduce emissions by roughly 20% to 40% below Reference Scenario levels.
- GHG emissions will lead to increases in global mean temperatures and a possible rise in sea levels. The climate is already changing, which in turn will almost certainly require adaptation measures.
- The most urgent policies needed to mitigate greenhouse gas emissions are: wide implementation of taxes on greenhouse gases; reform of subsidies that contribute to climate change; promotion of alternative transport fuels, vehicles and modes, and renewable energy technologies; and introduction and effective implementation of domestic and international emission trading systems.

13.1. Introduction

Recent scientific evidence indicates that global warming is a reality. Observed changes show that despite large variations from year to year, global mean temperature has risen significantly over the last hundred years. In this past century, the temperature rise was faster and lasted longer than at any period over the past ten thousand years, and the decade of the 1990s was the warmest on record for the Northern Hemisphere. A range of other indicators confirm the global warming trend, including increased precipitation, decreased snow cover and sea ice, and sea level increase.

Scientists believe that the observed global warming is mainly due to changes in human activities and related increases in greenhouse gas emissions. These changes are driven by worldwide population and economic growth, and the underlying production and consumption of fossil fuels, as well as the intensification of agricultural activity and land use changes. OECD countries are the main past and present contributors to human induced climate change and, as such, have a special responsibility in implementing policies to reduce emissions of greenhouse gases and enhance sinks. However, OECD countries will not be able to effectively combat climate change on their own – long-term stabilisation of greenhouse gas concentrations will require broadening the effort to include emission reductions from at least the world's major developing economies.

Reducing greenhouse gas emissions and enhancing sinks is the aim of the UN Framework Convention on Climate Change (UNFCCC), which has more than 170 signatories. Action under the Convention has advanced beginning with the targets that were established for industrialised countries (Annex I countries) under the Kyoto Protocol. The coming years will show whether or not OECD countries are able to demonstrate leadership and take action to ratify and implement the Protocol as a first step towards achieving the long-term objective of the

Convention – the stabilisation of atmospheric concentrations of greenhouse gases. The Sixth Conference of the Parties (COP6) in The Hague made progress on governance, financing, and technical elements related to the Protocol, but was unable to arrive at a final comprehensive agreement.

13.2. Pressures on the environment: emissions of greenhouse gases

Overall emission trends

Three greenhouse gases account for the majority of human induced global warming effects: carbon dioxide (CO_2), methane (CH_4), and nitrous oxide (N_2O) (see Figure 13.1).[1] CO_2 is the dominant greenhouse gas, accounting for 75% of global emissions and 81% of emissions from OECD countries (excluding land use and forestry uptake). This is followed by methane (10% in the OECD regions) and nitrous oxide (roughly 7% in the OECD regions). Hydrofluorocarbons (HFCs), perfluorocarbons (PFCs) and sulphur hexafluoride (SF_6) account for a small share (about 2%) of total emissions in OECD regions, but they are growing quickly and thus are subject to management under the UNFCCC (Burniaux, 2000; UNFCCC, 2000).

Figure 13.1. **Greenhouse gas emissions by gas in OECD countries, 1998**

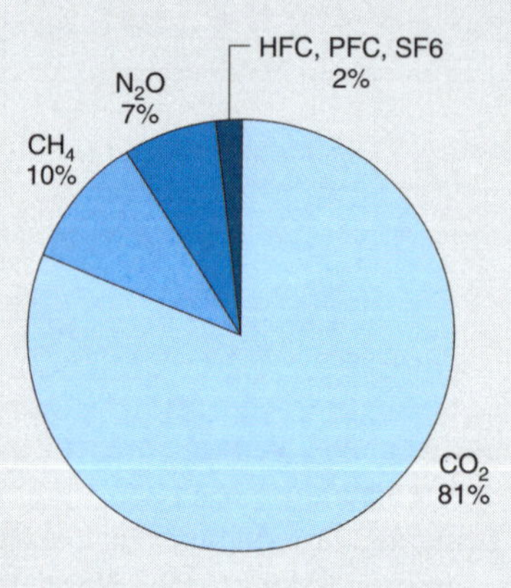

Note: Gases are in CO_2 equivalents.
Source: UNFCCC (2000).

CO_2 emissions from fossil fuel combustion can be broken down into emissions related to energy production and transformation (to a large extent power generation) and emissions related to other energy end-use sectors (industry, transport, residential and commercial). The largest CH_4 emission sources in OECD countries are the natural gas and oil industry (leaks and deliberate releases), livestock enteric fermentation (the normal digestive process of livestock) and landfilling of solid waste (the decomposition of organic matter in landfills) – accounting for 75% of total methane emissions. The largest source of N_2O emissions in OECD regions is agricultural soil (over 60%), with industrial processes accounting for another 15%. N_2O emissions from transport grew rapidly (over 20%) during the last decade and are closely linked to air-fuel mixes and combustion temperatures, and to the use of pollution control equipment (catalytic converters) on vehicles.

1. Unless otherwise noted, the figures used in this chapter are based on UNFCCC "official" national data and refer to emissions of the 3 main gases – CO_2, CH_4, and N_2O – and do not include emissions from the land use change and forestry sector. Data may not be completely consistent across time or countries

Recent trends in greenhouse gas emissions for OECD and other Annex I countries[2] are carefully monitored under the UNFCCC. Total OECD country emissions (excluding Mexico and Korea) of all greenhouse gases increased by about 4% over the 1990-1998 period, reaching 13 902 Mt CO_2 equivalent in 1998. All but five OECD countries have reported emission growth during this period, partly due to robust economic growth. Emissions are rising, although the pace of emission growth has tapered off in the last few years. OECD country emissions in 1996 alone grew more than the first five years of the decade combined,[3] but this rapid growth appears to have levelled off in 1997 and 1998. The five largest emitters in the OECD regions (the US, Japan, Germany, PCanada and the UK) account for around 75% of OECD country greenhouse gas emissions, which means that OECD trends are dominated by changes in these countries.

OECD countries currently account for about 50% of global CO_2 emissions from energy use, but the share of emissions from OECD countries is gradually falling as other parts of the world show higher economic growth. For the other greenhouse gases, the OECD area share is smaller and is likely to decline in the period to 2020.

Total emissions of CO_2 in OECD countries increased by about 6% in the 1990-1998 period, reaching 11 026 Mt in 1998. In the OECD regions, the highest rates of growth in CO_2 emissions have occurred in North America (11% from 1990 to 1998) and Asia Pacific (12%), while emissions in Europe actually decreased by 5%[4] (UNFCCC, 2000). From 1996 to 1998, the emissions growth rate from fuel combustion has slowed considerably in the US, from 3.5% to 0.5%. During this period, Annex I emissions as a whole declined by 7%, due to the economic slowdowns in countries with economies in transition.

Emissions of CO_2 in OECD countries are projected to increase by approximately 33% from 1995 to 2020.

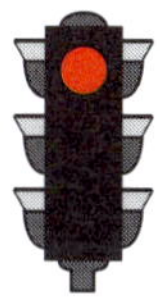

The Reference Scenario and other scenarios show that CO_2 emissions are likely to increase substantially to 2020, both in OECD countries and in other regions of the world (see Figure 13.2). In the Reference Scenario, emissions in OECD countries are projected to increase by approximately 33% from 1995 (38% from 1990) to 2020 and almost 100% in non-OECD countries. It projects average annual OECD emission growth rates of 1.2% from 1995 to 2010, and 1.1% from 2010 to 2020. The projections are within the range of results from other major modeling exercises, as shown in Figure 13.2. CO_2 emissions in Central & Eastern Europe are projected to grow by 78% from 1995 to 2020. In Australia & New Zealand the projected increase is 38%; in North America it is approximately 35%; and in the regions of Western Europe and Japan & Korea, growth is estimated at 23% and 17% respectively. In all OECD regions, the growth rates for CO_2 emissions are projected to be lower from 2010 to 2020 than in the period 1995 to 2010, mainly because economic growth is expected to slow down after 2010.

The IPCC A2 scenario, which assumes regionally oriented economic development with fragmented and relatively slow per capita economic growth and technological change, projects a 40% increase in emissions to 2020, slightly above the Reference Scenario. The B1 scenario of the IPCC – which assumes global solutions to economic, social and environmental sustainability, including improved equity, but without additional climate initiatives – projects an increase of 13%. The real difference in the environmental outcomes of the two IPCC scenarios is due to the emissions intensity of the energy used, with scenario B1 showing a rapid decline in emissions intensity.

The *World Energy Outlook* scenario by the International Energy Agency (IEA) projects a 34% increase in emissions from 1990 to 2020, slightly lower than the Reference Scenario used here (IEA, 2000). The OECD GREEN scenario shows a higher rate of CO_2 emissions growth, largely because the model assumes that as oil and gas reserves are depleted, energy demand will shift toward coal and a carbon-based fuel substitutes, so there is an increase in the carbon intensity of fuel use. The model also assumes no further reductions in energy subsidies, which would be a change from trends in recent years.

2. Under the UNFCCC, Annex I (industrialised) countries agreed to an aim of limiting their greenhouse gas emissions. Annex B of the Kyoto Protocol established legally binding greenhouse gas limitation commitments for almost all the Annex I countries.
3. For an overview of national emission trends, Kyoto objectives and EU burden sharing, see Newman *et al.* (2001).
4. The number for Europe includes some non-OECD eastern European countries with economies in transition.

Figure 13.2. **Projected CO_2 emissions from OECD countries, 1990-2020**

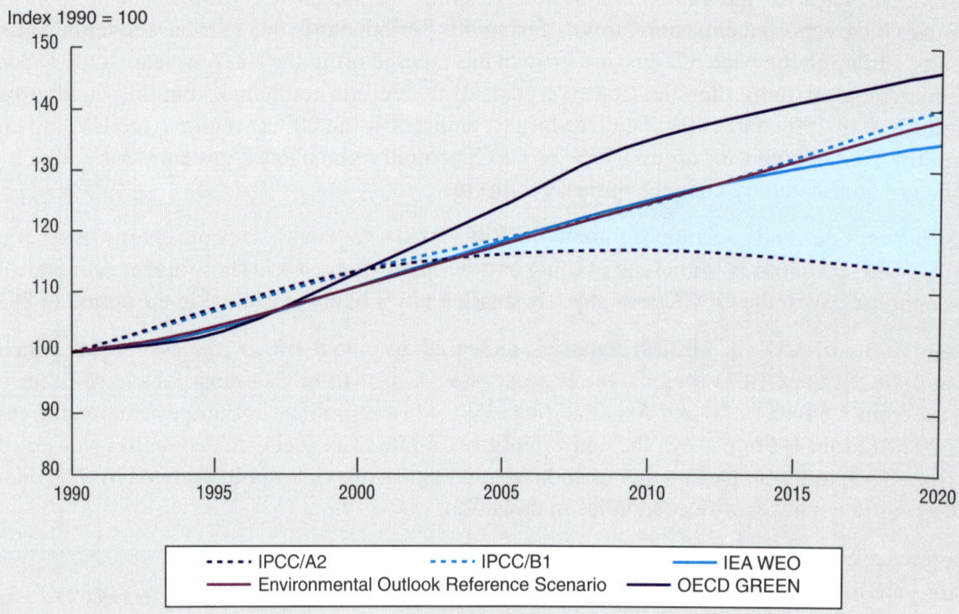

Notes: 1990 data for the GREEN and WEO scenario's are calculated from IEA (1999).
Sources: Reference Scenario, Burniaux (2000), IEA (1999), IEA (2000) and IPCC (2000*b*).

Meeting Kyoto targets will require reducing greenhouse gas emissions in OECD countries by roughly 20% to 40% in 2020 compared with Reference Scenario projections.

Methane and N_2O emissions in OECD regions are likely to grow more slowly than CO_2 emissions, and in some cases even decline slightly. During 1995-2010, emissions of N_2O in Annex I countries are projected to increase by 0.9% on average annually (14% in total), while emissions of methane are projected to decrease by 0.8% annually and 13% in total (Burniaux, 2000). Over the same period, total emissions of the three greenhouse gases are projected to increase by 1.4% annually (23% in total) in Annex I countries. The projected increases in greenhouse gas emissions indicate the challenge of achieving Kyoto targets, which could require the OECD countries to reduce emissions by as much as 40% below the Reference Scenario should they choose to meet targets unilaterally. If OECD countries work with countries with economies in transition, the gap could be reduced to approximately 18%.

Emission trends by sector

Stationary fuel combustion – mostly for power generation – is the largest single source of greenhouse gas emissions (59% in 1998) (see Figure 13.3). CO_2 emissions from fossil fuel combustion can be broken into emissions related to energy production and transformation (34%), and emissions related to other energy end-use sectors (industry 17%, transport 27%, residential and commercial 13%), with other sectors contributing 9% in 1998. Presently, coal is responsible for 34% of total CO_2 energy-related emissions, while oil is responsible for 45% (IEA, 2000). Due to market reform policies and changes in technology, recent shifts away from coal towards natural gas have led to a drop in carbon intensity from electricity production and other industry sectors. However, this decline may not continue if, or when, nuclear power is phased out (see Chapter 12).

Figure 13.3. **Greenhouse gas emissions by sector in OECD countries, 1998**

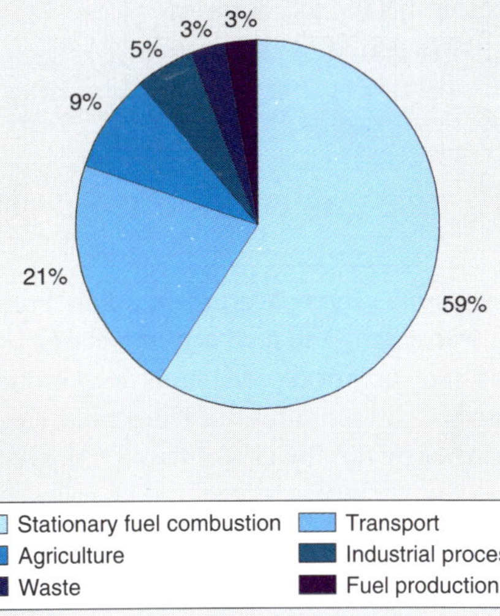

Note: Gases are in CO_2 equivalents.
Source: UNFCCC (2000).

In many countries, increasing demand for electricity production is a main driver for CO_2 emission trends. The IEA estimates that CO_2 emissions from the power sector in OECD countries will rise to 21% above their 1997 level in 2010 and 33% in 2020. Growth in CO_2 emissions is projected to remain slightly below the growth in electricity demand because of improvements in fossil-fuel power generation efficiency (IEA, 2000).

In most OECD countries, emissions from transport have risen faster than those from all other energy-related sources since 1990 (Schipper *et al.*, 2000). Emissions from transport rose from 19% of total GHG emissions in 1990 to about 21% in 1998, the absolute increase for the sector being 15% (UNFCCC, 2000). Within the transport sector, emissions from aviation are the most rapidly growing, although they still account for a relatively small share of total transport emissions.[5] CO_2 accounts for the overwhelming majority of increasing emissions from transport, but N_2O emitted by that sector is increasing even more rapidly. It is projected that the contribution of the transport sector to total CO_2 emissions in OECD regions will increase from approximately 20% in 1995 to 31% in 2020. Other main sectoral sources of greenhouse gas emissions are agriculture (9% in 1998), industrial processes (5% in 1998), waste (3% in 1998), and fuel production (3%).

Land use and forestry activities are accounted for separately from other greenhouse gas sources. The establishment and harvest of plantations, commercial forest management, and fuel-wood gathering lead to changes in forests and other woody biomass stocks. Converting forests for agricultural use and abandoning managed lands affects not only the amount of carbon stored in above-ground biomass, but also the amount of carbon stored in the soil. Depending on the overall change in forest coverage and density, and change in agricultural land use, these activities may result in either net emission of greenhouse gases to the atmosphere, or net removal.

5. International aviation emissions are not covered under national mitigation obligations in the UNFCC or the Kyoto Protocol, as these emissions are considered to be "international" rather than "national" in origin. Countries report the emissions but they are not included in national inventory totals.

As of 1998, changes in land use and forestry are estimated to be a net CO_2 sink for most OECD countries. Across countries that reported data, the aggregate sink is equivalent to about 10% of total annual CO_2-equivalent emissions in 1990.[6] According to current national inventory accounting under the UNFCCC, absolute carbon sequestration in OECD countries declined significantly over the 1990 to 1998 period, with a drop of about 25% (IPCC, 2000*a*).

Changes in land use and forestry are estimated to be a net CO_2 sink for most OECD countries.

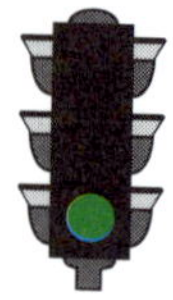

13.3. The effects of greenhouse gas emissions on climate change

Climate change scenarios prepared by the Intergovernmental Panel on Climate Change (IPCC) show a wide range of possible impacts from global warming in the next century. Based on the range of climate sensitivities and the plausible ranges of greenhouse gas and sulphur dioxide emissions reported by the IPCC, a number of climate models project that the global mean surface temperature could increase by about 1.5 to 6.0 degrees C by 2100. This range is comparable to that reported earlier by the IPCC of 1.0 to 3.5 degrees C (IPCC, 2001*a*).

Worldwide precipitation is likely to increase, but projections vary by region (*e.g.* decreases in precipitation levels are expected in subtropical areas). Changes in precipitation and temperature levels are expected to increase the risk of more extreme weather events in some regions, such as flooding. In some regions, the frequency and severity of droughts could increase with drier summers and lower precipitation overall as the result. A continued decline in glacier mass is foreseen, as well as a decline in snow cover and sea ice, especially in the Northern Hemisphere. For the next 100 years, the scenarios also show a wide range of possible sea level rise, from 15 to 95 cm.

Climate change will have an impact on human life and the ecology of the planet in a variety of ways. Some of the major impacts will affect agriculture, water supply and quality, human settlements, and human health, and will also affect the ecosystem, including biodiversity and migratory patterns. Our understanding of the impacts resulting from long-term climate change is limited, although in-depth regional studies have recently yielded important insights.

Projected emissions of greenhouse gases will lead to increases in global mean temperature and a rise in sea levels.

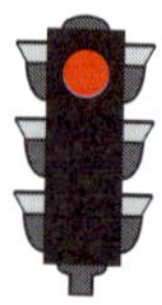

In the Northern Hemisphere, potentially significant shifts in agricultural productivity could occur due to changes in growing seasons and pest or plant disease vectors from climate change. Water supply and quality could also be affected, intensifying existing pressures on water supply. Another impact of climate change in both the Northern and Southern Hemispheres is likely to be the spread of vector-borne diseases, including dengue fever and malaria (WHO, 2000). Adverse impacts could partly be offset by some positive impacts, such as increased agricultural productivity in some regions of the Northern Hemisphere, and lower energy use due to warmer winter seasons. Effects will be varied. Forest productivity is expected to increase in northern Europe and decrease in southern Europe. Overall studies show that some of the most adverse impacts are bound to occur in the Southern Hemisphere where countries are most vulnerable and least likely to easily adapt to climate change. Finally, scientists indicate that unprecedented rates of global climate change raise the potential for surprises and catastrophic or irreversible developments to happen, making it difficult to adopt successful adaptation strategies.

Humankind has throughout history adapted to normal climate variability, and has absorbed the economic losses that result from weather instability and shifting climatic conditions. Decisions about which adaptation measures to take with regard to climate change involve choices about the level of expenditure and the benefits to be gained. While an over-expenditure on adaptation would result in high costs from using resources better spent elsewhere, under-investment would have more severe consequences. Under-investment on land-use planning and flood protection may result in significant damage to buildings and infrastructure caused by flooding, and high produc-

6. With all OECD countries reporting except Greece, Iceland, and Luxembourg.

tivity losses in the agricultural, commercial, and industrial sectors. Under-expenditure on public health programmes and in disaster preparedness (for fires, flooding, and other natural catastrophes) may lead to high welfare losses in the event of catastrophic weather events or climatic conditions.

Since climate change poses a risk of uncertain (and perhaps very serious) consequences, it would seem appropriate to extend the precautionary approach to the improvement of adaptation capacity. Moreover, it seems clear from the increased losses due to extreme weather events over the last decade that current expenditure on adaptation falls far short of what is required to prevent significant economic losses.

13.4. Policy options and their potential effects

OECD, IEA and IPCC projections of greenhouse gas emissions show that significant policy action and changes in investment and consumption patterns will be required to meet Kyoto commitments and, more importantly, to stabilise long-term emissions and mitigate human induced climate change. IPCC suggests that in order to stabilise greenhouse gas concentrations at "safe" levels, reductions of global emissions in the range of 50-60% compared to present levels will be required by around 2050 (IPCC, 1996).

Policy instruments for addressing greenhouse gas emissions at the sector level have been analysed in the sectoral chapters of this report. Policy instruments recommended in those chapters are mainly economic instruments, such as taxes and removal of subsidies, combined with regulatory and other instruments, and the promotion of more environmentally friendly technologies. Instruments such as carbon taxes, tradable permits and voluntary agreements are receiving far more attention in the climate change context than before and are briefly reviewed here.

The management regime of the Climate Change Convention and the Kyoto Protocol cover emissions of all direct greenhouse gases.[7] This provides the flexibility to mitigate emissions where it is least costly to do so. While non-CO_2 emission sources may appear relatively minor compared to CO_2 emissions from energy use, they appear to offer cheaper mitigation opportunities in the short term, and may thus allow more time for making the transition in the energy sector to cleaner technologies and fuels.

Box 13.1. Ancillary benefits of greenhouse gas mitigation

Greenhouse gas mitigation has benefits beyond those directly associated with climate change – the so-called ancillary benefits. In particular, reductions in emissions of greenhouse gases also have positive effects on urban air pollution, and thereby on human health. Estimates of the magnitude of such ancillary benefits per tonne of carbon reduced in OECD countries vary widely. Nevertheless, even the most conservative estimates suggest that they are substantial, and may offset a third or more of the abatement costs for modest mitigation efforts. A study in Hungary estimated that a 7.7% reduction in CO_2 emissions would result in health benefits of US$650 million – enough to cover the investment required to implement the CO_2 reduction measures (Aunan *et al.*, 2000). A second example of ancillary benefits is greenhouse gas reduction policies that lower demand for transport and lead to social benefits, such as those that result from less traffic congestion.

The prospect of large ancillary benefits has the potential to lower the net social costs to society of reducing greenhouse gases. It is therefore important to identify the major areas where these benefits are present, to develop methods to quantify them and find the best ways of injecting this information back into the regulatory and decision-making process.

7. In addition to carbon dioxide, methane, and nitrous oxide, direct greenhouse gases are perfluorocarbons (PFCs), hydrofluorocarbons (HFCs) and sulphur hexafluoride (SF_6). The indirect greenhouse gases are carbon monoxide (CO), nitrogen oxides (NO_x), and non-methane volatile organic compounds (NMVOCs). While Parties are encouraged to provide emission projections for these gases, they are not required for inclusion in Parties' National Communications.

National climate change policies are still in their initial stages and it is too early to assess overall costs and effects. Present policies are fragmented rather than strategic and comprehensive. With these policies alone, emissions across OECD regions are expected to grow steadily in coming years. In order to meet Kyoto objectives on time, more comprehensive and consistent policy-making across OECD countries will be needed (OECD, 1999*c*). A basic framework for national climate action emerges from OECD analyses of environmental policy experience. The elements for a framework of good practices include: getting prices right; creating and using markets (*e.g.* through tradable permit schemes); a mix of other complementary policies; improved monitoring of emissions at end-use; improving institutions for decision-making, and international co-operation.

International co-operation

Impacts and costs of climate change are likely to be unevenly distributed among major regions of the world and among sectors of the economy within a country. In addition, countries that have caused the bulk of emissions in the past differ from those most likely to suffer the worst impacts. The majority of harmful impacts will occur in the developing countries, possibly aggravating current economic disparities between different regions of the world (IPCC, 2001*b*). Thus, devising equitable sharing of costs to mitigate and adapt to climate change will be key to global agreement on an effective policy response. The UN Framework Convention on Climate Change (UNFCCC) establishes an initial framework for international action and basic principles for burden sharing.

The objective of the UNFCCC is to stabilise greenhouse gas concentrations in the atmosphere at a level that would prevent dangerous anthropogenic interference with the climate system. Achieving this objective clearly requires extensive international co-operation, contributions from different levels and areas of government, as well as stakeholder engagement to change current production and consumption patterns.

The Kyoto Protocol emerged in 1997 as a first step towards implementing the UNFCCC. The Sixth Conference of the Parties (COP6) in The Hague in November 2000 produced some advances on governance, financing, and the design of specific elements of the Protocol. Although there was a narrowing of views among countries on technical and political points, delegates were unable to arrive at a final comprehensive agreement. It is hoped that such an agreement will be reached in early to mid-2001 and lead to rapid ratification so that the Protocol could enter into force in 2002. The Protocol establishes binding, greenhouse gas emission reduction targets for individual developed countries with a view to achieving at least a 5% aggregate reduction from 1990 emissions levels in the period 2008 to 2012.

Box 13.2. The Kyoto Protocol and market mechanisms

The Kyoto Protocol also calls for the establishment of three types of market mechanisms to help Parties to the Protocol achieve their national emission targets at lowest cost :

- *International emissions trading*: Allows any Annex I Party or authorised legal entity to trade a portion of its allowable emission level; any additional emission reductions below a Party's target level may be traded to another party or any legal entity to another Party. (Article 17 of the Protocol)
- *Joint Implementation*: Allows any authorised legal entity in any Annex I Party to achieve emission reductions through specific projects and to transfer them to another Annex I Party. The Party acquiring a certain amount of project-level emission reduction units is permitted to increase its level of allowable emissions, while the transferring Party would decrease it. (Article 6 of the Protocol)
- *Clean development mechanism*: Allows developing countries (*i.e.* non-Annex I Parties) to transfer certified emission reduction units from projects to Annex I Parties. The Article allows Annex I Parties to count such project-level emission reductions achieved from the year 2000 towards their compliance in the first commitment period (2008 to 2012). (Article 12 of the Protocol)

The specific design and rules governing these mechanisms are still the subject of international negotiations. A priority is to maintain a balance among key criteria, such as environmental effectiveness, economic efficiency and equity.

Economic analyses indicate that the cost of meeting Kyoto targets for Annex I countries would be low (0.2% of GDP per year in the 2010 time frame), and even lower with an effective international emission trading system (0.1% of GDP or lower) (Burniaux, 2000). Yet adjustment or transition costs are expected to be significant in some sectors. Results assume multiple gas mitigation policies (CO_2, CH_4 and N_2O) as this is estimated to lower the marginal and average costs of achieving Kyoto targets by 20-30% compared to CO_2 only policies. Thus the main policy challenge is to identify and implement an effective method of achieving greenhouse gas emission reduction and abatement. Policies targeting CH_4 and N_2O emissions, as well as landuse and forestry options, may help to ease the transition in the energy sector.

Economic analyses indicate that the costs of meeting Kyoto targets for Annex I countries would be low, though adjustment costs may be significant in some sectors.

In the case of forestry and land use, policies could be designed to either augment the storage and accumulation of carbon in forest ecosystems and agricultural lands, or reduce the release of stored carbon. These policies could enhance the use of forests as net carbon sinks, and maintain or expand pools of carbon stored in soils and vegetation. To offset national emission reduction obligations, the Kyoto Protocol allows accounting for changes in sinks that have occurred due to human intervention after 1990. Even though the details of sink accounting have not yet been agreed, they could influence the level of emission reductions otherwise required under the Protocol.

Technological development and diffusion

While the energy technologies necessary to meet the Kyoto targets are largely "on the shelf" today, diffusion is a problem due to their relatively high costs (OECD, 1999*a*). Programmes supporting clean technology diffusion are a traditional focus of energy and environmental policy in OECD countries and are becoming an important part of national climate programmes. Although such programmes need to be designed carefully to ensure cost-effectiveness and to avoid subsidising certain technologies or firms, they can be effective in accelerating the development and use of environmentally sustainable technology.

Besides providing direct government spending, governments can also foster industry research and development investment through a variety of fiscal instruments, such as tax incentives. These policies may accelerate the pace of clean technology diffusion by speeding up capital turnover in targeted areas. Under normal conditions, companies only replace old technologies with new ones when the former have reached the end of their useful lives. Fiscal policies can allow companies to accelerate depreciation schedules or provide tax credits for investment in clean technologies, although at the cost of premature scrapping of productive equipment. Experience with a limited number of these programmes suggests that they may be valuable, but more studies are needed on their effectiveness and efficiency. At a minimum they may be useful if used in a temporary manner to help create niche markets for new, clean technologies. This in turn can bring down the cost of new technologies, allowing them to compete with conventional alternatives.

Legal and regulatory instruments

Standards (voluntary and regulatory) can be useful in encouraging the development, marketing and purchasing of more energy efficient products, vehicles and buildings (see also Chapter 12). The gap between real investment decisions and opportunities for cost-effective investment in more energy efficient products reflects high transaction costs, conflicting incentives (*e.g.* between tenants and building owners, or between building owners and developers), and information failures. Performance standards, when introduced with enough lead-time for manufacturers and builders to adapt, can overcome these barriers. They can help shift investment and narrow the efficiency gap, making increased energy efficiency into a viable alternative to increased energy supply, and improving its competitiveness (OECD, 1998).

Economic instruments

An important element in a policy mix to mitigate climate change is getting the prices right. Policies should aim to include price signals that reflect the full costs of different courses of action. Carbon taxes, removal of subsidies that have adverse effects on the climate, and the creation and implementation of domestic and international tradable permit systems are the most important economic instruments to mitigate climate change.

In OECD countries, significant subsidies are in place in most sectors that emit greenhouse gases. Since many of these subsidies contribute to increases in greenhouse gas emissions and climate change, OECD countries should consider their removal or reform as an important element in achieving Kyoto targets. While many OECD countries have started to reform or remove environmentally damaging subsidies, the process has been slow and significant levels of support remain in place. Various OECD studies have found a significant potential for greenhouse gas emission reductions through subsidy removal. Thus, a series of OECD case studies that simulated the effects of removing coal and other energy subsidies, either at the world or country level, found that in 2010 CO_2 emissions in these sectors would be reduced by 1 to 8% compared to the business as usual scenario (OECD, 1997). Where the economic effects were analysed, most studies suggested real increases in GDP, indicating that subsidy removal may be a "no regrets" policy (IEA, 1999).

OECD countries can also introduce energy taxes to encourage reductions in total fuel use or – where the tax rates are linked to fuel carbon content – substitution with fuel sources that emit less greenhouse gases. A policy simulation was undertaken to examine the effects of removing subsidies to energy production and use in OECD regions combined with the application of an annually increasing *ad valorem* tax on fuel use in OECD countries, with the increase linked to the carbon content of the fuels (see Chapter 12).[8] The tax was thus set to increase by 2 percentage points per annum for coal, 1.6 percentage points for crude oil and 1.2 percentage points for natural gas, reaching a total tax rate of 50%, 40% and 30% of pre-tax prices respectively in 2020. The effects of this simulation were substantial: CO_2 emissions were almost 25% lower in 2020 for OECD regions than projected under the Reference Scenario, and 11% lower worldwide. Applying an energy tax of this magnitude, in combination with subsidy removal, would not enable countries to reach Kyoto targets (CO_2 emissions in OECD would still increase by 3% from 1990 to 2010), but would move OECD countries much closer to that objective.

Other OECD and external modeling work has studied the level of carbon tax that would be required to achieve the Kyoto targets through domestic action alone. The OECD has estimated that meeting the Kyoto targets in OECD countries could be achieved with a multigas GHG tax at a cost of 0.2% of GDP (Burniaux, 2000). While a number of OECD countries already have carbon taxes in place as a key element of their climate change policies, their success to date has been limited because of the preferential low rates or large exemptions offered to large industrial energy users due to concerns about effects on their international competitiveness. Countries that have such taxes in place should make an effort to strengthen them, and those that have none, should introduce carbon taxes with no exemptions for large energy users. Taxation may also be appropriate for a number of other greenhouse gas sources, such as landfill methane, or nitrous oxide from industrial sources (OECD, 2000).

Another option that has become important and is gaining recognition is domestic and international tradable permit systems. As trading permits are likely to play a major role in the international regime, many OECD countries are currently examining them for use at the domestic level, with the intention to join an international regime over time. Unresolved issues, such as the allocation of permits, inclusion of non-energy emissions and other factors, have meant that no country has implemented a comprehensive national regime so far. An international regime has not yet been established either, due to delays in concluding the Kyoto Protocol negotiations. However, a number of OECD countries already have, or are planning, pilot national emission trading systems as part of their climate strategies. OECD modeling (using the GREEN model) shows that with emissions trading in an international regime, the costs of achieving Kyoto targets in OECD regions could fall from 0.2% to 0.1% of GDP. OECD countries should implement comprehensive national permit schemes to realise this potential to lower the overall costs of achieving Kyoto targets.

8. Because of the limited number of energy-related subsidies listed in the GTAP database used in the modeling exercise, the effects of simulating the removal of subsidies included in the database were found to be insignificant.

It is important to recognise that greenhouse gas mitigation policies can be closely linked to resource saving projects in other sectors. In the transport sector, investment in public transportation and effective transport planning policies reduce local pollution and congestion in addition to greenhouse gas emissions. The promotion of sustainable forestry and land use could have positive consequences for biodiversity protection and welfare gains for farmers and smallholders, as well as for enhancing CO_2 uptake. If policy incentives for sustainable forestry are to be established for climate change, care must be taken to ensure coherence with sustainability criteria, *e.g.* the preservation of old-growth forest. Energy efficiency and materials use policies may also interact in positive ways, both leading to CO_2 reductions.

Voluntary agreements

Voluntary agreements figure prominently in OECD national climate programmes and target mainly the industrial sector, although they take many different forms (OECD, 1999*a*). The distinguishing features of these measures are whether they are enforced or closely monitored; the rigidity, clarity and magnitude of the emission reduction objectives; and the nature of the participants (*e.g.* whether industry associations or individual industry players have signed the agreements). Voluntary agreements also vary by the degree of government involvement. Unilateral commitments, initiated by industry without involvement of a government authority, have also emerged in some countries, most notably in Japan (OECD, 1999*a*). Recently, unilateral commitments by multinational corporations have emerged. It is unclear how these unilateral actions by multinationals can be taken into account in national policy frameworks.

Nevertheless, voluntary agreements could be an important element in the policy mixes used by governments to achieve climate change objectives. They provide a flexible policy instrument and a means to fully engage industry in greenhouse gas mitigation. In addition, voluntary agreements may be particularly effective for exploring mitigation action in areas where learning about technical mitigation options and costs is essential, as with the industrial process emissions and SF_6, HFC and PFC (OECD 1999*b*). While the effects of voluntary agreements are difficult to measure, there is a broad consensus that they can influence the environmental culture of industries and businesses by raising the profile of greenhouse gas performance or energy efficiency objectives (OECD 1998; 1999*b*).

Information and other instruments

Education and public awareness policies are an essential complement to other greenhouse gas mitigation policies. Attitudes in business and industry, and among consumers, will ultimately influence the acceptance and pace of uptake of new "green" products and technologies throughout national economies. Green power schemes, where consumers may choose to pay more for "green" electricity, are an interesting example of combining information with real choice for the customer (see Chapter 12). Overall, education and information instruments should be seen as complementary to other policies that provide more direct and continuous incentives to think, act and buy "green".

Monitoring greenhouse gas emissions, and the effects of policies and measures in particular, requires significant government administrative capacity. OECD countries have varying capacities to monitor emissions and policy performance and could benefit from stronger networking to share experience. Goal setting and comprehensive monitoring can also be effective at the local level to raise awareness and stimulate mitigation initiatives. OECD and IEA are working to extend and reinforce the monitoring framework established under the UNFCCC. OECD Performance Reviews in the energy, transport, agriculture, environment and economic areas cover some aspects of climate policy. Over time, the OECD reviews should provide insights about the driving forces behind emission trends in OECD regions, identify and examine good practices, assess current policies and promote better policy performance.

REFERENCES

Aunan, K., H.A. Aaheim, H.M. Seip (2000), "Reduced Damage to Health and Environment from Energy Saving in Hungary", in OECD (2000), *Ancillary Benefits and Costs of Greenhouse Gas Mitigation* OECD, Paris.

Burniaux, J.-M. (2000), "A Multigas Assessment of the Kyoto Protocol", Economics Department Working Paper No. 270, OECD, Paris.

IEA (International Energy Agency) (1999), *World Energy Outlook – 1999 Insights: Looking at Energy Subsidies: Getting the Prices Right*, OECD, 1999.

IEA (2000), *World Energy Outlook* 2000, OECD, Paris.

IPCC (Intergovernmental Panel on Climate Change) (1996), "Impacts, Adaptation and Mitigation of Climate Change: Scientific Technical Analysis", in Watson, R. T., M.C. Zinyowera and R.H. Moss (Eds): *Contribution of Working Group II to the Second Assessment Report of the* IPCC, Cambridge University Press, Geneva.

IPCC (2000*a*), *Land Use, Land-Use Change, and Forestry*, IPCC, Geneva.

IPCC (2000*b*), "Special Report on Emissions Scenarios, A Special Report of Working Group III of IPCC", Cambridge University Press, Geneva.

IPCC (2001*a*), "Climate Change 2001: The Scientific Basis", Working Group I Third Assessment Report, IPCC, Geneva.

IPCC (2001*b*), "Climate Change 2001: Impacts, Adaption, and Vulnerability", Working Group II Third Assessment Report, IPCC, Geneva.

Newman J., N. Beg, J. Corfee and G. McGlynn (2001): "Energy and Climate Change: Trends, Drivers, Outlook and Policy Options", Background document for the OECD *Environmental Outlook*, OECD, Paris.

OECD (1997), *Reforming Energy and Transport Subsidies: Environmental and Economic Implications*, OECD, Paris.

OECD (1998), "Energy Efficiency Standards for Traded Products", Annex I Expert Group on the United Nations Framework Convention on Climate Change, Working Paper No. 5, OECD, Paris.

OECD (1999*a*), *Voluntary Approaches for Environmental Policy: An Assessment*, OECD, Paris

OECD (1999*b*), *National Climate Policies and the Kyoto Protocol*, OECD, Paris.

OECD (1999*c*), *Action Against Climate Change: The Kyoto Protocol and Beyond*, OECD, Paris.

OECD (2000), "The Potential for Using Tax Instruments to Address Non-CO_2 Greenhouse Gases: CH4, N20, HFCs and SF6", COM/ENV/EPOC/DAFFE/CFA(99)110/FINAL, OECD, Paris.

Schipper, L. C. Marie-Lilliu, and R. Gorham (2000), *Flexing the Link Between Transport and Greenhouse Gas Emissions: A Path for the World Bank*, World Bank, Washington DC.

UNFCCC (United Nations Framework Convention on Climate Change) (2000), "UNFCCC database (*www.unfccc.de*) drawn from official national inventory submissions", UNFCCC, Bonn.

WHO (World Health Organisation), (2000) *Climate Change and Human Health: Impact and Adaptation*, WHP/SDE/ODE/00.4, WHO, Geneva.

14
Transport

KEY SIGNALS

- Growth in the transport sector continues to be strong, with road transport (passenger and freight) and aviation showing the highest growth rates.
- Motor vehicle kilometres travelled are projected to increase by 40% in OECD regions and by 86% worldwide to 2020. Over the same period, air travel is projected to triple worldwide.
- Emissions from transport in OECD countries have been reduced for some major air pollutants, notably lead, carbon monoxide (CO), volatile organic compounds (VOC), and more recently for nitrogen oxides (NO_x). At the same time, emissions of greenhouse gases, in particular carbon dioxide (CO_2), from transport have increased in OECD countries.
- Further reductions of 15% to 55% compared to present levels are expected for the above air pollutants for transport in OECD countries by 2020. However, CO_2 emissions from motor vehicles in OECD countries are projected to increase by approximately 44% from 1995 to 2020.
- Projected worldwide increases in transport will have serious impacts on public health and the environment. Issues causing the most concern are climate change, air pollution, noise and land use.
- More focused transport policies are needed which address growth in transport and integrate measures to support technological improvements and transport demand management. The available instruments are well known and have been thoroughly assessed, but there is a significant implementation gap. In the future, economic instruments need to be strengthened, better combined and more consistently deployed.

14.1. Introduction

A striking feature of economic development in the 20th century was the tremendous growth in the number of motor vehicles, both for passenger and freight transport. Transport demand has been growing rapidly in OECD countries and worldwide – it more than doubled over the last 25 years in OECD regions. This growth has been almost entirely in the form of road transport and, more recently, aviation. Without major changes in policies and practices, these trends seem set to continue for several decades.[1]

Transport activities have major adverse impacts on the environment and human health, in particular on climate change and local air pollution. Other effects include noise, deterioration of ecosystems through acidification and eutrophication, habitat-disrupting land use, water pollution and waste generation.

Although important fuel efficiency gains have been achieved for individual vehicles, overall growth in transport in OECD countries and worldwide has largely outweighed these improvements. De-coupling environmental impacts of transport from overall economic growth worldwide has not yet taken place, except for some air pollutant emissions in OECD regions.

1. To produce detailed sector-specific projections for the transport sector, the modelling framework used for the Reference Scenario in this chapter differs from that used for other chapters (see Annex 2). Thus, the regions used here differ slightly from those of other chapters.

14.2. Developments in the transport sector

Table 14.1. **Key transport sector statistics and projections**

		1980	1998 (or latest available year)	2020 projected	Total change 1997-2020
Road motor vehicle stocks (million)	OECD	348	552	730	32%
	World	411	688	1 200	74%
Distance travelled in motor vehicle kilometres (billion km)	OECD	4 924	8 472	11 900	40%
	World	..	11 000	20 500	86%
Air travel (billion passenger km)	*World*	1 310	3 000	9 000	200%

Sources: OECD (1999*b*; 2001*b*) and Reference Scenario.

Transport supply and demand

Transport is a sector that contributes 4-8% of GDP and 2-4% of the labour force in OECD countries (OECD, 2001*a*). The supply of transport includes the production, maintenance and use of transport infrastructure and mobile equipment. Transport also plays a substantial role in international trade, in business operating and household consumption expenditures, and in public expenditure (see Chapters 5 and 16).

Demand for transport derives from the demand for access to people, places, goods and services in modern society. The demand for personal mobility, and thus demand for passenger transport, is closely related to income levels, location, distance from home to employment and educational services, and shopping and recreational opportunities. In the last few decades, growth in income, improvements in technology and infrastructure, and more leisure time have allowed people to travel more and longer distances, and have given them more freedom to choose their mode of travel. This choice depends largely on all of the above factors, but also on personal preferences and priorities.

The demand for freight transport is closely linked to economic growth and international trade, to the overall development of the various sectors of the economy, and to land-use planning and infrastructure. Freight transport has been growing faster than passenger transport in the past decades, but total vehicle kilometres travelled for freight transport is still less than for passenger transport.

Historically, there has been a strong correlation between economic growth and the demand for passenger and freight transport in OECD countries. Growth of GDP has been accompanied by slightly higher rates of growth in overall transport demand, particularly for road transport. While GDP in OECD countries has grown by 46% from 1980 to 1995, the number of motor vehicles has increased by 59% and vehicle kilometres travelled by 72% from 1980 to 1997. Growth rates in air traffic have been much higher than GDP growth rates in OECD countries in the past decades, increasing by around 10% annually.

Road transport, both passenger and freight, has experienced tremendous growth over the past decades, having more than doubled in the last 25 years in OECD countries. Road transport is the dominant transportation mode, representing over 90% of passenger travel and 75% of goods transported. Today, over 550 million motor vehicles (75% of which are passenger cars) are registered in OECD countries, and almost 700 million worldwide (OECD, 1999*b*).

It is projected that the total motor vehicle stock in OECD countries will increase from 552 million vehicles in 1998 to approximately 730 million vehicles in 2020, a total growth of 32%. At the world level, the stock is projected to increase by 74% in the same period. Growth will take place across all vehicle categories with higher growth for passenger cars and light trucks in OECD countries. Cars will continue to be the dominant vehicle category worldwide.

Motor vehicle kilometres travelled are projected to be at substantially higher levels in 2020 than today (Figure 14.1). Total distances travelled are projected to grow by 40% in OECD countries. Within OECD regions, growth rates are expected to be very high in Central & Eastern Europe, but more moderate in the other OECD regions. Growth in vehicle kilometres travelled is expected to be higher at the world level, with a total increase of approximately 86% from 1997 to 2020.

Transport activity by heavy trucks is projected to become much more important worldwide, with distance travelled by heavy trucks expected to almost double from 1995 to 2020. Total distance travelled by passenger cars is projected to increase by 79%.

Figure 14.1. **Motor vehicle kilometres travelled (VKT), 1990-2020**

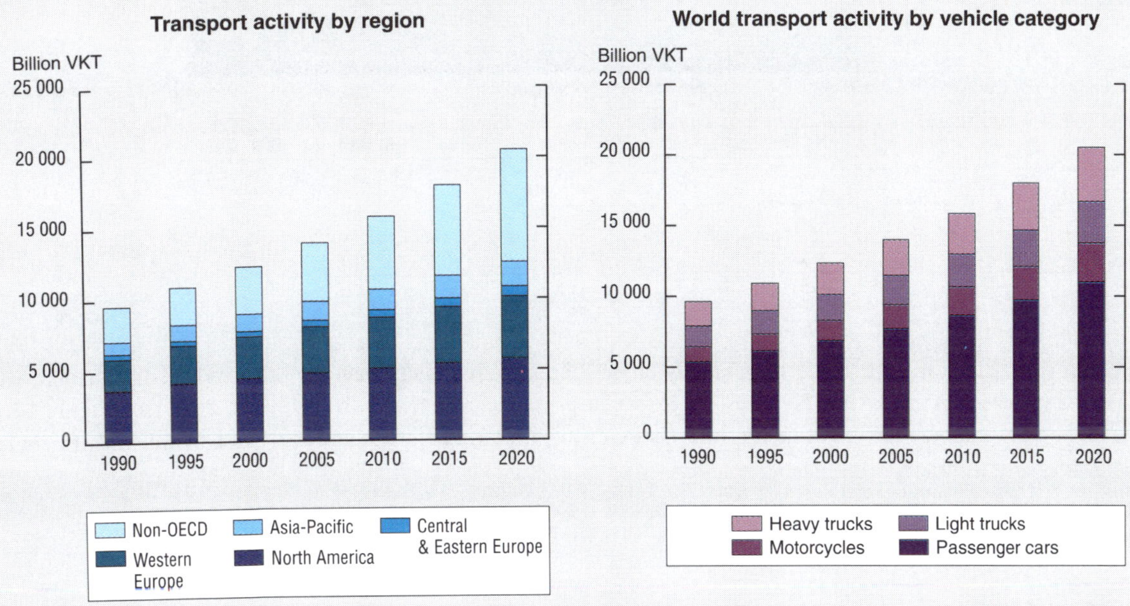

Source: Reference Scenario.

The outlook for rail transport is completely different from that for motor vehicle traffic. The share of railways, both for passenger and freight transport, has declined in recent decades in all OECD countries, except in the US where rail freight has been growing rapidly. While total passenger transport by rail has experienced a growth of 10% over the past fifteen years, it accounts for only 6% of passenger travel in OECD countries. It is projected that rail freight will increase by 40% in the period to 2020, although it is expected that rail will continue to lose market share in freight transport (OECD, 1997*a*).

Water-bound transport has also experienced a declining market share, both for inland waterways and seaborne shipping, although overall freight transport by these modes has been growing steadily in recent years in many OECD countries. Maritime transport is projected to increase by approximately 90% from 1990 to 2020, and thus increase its market share (OECD, 1997*a*).

Air transport's share of global passenger travel is currently less than 3%. However, in OECD countries, the volume of passenger travel by air in 1997 was approximately 10% of total passenger travel. The international aviation industry has grown tremendously over the past 30 years as a result of economic growth, higher disposable incomes and increased leisure time on the demand side, and falling airline tariffs and technical change on the supply side. Passenger traffic has expanded at an average rate of 9% annually since 1960, and air cargo by 11% annually.

Figure 14.2. **Global passenger air travel, 1990-2030**

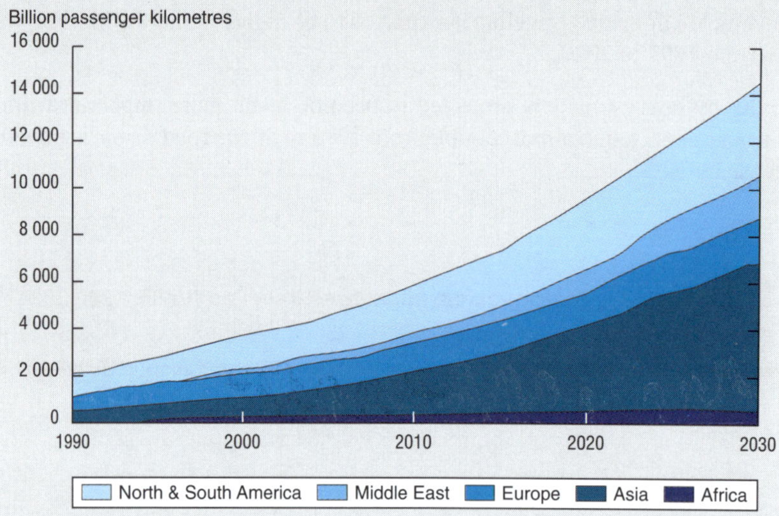

Source: IPPC (1999).

Aviation is projected to experience significantly larger increases in activity than the other modes of transport, with projected increases of approximately 200% in global air passenger kilometres from 1997 to 2020 (see Figure 14.2). The highest growth rates are expected for intra-Asia air travel, between Europe and Asia, and on trans-Pacific routes.

14.3. Environmental effects of transport

Considerable progress has been made in reducing some traditional air pollutants from transport in OECD countries. However, transport activity continues to cause large adverse impacts on the environment and human health, in particular on climate change, local air pollution and noise. It also results in significant depletion of non-renewable resources.

In OECD countries, road transport is responsible for most of the transport sector's impact on the environment. It accounts for over 80% of all transport-related energy consumption (OECD, 1999*b*). More recently, there has been a growing concern over environmental impacts from air traffic, which have been increasing rapidly, particularly for private and leisure trips. Air transport currently represents about 11% of transport-related energy consumption (OECD, 1999*b*). Rail transport – including high-speed rail – causes considerably less environmental impacts than road and air traffic. As for maritime transport, shipping is generally associated with lower overall environmental impacts per tonne-km transported than aviation, road or rail transport.

Air pollution and climate change

Transport contributes significantly to climate change and to air pollution at the local, regional and global level. Air emissions from transport are directly related to the consumption of fossil fuels, and represent a high proportion of overall man-made emissions in OECD countries (OECD, 1995). Carbon monoxide (CO), carbon dioxide (CO_2), nitrogen oxides (NO_x), particulate matter (PM), lead (Pb) and volatile organic compounds (VOC) are the main pollutants emitted from transport.

At the local level, road transport is the main contributor to air pollution in urban areas where road traffic and congestion concentrate, degrading the urban environment. Human health effects cause the most concern, particularly from nitrogen oxides and fine particles emitted by diesel vehicles. For a number of these pollutants, motor vehicles are the single largest source of total emissions in OECD countries. In 1997, motor vehicles contributed 89% of CO emissions, 52% of NO_x emissions and 44% of VOC emissions (OECD, 1999*a*).

Stringent emission standards and targets for motor vehicles to 2005 and beyond have been adopted in Europe, Japan and the US. With measures in place in OECD countries, emissions of nitrogen oxides (NO_x), volatile organic compounds (VOC), carbon monoxides (CO) and fine particulates are expected to decline by 15% to 55% between now and 2020, and possibly stabilise thereafter at much lower levels. However, a similar development for all air pollutants is not expected for the rest of the world, where CO, NO_x and VOC emissions from transport are expected to increase by 70%, 20% and 55% respectively to 2020 (Wiederkehr, 2001).

The primary energy sources of transport are fossil fuels. In OECD regions, their use for transport increased by more than 45% from 1980 to 1997, while for rail transport it decreased by 17% over the same period. Total transportation fuel use continues to increase along with growth in vehicle stock and distances travelled. This has resulted in growing emissions of greenhouse gases, in particular CO_2, worldwide, and in an increasing contribution to global warming. Emissions of greenhouse gases from transport rose from 19% of total greenhouse gas emissions in OECD countries to 21% in 1998, faster than emissions from any other source (see Chapter 13).

Global CO_2 emissions from motor vehicles are projected to increase by approximately 83% from 1995 to 2020, and almost double from 1990 levels (Figure 14.3). In OECD countries, CO_2 emissions from motor vehicles are projected to increase by approximately 44% from 1995 to 2020 (59% from 1990). The contribution of the transport sector to total CO_2 emissions in OECD countries is projected to increase from approximately 20% in 1995 to 30% in 2020 (see Chapter 13).

The contribution of the transport sector to total CO_2 emissions in OECD countries is projected to increase from approximately 20% in 1995 to 30% in 2020.

Figure 14.3. **Global CO_2 emissions from motor vehicles, 1990-2020**

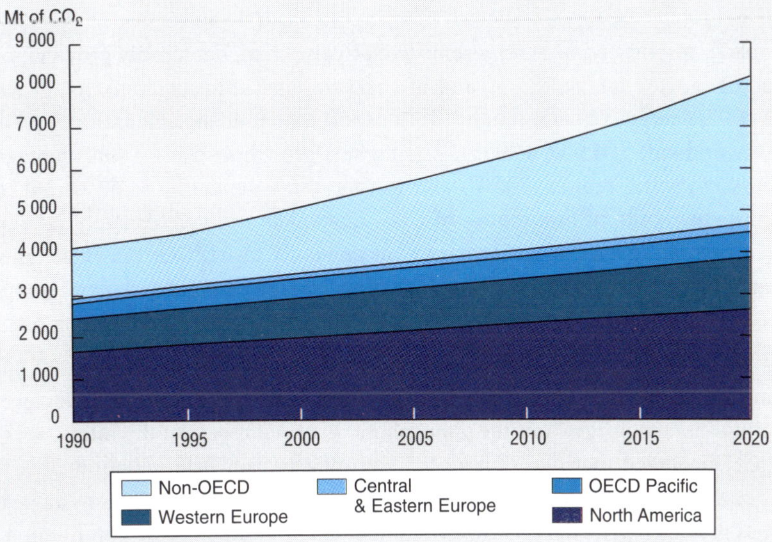

Source: Reference Scenario.

A de-coupling of environmental impacts from overall growth in transport activity worldwide is not expected. As mentioned above, emissions of major air pollutants (*e.g.* VOC, NO_x and CO) are expected to significantly decrease in the OECD region in a period of continued growth of vehicle stock and transport activity (Figure 14.4). Yet, CO_2 emissions from transport will continue to grow rapidly due to the absence of strong policies that target fuel efficiency improvements.

Figure 14.4. **Motor vehicle emissions, GDP and population, 1995-2020**

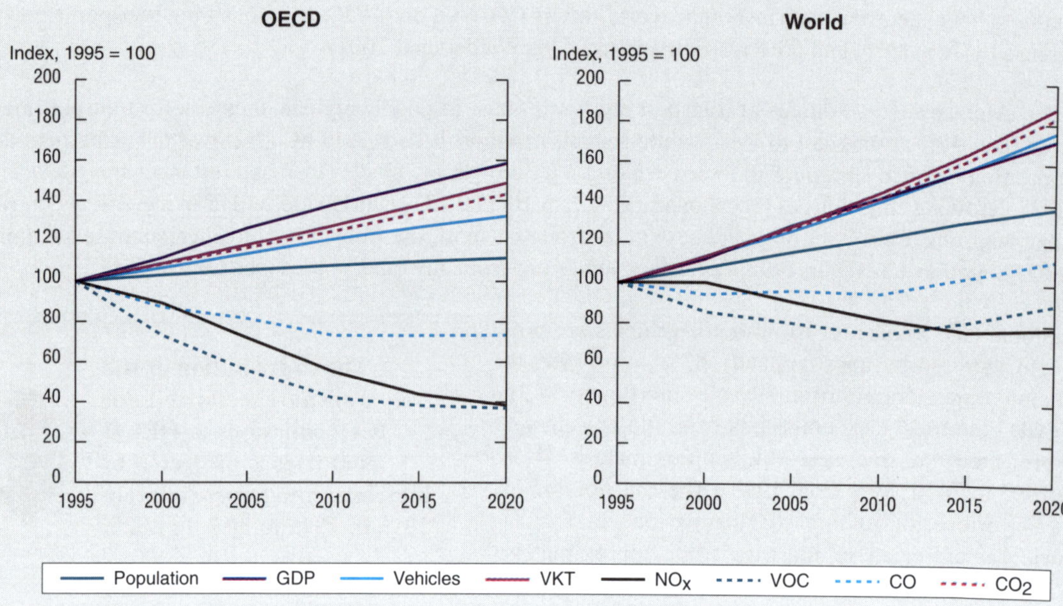

Source: OECD (2000*b*) and Reference Scenario.

Compared to road transport, aviation represents a relatively small, but rapidly growing, source of environmental impacts, including noise, air pollution and global climate impacts from its growing CO_2 emissions. Between 1980 and 1997, energy consumption by air transport increased significantly, by 65% in OECD countries, and by almost 50% worldwide (OECD, 1999*b*). The highest growth in energy consumption by air transport has been noted for the Asia-Pacific region (158%) and Europe (86%) over the same period. Commercial aviation releases annually more than 500 million tonnes of CO_2, equivalent to approximately 2.5% of global greenhouse gas emissions, and representing 12.4% of transport emissions of CO_2 (IPCC, 1999). Aviation related CO_2 emissions have accelerated in recent years as a consequence of increasing demand and increasing aviation services, and continuing rapid growth in air traffic is expected to generate further upward pressure on future emissions.

When assessing the environmental impacts from aviation, it is important to take into account the emission of all pollutants to the mid- to upper- levels of the atmosphere. A recent report of the Intergovernmental Panel on Climate Change (IPCC) concluded that the total global warming impact from aviation emissions of all pollutants amounts to two to four times the global warming impact from aviation CO_2 emissions alone (IPCC, 1999). Based on these findings and projected growth rates for aviation and motor vehicles, the contribution of aviation to global warming could surpass the impact of passenger cars and heavy trucks by 2020 and beyond, despite aviation's comparatively low CO_2 emissions at present (OECD/BMFLUW, 2000).

Land use

Land use for transport is both a factor in generating transport activity (infrastructure induced mobility) and a contributor to environmental stress. Transport infrastructure, mainly in terms of roads, consumes about 25-30% of land in urban areas, and just under 10% in rural areas in OECD countries. The road network occupies 93% of the total land area used for transport in the EU. Rail is responsible for 4% of land take, and airports for less than 1%. Railways require the lowest land take per transport unit, being 3.5 times lower than for passenger cars (EEA, 2000). Due to the expected increase in economic and transport activity, the area of land used for transport infrastructure (roads and parking, rail corridors, airports and harbours) is likely to grow to 2020. This will put growing pressure on biodiversity because of habitat fragmentation and destruction, and the contribution of transport to acidification and eutrophication (see Chapter 11). As infrastructure may act as barriers to movement and interchange between animals, this can clearly affect habitat and species.

Waste

The production, maintenance and use of transport infrastructure and mobile equipment including cars, trains and airplanes, contributes to the generation of solid and hazardous waste in OECD countries (see Chapter 20). Some 50 million vehicles are scrapped annually generating large waste streams of metals, plastics, tyres and used oil. Requirements for recyclability and final disposal of most of the vehicle parts in OECD countries aim to close the loop for material flows in order to save energy, materials and scarce resources. The flow of municipal and hazardous wastes to land fills, incineration plants, recycling and other treatment facilities also increases the demand for transport. Studies in France have shown that the transport of waste accounts for approximately 5% of the total energy consumption by the transport sector, and that 15% of freight movements involves the transport of waste (EEA, 2000).

Noise

Noise levels seriously affecting human health are still recorded in many OECD countries.

Transport noise, particularly from road traffic and aircraft movements, is the major source of external acoustic nuisance in urban areas. Low-level noise affects the well-being of people, while at levels higher than 65db(A) it is detrimental to health. Noise standards for cars and commercial vehicles and airplanes have been progressively improved over the last 20 years in the majority of OECD countries. However, excessive noise levels of 65dB(A) and above are still being recorded in many countries, seriously affecting health by contributing to high blood pressure and cardiovascular diseases. In European OECD countries, about 30% of the population are exposed to road traffic noise levels above 55 dB(A), and some 13% above 65 dB(A) (EEA, 2000). Low-noise technologies for vehicles and road surfaces together with noise protection measures could reduce current levels significantly. There is also a large potential for reducing railway noise. The use of disc-brakes, reducing noise characteristics of wheels and fitting noise absorbers, and the use of noise reduction aprons and low-noise protection walls could bring down railway noise to levels where exposure would not exceed 55 dB(A). It is estimated that some 10% of the population in the EU are seriously disturbed by aircraft noise, *i.e.,* above 55 dB(A).

Despite technological progress to reduce noise at the source and the introduction of low-noise technologies (road surfaces, tyres, wheels, protection walls, etc.), future prospects are less promising. Noise nuisances are expected to worsen near roads due to projected increases in vehicle traffic, and more aviation will generate considerably higher noise emissions around airports and along airways.

Environmental and social externalities from transport activities

Negative environmental and health externalities arising from transport impose a high cost on society. A recent study undertaken by the World Health Organization (WHO) examined the effects of exposure to very small particles (PM_{10}) on human health in Austria, France and Switzerland (WHO, 1999). The study found that long-term exposure of adults to air pollution from motor vehicles caused 21 000 premature deaths annually from respiratory or heart diseases in these countries – more than twice the number of deaths caused by traffic accidents. The study confirmed that road transport is the most important source of human exposure to air pollution and noise. A similar study done on additional cancer risk from exposure to diesel particulate matter has been carried out by the South Coast Air Quality Management District in California (SCAQMD, 1999). It concluded that mobile sources (primarily motor vehicles) were responsible for 90% of the cancer risk due to air pollution in the area, and that 70% of that was attributable to diesel particulates; for all the metropolitan areas across the US, the estimate was that 125 000 cancer cases were caused by diesel particulates.

Total environmental and health costs from transport – including impacts from air pollution, noise, accidents and climate change – have, in a recent study, been estimated to amount to almost 8% of GDP in European countries (excluding congestion costs). For 1995, this represented some 530 billion Euros, primarily from accidents (21%), and air pollution and climate change (48%) (INFRAS/IWW, 2000). Road transport is responsible for 92% of these costs, followed by aviation with 6%, and rail with approximately 2% (Figure 14.5). Two-thirds of the costs are caused by passenger transport, one-third by freight.

Figure 14.5. **Externalities from transport in Europe, 1995**

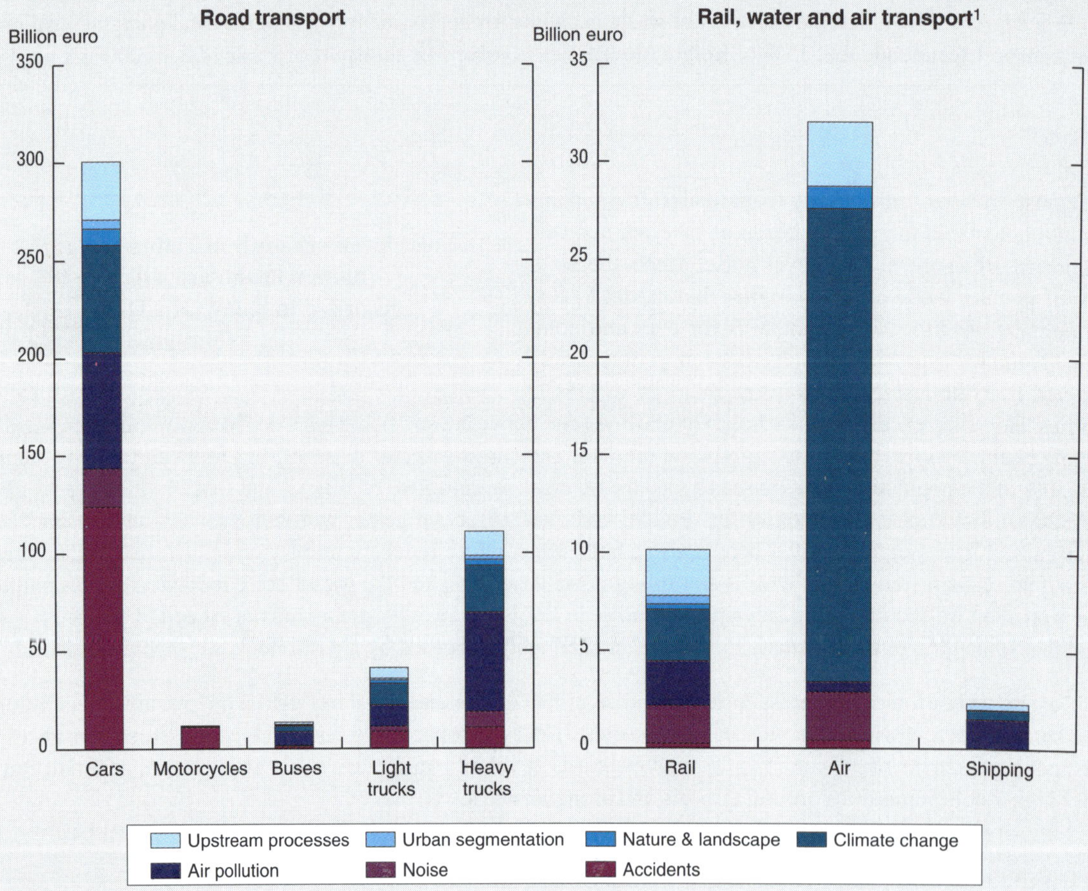

Note: Includes passenger and freight.
Source: INFRAS/IWW (2000).

14.4. Policy options

Advances in technology are not likely to be sufficient to overcome increased environmental pressures and impacts stemming from growing transport demand in the period to 2020. The principal areas of concern will continue to be climate change, air pollution, noise, congestion, resource use (fuels, materials, etc.), land use and habitat destruction.

Numerous measures have been taken or are being discussed to reduce the negative environmental and health impacts of current transport systems. Through national and international regulations controlling vehicle emissions and fuel quality, there have been significant gains with respect to specific pollutants, notably carbon monoxide and lead. However, many of the measures proposed for the future, in particular those targeting structural changes in transport activity or those aiming to reduce CO_2 emissions, lack a framework for effective implementation. Moreover, many do not address the prospect of continuous transport growth which threatens to outweigh the gains likely to be achieved through technological improvements.

A more focused and target-oriented policy approach is needed at international, national and local levels, as well as across the sectors concerned, which places environmental and health goals higher on the political agenda. Such an approach has worked well in the field of air pollution control, for which air quality goals adopted at the national or international level have been used to set stringent control requirements in the form of standards, limits and target levels for emissions and fuel composition to meet these goals.

The feasibility of achieving agreed environmental and health goals in the long term by using a set of specific targets derived from these goals has been explored within the OECD project on environmentally sustainable transport (EST), and the EST Vienna guidelines (OECD, 1996; OECD, 1997b; OECD/BMLFUW, 2000). The EST project has concluded that ambitious, but appropriate, goals may be met through a combination of measures ensuring technology advancements and, especially, influencing transport demand through *i)* improved transport management to promote better use of existing infrastructure and modes; *ii)* inducing transport modes with lower environmental impacts; and *iii)* achieving efficiency improvements through integrated mobility services and logistics. However, there is currently an implementation gap between the potential offered by demand management measures and the level of public support and political will.

Technological development and diffusion

Barriers to the widespread use of more environmentally friendly transport technologies and good practices exist across OECD countries.

Technologies already exist which could significantly reduce the environmental impacts of motor vehicles. If the presently available improved technologies were introduced today for all new passenger cars and commercial vehicles, they could contribute to reductions in fuel consumption by around 30% by 2010. The development and wider use of alternative fuel technologies, *e.g.* hybrid fuel cell vehicles, could further reduce emissions. However, a range of barriers across OECD countries has prevented their wide-scale use. These include their high costs and consumer preferences for large vehicles, but also the reluctance of industry to market more fuel efficient vehicles which could counter current trends. Market forces have so far failed to generate widespread demand for vehicles with best available technologies. It is therefore important to introduce pricing and fiscal policies that would encourage technological development and facilitate market penetration of vehicles with less impact on the environment.

At present, combinations of advanced emission control technologies are available that could allow cars and other light-duty vehicles to meet tailpipe emission standards 50% to 80% tighter than the most stringent values currently in effect in OECD countries. Existing fuel efficiency technologies and the downsizing of vehicles could generate further fuel and resources savings of at least 50%, compared with today's average fuel consumption in OECD regions. Technologies to substantially reduce CO_2 emissions from motor vehicles, however, will not saturate the market in the short term, as their deployment and market penetration will have their full effect only in about

20 years, when the entire vehicle stock has been completely renewed. To ensure faster market penetration of these new technologies, incentives will have to be provided and gradually increasing performance standards will have to be introduced.

Alternative fuel technologies do exist and, if competitively priced and used more broadly, they could bring about significant decreases in environmental pressures. Natural gas and bio-diesel could be promoted for different applications, in particular for urban buses and corporate vehicle fleets with central fuelling facilities. To reduce fossil fuel use and environmental impacts, OECD governments should encourage and provide incentives for the development and deployment of hybrid and electric vehicles, as well as fuel cell vehicles. The development of clean fuels, particularly very low-sulphur or sulphur-free fuels, should also be promoted, as they are necessary for introducing advanced control technologies that will significantly reduce pollutants.

Legal and regulatory instruments

Standards, limits and restrictions have worked most effectively in reducing air pollution when based on clear health and environmental goals and objectives. Other successful policies include the recent ban on leaded petrol in many OECD countries, as well as the introduction of lower speed limits. In urban areas, limiting access of motor vehicles to city centres by parking restrictions and charges will continue to be efficient in restraining traffic, although this should be combined with economic measures such as city road tolls.

National and international efforts should focus on using legal and regulatory instruments to introduce clean and very low-sulphur fuels to reduce the formation of particulate matter. These fuels have multiple benefits. Compared to current fuels, they will not only reduce traditional pollutant emissions considerably, but also CO_2 emissions. In addition, they make possible the use of advanced catalysts to reduce NO_x emissions, and high efficiency filters to eliminate particulates from the exhaust. Efforts should continue to reduce sulphur levels in fuels to 30 ppm and possibly to 10 ppm or less, as proposed by several OECD countries. International co-ordination and clear regulatory requirements for fuel quality, together with longer-term targets and deadlines, will provide the necessary information and time for industry to prepare and adjust their business plans accordingly.

Economic instruments

In OECD countries, subsidies are still commonly used for all modes of transport. These include wide-spread tax deductions for transportation (*e.g.* for employer provided motor vehicles), and fuel subsidies which contribute to higher overall transport levels, and thus increased negative effects on the environment. A gradual removal of environmentally harmful subsidies across OECD countries would result in significant decreases in environmental pressures from transport, in particular on air pollution and climate change.

A number of OECD countries, particularly in Europe, are reforming their tax systems to provide incentives to reduce external environmental costs of transport (OECD, 2001*a*). Changing the tax structure for motor vehicle use in OECD countries can substantially reduce environmental effects caused by increased transport levels. However, the impact on demand of a changing tax structure could differ from region to region depending on, for example, the existing tax structure and geographical conditions. Fully applying the Polluter Pays Principle and the User Pays Principle would mean greater reliance on road pricing, vehicle and fuel taxes, as well as removal of subsidies, across OECD countries. Such changes in taxation can ensure that prices fully reflect the social and environmental costs of growing motor vehicle use, and should be structured so that they encourage optimal changes in behaviour.

A shift in emphasis from taxation based on ownership to emission-related and use-based vehicle taxes would affect the level of transport use and contribute to environmental improvements and less resource use, in most cases without reducing mobility and choice of transportation. However, resistance can be expected and equity concerns would need to be addressed. Fiscal measures, such as road or user charges, should be applied to rapidly increasing road freight to encourage greater efficiency and more completely internalise its high external costs.

To promote the use of public transport, which in dense areas is more energy efficient and less environmentally damaging than individual motorised transport, OECD countries need to counter the factors that have contributed to its decline. These include the current trend of rising prices for public transport. Alternatives need to be explored to increase its attractiveness, reliability and performance, while balancing real costs and prices. The increased use of private passenger cars to a large extent reflects their declining operating costs primarily due to lower fuel prices which decreased in real terms in the 1980s and 1990s. Action could focus on bringing diesel fuel taxes in line with petrol taxes in order to better account for their higher externalities. However, recent protests against the increasing fuel prices and the levels of diesel taxes in many European countries indicate that this will be very difficult to achieve without accompanying measures that provide attractive mobility alternatives.

Attempts are being made to introduce taxation on aviation fuels, at least for domestic flights in some European countries, while internationally this sector enjoys full tax exemptions. Due to the very high growth rates of air traffic, which are expected to continue to 2020 and beyond, such taxes, if effective in changing demand, could help reduce environmental impacts and would also be consistent with the Polluter Pays Principle.

Voluntary agreements

Voluntary agreements can provide opportunities for reducing environmental effects from transport. While such agreements are not considered to be effective as a stand-alone instrument in the transport sector, in combination with and as a supplement to other policy measures they have proven to be useful for achieving improvements beyond the minimum requirements of regulations. Voluntary agreements – possibly long-term agreements on fuel efficiency, negotiated between government and vehicle and aircraft manufacturers, industry, and airline or rail operators – could prove to be effective in increasing fuel efficiency and reducing total fuel use if combined with other instruments.

Information and other instruments

The importance of information and other instruments is likely to increase in the future since they could be crucial in changing consumer behaviour and transport demand. Some policy options have already proved effective in OECD countries and should be explored and expanded further. These include: provision of integrated mobility services by combining car-sharing schemes with public transport services; comprehensive and real-time passenger information and education services; public transport incentives for employees; awareness raising and experimental programmes; and support for local initiatives, as well as marketing campaigns, to promote walking, cycling and using public transport.

However, experience with fuel efficiency labelling of vehicles has shown that simply providing consumer information has a limited influence on customers when they are buying a car because of the low priority they give to fuel efficiency in the face of other criteria, such as vehicle size, power, comfort, and accessories. In order to be effective, consumer information campaigns and education have to be supplemented more consistently with other policy instruments, in particular financial and fiscal incentives.

Integrated urban and transport planning for demand-side management

Carefully conceived and integrated urban land use and transport planning is a potentially effective approach for significantly modifying demand for travel in the long term, especially in favour of public transport, and for reducing adverse impacts on habitat. Such planning should make use of the different types of instruments discussed above.

Household and business location preferences are major drivers of land use development and transport demand, and of environmental impacts from transport. Better planning for land use and development should improve the accessibility to jobs, shops and other facilities without the need to travel by car. Zoning, in particular the introduction of low-traffic zones, has proven to be effective in changing urban travel patterns and reducing environmental impacts. Other policy options include the establishment of more car-free zones, greater traffic control, improving and extending railway and other public transportation networks, providing terminals and logistics centres for combined and optimised transport, and establishing roadway priority for public transport, bicycles and pedestrians.

Measures for influencing transport demand generally fall into three categories: incentives for using (or not using) a particular mode of transport; providing an alternative mode (*i.e.* inducing modal shifts); and introducing user charges to influence transport organisation. Success has already been achieved for certain areas through restricting access to historic city centres by motor vehicles, traffic calming zones, restrictive parking policies, etc. A more effective use of existing infrastructure, as well as more innovative approaches for passenger and freight transport, such as the ones listed below, could result in considerable cost savings and reduce environmental impacts. They should be further explored by OECD countries:

- Better use of existing infrastructure: introducing road user charges to increase vehicle occupancy and load factors and reduce the number of empty freight runs (some 30% in Europe); bundling freight movements to improve efficiency and reduce dead weight; enforcing speed limits to harmonise traffic flows, reduce congestion and save fuel; and promoting non-motorised transport by expanding walking and cycling infrastructure.
- Increasing the use of public transport (the mode with the least environmental impacts and external costs) in locations where it can serve passenger demand efficiently as well as environmentally by increasing frequency, reliability, comfort and safety.
- Encouraging dual or hybrid systems that combine public-private partnerships and organising transport to provide better access to both public transport and individual car use.
- Developing combined road-rail transport where this can help serve markets in environmentally friendly and efficient ways by applying transport chains and total time management for ordering, producing and delivering products.

OECD countries should further explore, develop and apply demand-side management tools. They can be effective in encouraging transport that both meets mobility needs and minimises negative environmental effects. Experience in a number of OECD countries suggests that comprehensive packages of instruments ensuring technological advancements and transport demand management offer the most promising prospects for significant environmental improvements for the sector.

REFERENCES

EEA (European Environment Agency) (2000), *Environmental Signals* 2000, EEA, Copenhagen.

INFRAS/IWW (2000), *External Costs of Transport: Accident, Environmental and Congestion Costs of Transport in Western Europe*, INFRAS and IWW, Zurich and Karlsruhe.

IPCC (Intergovernmental Panel on Climate Change) (1999), "Aviation and the Global Atmosphere, Summary for Policy Makers – A Special Report of the Working Groups I and III of the Intergovernmental Panel on Climate Change", IPCC, Geneva.

OECD (1995), *Motor Vehicle Pollution: Strategies Beyond* 2010, OECD, Paris.

OECD (1996), "Environmental Criteria for Sustainable Transport: Report on Phase 1 of the Project on Environmentally Sustainable Transport (EST)", OCDE/GD(96)136, OECD, Paris.

OECD (1997*a*), "Freight and the Environment: Effects of Trade Liberalisation and Transport Sector Reforms", OECD/GD(97)213, OECD, Paris.

OECD (1997*b*), "Environmentally Sustainable Transport: Synthesis Report on Phase 2 of the EST Project", ENV/EPOC/PPC/T(97)1/Final, OECD, Paris.

OECD (1999*a*), "Indicators for the Integration of Environmental Concerns into Transport Policies", ENV/EPOC/SE(98)1/Final, OECD, Paris.

OECD (1999*b*), OECD *Environmental Data: Compendium* 1999, OECD, Paris.

OECD (2001*a*, forthcoming), *Analytical Report on Sustainable Development*, OECD, Paris.

OECD (2001*b*), *Policy Instruments for Environmentally Sustainable Transport: Report of Phase 3 of the* EST *Project*, OECD, Paris.

OECD/BMLFUW (OECD/Federal Ministry for Agriculture, Forestry, Environment and Water Management, Austria) (2000), "est! Environmentally Sustainable Transport – Futures, Strategies and Best Practices, Synthesis Report of the OECD project", presentation at the International est! Conference, Vienna, 4-6 October 2000.

SCAQMD (South Coast Air Quality Management District) (1999), "Multiple Air Toxics Exposure Study in the South Coast Air Basin (MATES II)", SCAQMD, Los Angeles.

WHO (World Health Organisation) (1999), "Health Costs Due to Road Traffic-Related Air Pollution – An Impact Assessment project of Austria, France and Switzerland: PM10 Population Exposure, Technical Report on Air Pollution", WHO Regional Office for Europe, Copenhagen.

Wiederkehr, P. (2001), "Environmental Outlook to 2020 for Transport", background document for the OECD *Environmental Outlook*, OECD, Paris.

15
Air Quality

KEY SIGNALS

- Efforts in OECD countries to meet air quality objectives have resulted in significant reductions in emissions of sulphur oxides, carbon monoxide and lead, and are likely to result in further reductions in emissions of these pollutants in the period to 2020.
- Emissions of nitrogen oxides, particulate matter and volatile organic compounds have not been reduced to the same degree, although reductions have been observed in some countries. This is primarily due to the strong growth in motor vehicle traffic, which has largely outweighed the efficiency gains from new technological development, and contributed to photochemical smog (tropospheric ozone), acid deposition and eutrophication.
- Air quality standards for nitrogen dioxide, tropospheric ozone and fine particulate matter, as well as critical loads for acid rain and eutrophicating depositions, are still exceeded in many OECD countries and it is projected that this will continue to 2020.
- Important policy options to improve air quality are: further developing and implementing regional and global agreements and air quality action plans; strengthening of legal and regulatory instruments, including combinations of regulations and standards on emissions and fuel use; applying economic instruments to promote clean fuels and fuel substitution, and to encourage energy saving measures; and the use of voluntary agreements.

15.1. Introduction

Emissions of air pollutants reduce air quality in OECD countries and have significant effects on human health, in particular in urban areas where the majority of people in OECD countries live. Pressures on urban air quality in OECD countries are caused by emissions of sulphur dioxide (SO_2), carbon monoxide (CO), lead (Pb), particulate matter (PM), nitrogen oxides (NO_x), volatile organic compounds (VOC) and ozone (O_3) generated primarily by the transport, energy and industry sectors. Levels of SO_2, CO and PM – which cause winter smog – have been significantly reduced in OECD countries. However, summer smog (tropospheric ozone), which mainly comes from emissions of NO_x and VOC, has increased. Toxic trace pollutants – including particle-bound heavy metals (*e.g.* cadmium, lead, zinc) and hazardous volatile organic pollutants (*e.g* benzene and halogenated organics) – emitted from specific industries (*e.g.* smelters, coke ovens and waste treatment facilities) further reduce air quality in urban areas in OECD countries.

Air emissions also cause acid rain and eutrophication of soil and water. Deposition of SO_x, NO_x and ammonia, which are mostly produced by transport and agriculture, contribute to acidification of surface waters and damage to forests in large parts of Europe, North America and Asia. Furthermore, emissions of NO_x contribute to nitrogen deposition and eutrophication of terrestrial and marine ecosystems, which remain serious problems in OECD countries.

15.2. Pressures on air quality

Emission trends and outlooks

Significant reductions in emissions of sulphur oxides, carbon monoxide and lead have been achieved in OECD countries.

In the 1990s, significant reductions in emissions of SO_x, CO, PM and Pb were achieved in many OECD countries. Emissions of NO_x and VOC remained broadly stable over the last two decades, but have decreased slightly since the early 1990s.

In the period to 2020, emissions of most air pollutants are projected to decline in OECD countries,[1] but high emission levels still raise serious health concerns in urban areas, and cause high ozone concentrations, acidification and eutrophication in OECD countries.

Sulphur oxide (SO_x) emissions contribute to winter smog and acidification. SO_x emissions declined significantly in all OECD countries over the last twenty years, with average decreases of between 30-50%. OECD countries contributed about 40% to global SO_x emissions in 1995, with North America contributing 25%, OECD Europe 13% and OECD Pacific countries 1% (Figure 15.1). Assuming that improvements in emission controls and the use of clean, low-sulphur fuels in OECD countries will continue, emissions of SO_x are expected to decline by a further 30% from 1995 to 2020 (Wiederkehr, 2001). In North America, a more than 20% reduction (compared to 1990 levels) seems realistic by 2010 as a result of full implementation of the acid rain programme in the US. For Europe, a decrease of 75% of total sulphur emissions between 1990 and 2010 is achievable and would reflect efforts to meet the target agreed by the EU and under the UNECE Convention on Long-Range Transboundary Air Pollution (LRTAP) to reduce acid deposition (UNECE, 1999). Emissions in the OECD Pacific region could stabilise or slightly decrease due to more stringent controls in Japan, but they may increase if stricter controls are not introduced in other countries of the region. At the global level, emissions of sulphur oxides are expected to considerably increase because of strong economic growth and the use of high sulphur coal in some – primarily non-OECD – countries.

Figure 15.1. **Air emission shares by region, 1995**

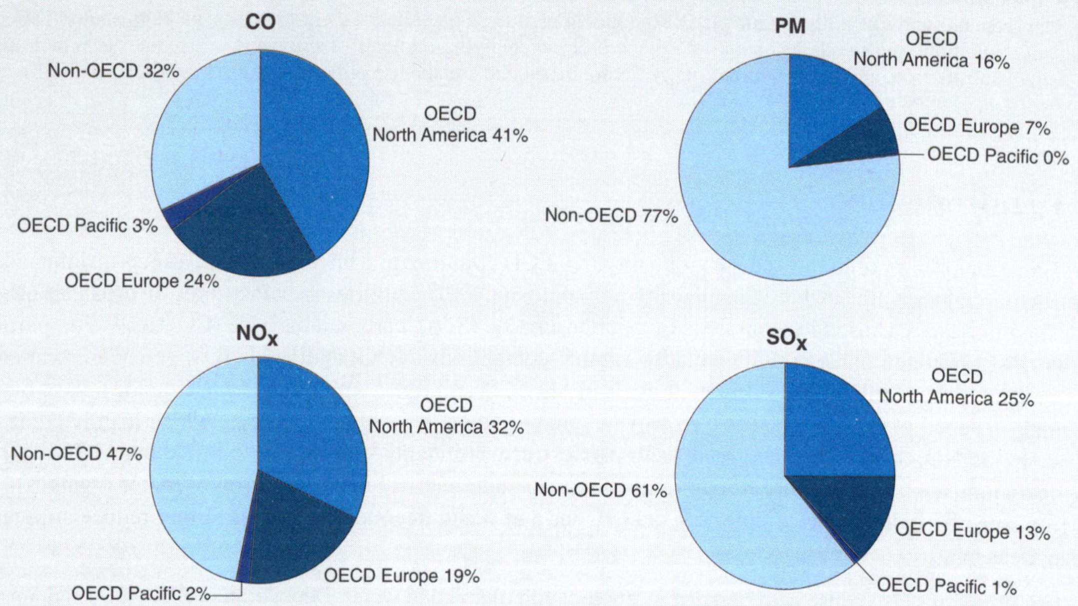

Source: OECD (1999a).

Carbon monoxide (CO) emissions contribute to winter smog. A downward trend has also occurred with respect to CO emissions across OECD regions, with emissions decreasing between 10-25% in OECD countries since 1980. In 1995, CO emissions from OECD countries contributed approximately 68% of worldwide emissions, with North

1. In order to expand the range of air emissions examined beyond those covered in the PoleStar model, the projections used in this chapter are based on the MOVE II model for mobile sources (see Annex 2) and on the background report of Wiederkehr (2001) for other sources. While the underlying assumptions remain the same as in the rest of the Reference Scenario, these projections differ in that they assume the implementation of some new and additional policies in order to ensure already agreed national and regional targets will be met. The regions used here also differ slightly from those of other chapters.

America contributing 41%, OECD Europe 24% and the OECD Pacific region 3%. Global emissions of carbon monoxide are likely to continue to increase, while for OECD countries they could decrease by a further 35% to 2020, with the highest decreases in OECD Europe and North America, if emissions standards for motor vehicles are further tightened (Wiederkehr, 2001).

NO_x emissions contribute to summer smog, the formation of tropospheric ozone and to acidification and eutrophication. NO_x emissions have either been stable, decreased or increased only slightly in most OECD countries over recent decades. The contribution to global emissions from OECD countries in 1995 was 53%, with North America contributing 32%, OECD Europe 19%, and the OECD Pacific region 2%. Emissions are likely to decrease in OECD countries to 2020, primarily because of continued efforts to tighten motor vehicle emissions. Large reductions are expected in OECD Europe (approximately 60%) and in North America (approximately 30%) (Wiederkehr, 2001). For other parts of the world, emissions of nitrogen oxides are likely to increase substantially as a result of economic growth and a rise in motor vehicle traffic.

Emissions of volatile organic compounds (VOC) have steadily decreased in North America and in the OECD Asia-Pacific region, while in OECD Europe they increased in the late 1980s and then started to decline. The large amount of emissions from motor vehicles have led to a greater formation of ground-level ozone (summer smog) in most OECD countries (OECD, 1995*b*). Emissions of VOC in the OECD region are expected to decrease by approximately 40% to 2020, primarily due to emission controls for motor vehicles (Wiederkehr, 2001). In the rest of the world, the emissions are likely to substantially increase, especially in those parts with modest or no emission controls.

Emissions of fine and ultra-fine particulate matter have been increasing in OECD countries, with significant negative impacts on human health in urban areas.

Fine particulate matter emissions have significant negative impacts on human health in urban areas. Combustion of coal and oil-fired power plants, smelters, cars and diesel vehicles represent the largest sources of fine particulate matter. Total emissions of particulate matter in OECD countries have decreased over the past decades as a result of stringent dust control for stationary sources. However, emissions of fine and ultra-fine particles have been increasing, since conventional control systems have not been able to remove them from exhaust gas. The contribution of OECD countries to global emissions of particulate matter in the mid-1990s was 23%, with North America contributing 16%, OECD Europe 7%, and the OECD Pacific region less than 0.5%. It is very difficult to project future emissions from the OECD regions as the amount of fine and ultra-fine particles released into the air will depend on the potential for introducing high-efficiency filter equipment. In other parts of the world, emissions of particulate matter are likely to increase due to growth in energy and transport demand, as well as the lack of efficient emission controls.

During the 1980s and 1990s, lead (Pb) emissions were reduced very efficiently in most OECD countries. This was achieved mainly through the phase-out of leaded petrol. Emission reductions of more than 90% have been achieved in the US and Japan since the 1980s, and over 70% in the EU (EEA, 1999). Future lead emissions from motor vehicles in OECD countries are projected to gradually decrease to negligible levels. However, emissions may well increase in other parts of the world, as leaded petrol is not being phased out in all countries.

Ammonia emissions in Europe could decrease by 15% to 2010 if the target agreed within the multi-effects protocol to the UNCECE LRTAP convention to reduce acidification, eutrophication and the formation of tropospheric ozone from 1999 is met (UNECE, 1999; Wiederkehr, 2001).

Emissions of greenhouse gases, including the three major gases of carbon dioxide (CO_2), methane (CH_4) and nitrous oxide (N_2O), have increased in most OECD countries in the past decade and are projected to increase substantially to 2020. In addition to their impacts on climate change, they also contribute to air pollution (see also Chapter 13).

Tremendous progress has been made in OECD countries in phasing out the emission of ozone-depleting substances, which have now been stabilised at very low levels. In some non-OECD countries, emissions of chlouroflourocarbons (HCFCs) are increasing and are likely to remain in the stratosphere for a long time (see Chapter 19). Also, the global impacts of the remaining substances from past releases are not expected to have their maximum effect until a few years from now.

For most of the numerous toxic trace pollutants, there are no long-term emission trends available, except for some heavy metals (*e.g.* lead, cadmium, zinc and arsenic), some carcinogens (*e.g.* benzene and 1,3-butadiene) and some persistent organic pollutants (OECD, 1995*a*). Data from OECD countries (primarily Europe, Japan, and the US) show that emissions of particulate-bound heavy metals, including arsenic, cadmium and zinc, have declined by more than 65% since the 1970s. Benzene emissions are expected to decline in the future due to stringent emission control for motor vehicles.

Emission trends by sector

Energy, transport, industry, agriculture and the household sector contribute to a variety of pollutant emissions. The contribution of different sectors to total emissions varies among the different pollutants (Figure 15.2).

Emissions from the transport sector represent a high proportion of overall emissions in OECD countries: about 85% of CO emissions, 52% of NO_X emissions, 44% of VOC emissions, at least 50% of airborne lead emissions, but only 4% of SO_X emissions (OECD, 1999a; OECD, 1995*b*). Generally, the proportion emitted by transport – in particular road transport – is increasing, except for some countries where emission regulations have been stringent enough to counteract the total growth in traffic volume (see Chapter 14).

Energy production by public utilities is a major source of SO_X emissions (23%) and the second largest contributor to NO_X emissions (28%), but only a minor source of VOC emissions (5%) and CO emissions (2%). SO_X emissions have been reduced substantially in most OECD countries as a result of flue gas treatment (desulphurisation), use of fuels with lower sulphur content, and replacing coal by oil (and ultimately by gas) as the primary fuel (see Chapter 12).

In the mid-1990s, emissions from industry (including from industrial energy production) contributed about 65% of SO_X emissions, 29% of VOC emissions and 11% of NO_X emissions, but only 2% of CO emissions. NO_X emissions from stationary sources have been reduced effectively in many OECD countries. The industry sector is also an important source of methane emissions from oil production (*e.g.* gas flaring) and natural gas distribution. Industry is also the main source of emissions of ozone depleting substances.

The household sector contributes to CO, NO_X, VOC and particulate matter emissions through fuel combustion (see Chapter 16). However, its contribution is much less than that of other sectors, namely 11% to CO, 6% to SO_X and 5% to NO_X and VOC.

Agriculture is a main source of ammonia, nitrous oxide and methane emissions (see Chapter 7). Ammonia contributes to acidification and eutrophication, while nitrous oxides and methane are important greenhouse gases.

Figure 15.2. **Emission sources for selected air pollutants in OECD regions, 1997**

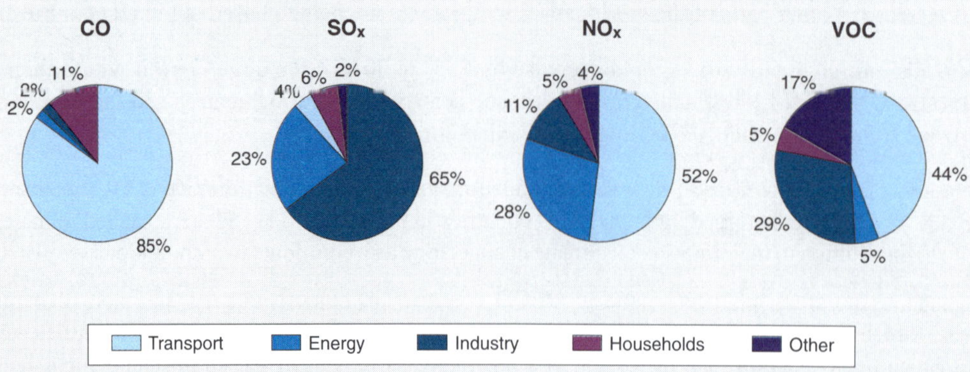

Source: OECD (1999*a*).

15.3. Changes in air quality

Air quality

Despite progress in reducing air pollution emissions in OECD countries, serious air quality and related human health problems persist.

In heavily industrialised and densely populated areas, substantial progress has been made in improving air quality related to emissions of sulphur dioxide, carbon monoxide, dust and lead. However, levels of fine and ultra-fine particulate matter, nitrogen dioxide and ozone have not decreased to the same extent. Data from thousands of monitoring sites in different locations – primarily in urban areas in Western Europe, North America and Japan – were analysed to determine which of these sites exceeded the WHO Air Quality Guidelines for major air pollutants: CO, lead, NO_2, O_3, PM and SO_2 (see Figure 15.3).

For SO_2, which causes winter smog and contributes to acidification, almost none of the sites exceeded the long-term annual average guideline, while about 12% of the reported sites still had concentrations higher than the daily short-term (24-hour) guideline value. Since the late 1980s, SO_2 concentrations have declined by 39% in the US, and up to 50% at highly polluted sites in Western Europe. Concentrations have generally remained at low levels in Japan. While it is rare that the chronic effect level (50μg m^{-3}, annual average) of SO_2 is surpassed in OECD regions, the acute health effect level (125μg m^{-3}, 24-hour average) is exceeded at more than 20% of urban residential sites in Western Europe and the US, but at only a small fraction of sites in Japan.

About 10% of the reviewed sites exceeded the short-term guideline for CO. Trends of CO concentrations show that significant reductions have been achieved in OECD countries, contributing to reducing the problem of winter smog. Despite this, approximately 30% of the sites in the largest cities of OECD Europe, and about 10% of the urban sites in the US, still exceed the WHO guideline which aims to prevent adverse health effects (10 mg m^{-3}, 8-hour average) in areas with heavy traffic.

For NO_2 – which contributes to high ozone concentrations (summer smog), acidification, eutrophication and climate change – 50% of the monitoring sites exceeded the annual long-term guideline, while about 10% were above the short-term guideline. Over 95% of the sites exceeded the 8-hour ozone guideline in Western

Figure 15.3. **Urban sites exceeding the WHO guidelines in the OECD region, 1993**

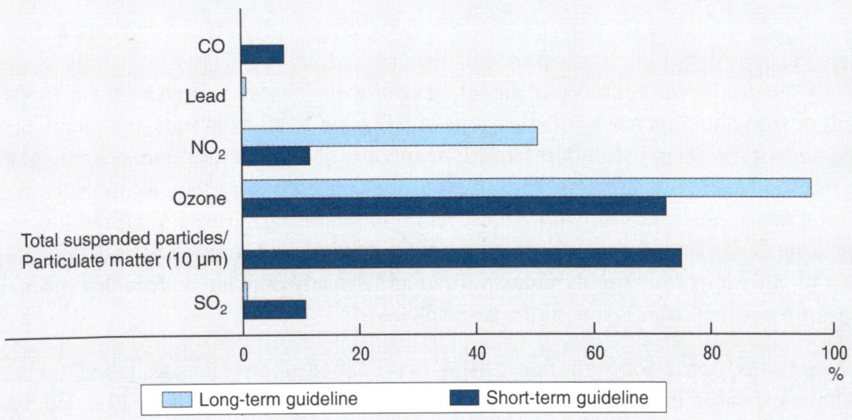

Note: Annual values, except for ozone.
Source: OECD (1999*b*)

Europe and the US. Since the late 1980s, NO_2 concentrations have declined by between 14-30% in both Western Europe and the US, while they have remained stable in Japan. The WHO guideline for maximum daily values of NO_2 (150µg m^{-3}) is widely exceeded in cities throughout OECD regions. In areas with heavy traffic, about 70% of sites in Western Europe, 60% in the US, and 65% in Japan exceed the long-term guideline (40µg m^{-3}, annual average).

Almost all lead monitoring sites were below the long-term guideline. Lead concentrations have declined by more than 55% in both Western Europe and the US since the late 1980s, predominantly due to reduced emissions from leaded petrol. Almost all OECD countries have lead concentration levels below the limit of the WHO guideline of the no-adverse-effect level (0.5µg m^{-3}), although a few European countries exceeded that limit in the early 1990s. In Japan, lead concentrations have remained unchanged since the 1980s at very low levels.

Applying the 24-hour guideline value for total suspended particles and tentative criteria for PM_{10} as a basis, about 75% of the sites were found to be above acceptable levels. Suspended particulate matter in the atmosphere consists of particles of different size and chemical composition. Particles smaller than 10µm and 2.5µm in diameter are referred to as PM_{10} and $PM_{2.5}$ respectively. PM_{10} can enter the upper respiratory tract, while $PM_{2.5}$ can penetrate the lungs. Increasing evidence suggests that health effects of particulates are primarily caused by fine particles, $PM_{2.5}$ and smaller. Recent research shows there is no "safe" threshold level for particulate matter. Since the late 1980s, total suspended particulate concentrations have declined in urban areas of Western Europe by some 35%. In the US, PM_{10} concentrations have been reduced by 35%, while in Japan they have remained stable over the same period, though at lower levels.

Concentrations of toxic trace pollutants and hazardous air pollutants

For many toxic trace pollutants – which contribute to chronic respiratory and other diseases – insufficient monitoring data are available to allow an assessment similar to that performed for major air pollutants. However, monitoring data in a few countries show that concentrations for several particle-bound heavy metals (*e.g.* cadmium, lead, zinc) have fallen considerably due to stringent dust controls of point sources of pollution and restrictions on the use of products containing these pollutants (*e.g.* leaded petrol). Ambient air quality levels for toxic trace pollutants have decreased by 20-50% since the 1980s, and by more than 75% since the 1970s in some OECD countries (OECD, 1995*a*). Concentrations of hazardous volatile organic pollutants, such as benzene and halogenated organics, have decreased by 30-50% during the 1980s in OECD countries as a result of control efforts to reduce health risks. However, benzene concentrations in Europe are still exceeding the limit values, exposing about half of the population in urban areas to excessive levels.

Tropospheric ozone concentrations

Summer smog, caused by high tropospheric ozone concentrations, occurs frequently in many OECD countries. Current daily mean concentrations of ozone (O_3) in Europe are twice as high as in the 1950s. These levels are mainly the result of substantial increases of emissions of NO_x and VOC from industry and transport. The complex formation processes for secondary pollutants like ozone involve numerous substances, sunlight and chemical reactions that need time and energy to develop. Thus, high ozone concentrations tend to occur far away from emission sources and during the summer season, with intense sunlight favouring ozone formation. It is increasingly recognised that large scale formation of photochemical oxidants (including ozone) is a global phenomenon caused by the large amount of emissions from highly industrialised and densely populated areas that substantially change the chemical composition of the lower layers of the atmosphere.

Since the late 1980s, ozone concentrations (using maximum 8-hour values as defined in the WHO guideline of 120µg m^{-3}) have increased in urban areas in Western Europe and Japan by about 15% and 5%, respectively. In some parts of the US, where concentrations were already much higher, peak levels have decreased by 4%. This difference is due to the relative changes in emissions of NO_x and VOC. The WHO ozone limit value is widely surpassed: 95% of the monitoring sites in the US and Japan, and about 90% of sites in Western Europe, exceed revised

WHO 8-hour guidelines aimed at protecting human health. In some urban areas, ozone concentrations are more than twice as high as the guideline limit.

About 330 million people in Europe are exposed to ozone levels above the threshold value, and more than 33 million people living in urban areas in Europe are exposed to excessive levels for more than 25 days per year (EEA, 1999). The WHO guideline for ozone is also used in Europe in relation to the critical levels concept as adopted by the UNECE Convention on Long-Range Transboundary Air Pollution (LRTAP) to estimate the accumulated ozone exposure of the population above the threshold value of 60 ppb (AOT60 indicator). Extensive modelling has been carried out for Europe to estimate the AOT60 in 1990 and 2010, taking into account the emission ceilings that have been agreed under the 1999 multi-pollutant protocol of the LRTAP Convention (50% for NOx and 53% for VOC emissions) The model projects that in 2010 the health relevant AOT60 levels will be reduced by 68% in Western Europe and by 74% in Central & Eastern Europe (see Figure 15.4). Nevertheless, levels will remain well above the targets and large proportions of the population will be exposed to ozone concentrations above the critical level for many days of the year. The forthcoming new EU ozone directive defines the target value to 25 days of exceedence in 2010.

An indicator for exposure of ecosystems has also been defined using the concept of accumulated ozone exposure above a threshold value of 40 ppb (AOT40 indicator) for assessing effects from ozone on vegetation during the relevant growing season. Most recently air quality targets have been agreed among the parties to the UNECE LRTAP Convention using the AOT40 indicators for determining the "critical levels" with a view to assessing effects on agricultural crops and forests. It is projected that for 2010 the AOT40 levels will be reduced by 45% in Western Europe and by 43% in Central & Eastern Europe (EEA, 1999). These results clearly demonstrate that significant improvements in ozone levels are to be expected from these emission reductions. The effect of these emission reductions seem more pronounced for peak values as indicated by the AOT60 levels, while lower ozone levels that protect agricultural crops and forest ecosystems will be much more difficult to achieve (less reduction of the AOT40 levels). Projections indicate that tropospheric background concentrations of ozone will continue to increase in both the North American and Euro-Asian continents (EEA, 1999). This steady growth is caused by increases in background levels of nitrogen dioxide, carbon monoxide and methane.

Figure 15.4. Exceedence of critical levels for ozone in Europe, 1990 and 2010

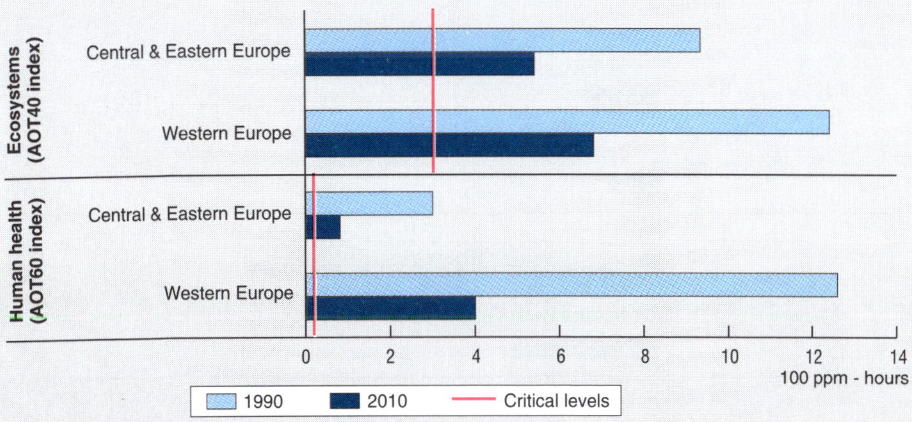

Notes: AOT40/10: cumulative vegetation exposure index (1 000 km² "excess" "ppm" hours).
AOT60: cumulative population exposure index (million persons "ppm" hours).
Source: EEA (1999).

In the US in the mid-1990s, some 90 million people lived in areas that violated the national ambient air quality standard for ozone. Projections for the next ten years indicate that excessive ozone levels will still occur in 28 non-attainment areas (US EPA, 2000). The situation in the US is similar to that described for Europe: in 1997 more than 110 million people lived in areas that exceeded one or several standards. Longer-term projections of air quality levels have also been performed for the US using emission projections and modelling of air quality trends to estimate exposure of the population to excessive pollution levels. In 2007, more than 129 million people are estimated to live in areas which still exceed air quality standards of one or several pollutants, *i.e.* 28 non-attainment areas for ozone and 80 areas with marginal exceedences are to be expected (US EPA, 2000). Estimates for other parts of the world are not available.

Acidification and eutrophication

Acid deposition originates from emissions of sulphur oxides (SO_x), nitrogen oxides (NO_x) and ammonia which are mostly produced by the combustion of fossil fuels for power generation, transport and agriculture. Deposition of these compounds and their reaction products causes acidification of surface waters and damage to forests in large parts of Europe, North America and Asia (see Chapters 8 and 10). While significant progress has been made in controlling sulphur compounds to reduce the acidification problem, nitrogen deposition has not been reduced to the same extent.

Current acid deposition levels in Northern Europe and parts of North America are at least twice as high as critical levels. Between 25% and 40% of the surface area of sensitive ecosystems in Europe receive higher depositions than can be tolerated, leading to continued acidification (Figure 15.5). This is expected to decrease to less than 10% in both Central & Eastern Europe and in Western Europe by 2010 (EEA, 1999). In other parts of the world, further substantial reductions of sulphur emissions beyond 2000 are unlikely, as the focus is on controlling SO_2 emissions from power plants rather than reducing total deposition loads. Thus, acid depositions are likely to continue to contribute to acidification of surface waters and soils in these areas and reduce the quality of the most sensitive ecosystems.

Eutrophication remains a serious problem in OECD countries. Emissions of NO_x contribute to nitrogen deposition and eutrophication of terrestrial and marine ecosystems. The areas of sensitive ecosystems experiencing nitrogen depositions above critical loads were almost 80% in Central & Eastern Europe and almost 40% in Western Europe in 1995. In the period to 2010, these values are expected to decrease to under 60% for Central & Eastern

Figure 15.5. **Exceedence of critical loads for sensitive ecosystems in Europe, 1995-2010**

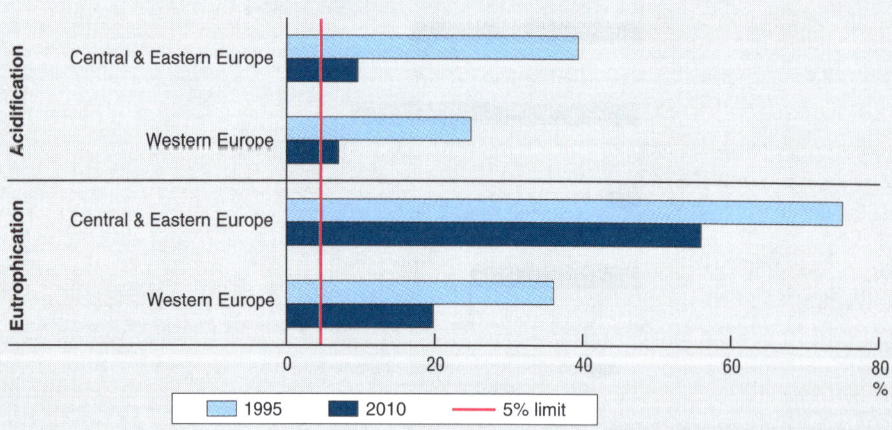

Source: EEA (1999).

Europe and about 20% for Western Europe (Figure 15.5) (EEA, 1999). Unfortunately similar studies and estimates are not available for other parts of the world.

15.4. Policy options

Adopted and planned control strategies are likely to lead to further reductions of air emissions over the next decade in OECD countries and substantial reductions by 2020 (Wiederkehr, 2001). Nevertheless, high emissions levels of NO_x and VOC remain a concern due to their contribution to the formation of photochemical smog (primarily ozone). Particulate matter emissions will continue to be a significant health impact in urban areas unless high-efficiency filter technology or sulphur-free fuels are introduced, and other sources of fine particles are controlled.

> **High emission levels of nitrogen oxides and volatile organic compounds remain a concern in OECD countries due to their contribution to the formation of photochemical smog.**

Existing air pollution control programmes have focused on various environmental media (air, water, land) and sectors (industry, energy, transport, agriculture and households) and addressed them separately, with an emphasis on end-of-pipe controls. Integrated approaches and upstream pollution prevention policies have been rather limited to date, as have area-wide air quality management approaches that simultaneously address stationary, mobile and diffuse sources.

A forward looking strategy for achieving lasting improvements in air quality and meeting stringent environmental standards and targets should strengthen existing regulatory instruments, better combine them with different economic instruments, and involve voluntary approaches in support of regulatory action. Such a strategy could make use of existing instruments, but they would have to be carefully combined, packaged and more consistently deployed in order to achieve environmental targets. This involves, first of all, the development of a comprehensive air quality and management action plan (as required, for example, by the EU directive) that combines regulations on emissions from all sources, including products and fuel use, economic instruments to enhance the development of clean technologies, vehicles and fuels, as well as voluntary agreements to supplement regulatory action.

Since some of the large-scale air pollution problems are caused by the long-range transboundary transport of air pollutants and the formation of secondary pollutants (such as ozone), a number of regional and global international agreements have been developed over the least two decades to address these pollutants. These regional agreements – in particular the UNECE Convention on Long-Range Transboundary Air Pollution (LRTAP), the regional sea conventions for the protection of the North Sea (OSPARCOM) and the Baltic Sea (HELCOM), and EU directives – have more recently focused on the concept of critical levels and loads as targets for setting longer-term emission reduction targets. Extensive modelling and evaluation suggest that this effect-oriented approach will result in much greater improvements than conventional approaches using available control technology.

Legal and regulatory instruments

Some OECD countries have set emission and performance standards for stationary and mobile sources that have resulted in significant improvements over time through continuous technological improvements. A precautionary approach is taken in a number of OECD countries, allowing the introduction of best available technology (BAT), and making continuous improvements in emission controls, even where the full environmental and health implications of the pollutants are not fully understood. Environmental permits for industrial facilities are important for efficiently implementing environmental requirements and ensuring compliance through appropriate monitoring and reporting.

Regulations of highly polluting fuels (*e.g.* heavy fuel oil, residual fuels, wastes) and limits on their composition (*e.g.* sulphur, heavy metals, toxic and aromatic content) should continue. Clean and low-sulphur, or sulphur-free, fuels that have numerous technological and environmental advantages should also be introduced.

Tighter emission standards for motor vehicles are needed, especially for diesel vehicles and motorcycles. In addition, monitoring and emission control programmes are necessary to ensure compliance with standards over time and prevent the use of highly polluting vehicles. Comprehensive and more stringent emission control programmes for motor vehicles (both light and heavy-duty) in Europe, Japan and North America may – over the next eight to ten years – lead to further significant emission reductions of CO, NO_x, VOC and particulate matter (notably from diesel vehicles).

There is also a need for better integrated pollution prevention and control, area-wide air quality action and management plans with goals, targets and deadlines covering all sources, and BAT-based performance standards for stationary combustion sources, including large industrial sources and power plants to control emissions of SO_x, NO_x, PM and heavy metals.

Regulations in most OECD countries require existing facilities to meet standards for new sources when retrofitted or major process changes have been made. There are a large number of facilities (up to 90%) that avoid this rule by declaring retrofitting and process changes as maintenance activities to improve the regular operation of the plant. This has significant implications on the environmental performance of these plants, as retrofitted plants usually operate with higher output capacity, but under more lenient emission controls than what is required and achievable by using state-of-the-art controls. Future control strategies and enforcement should address this issue and focus particularly on implementation.

Economic instruments

Economic instruments are increasingly used to supplement traditional regulatory approaches to air quality improvement in OECD countries. They include emission fees and charge systems, subsidies and accelerated depreciation schemes and, more recently, the trading of pollution credits. The main issue for the future application of economic instruments is to more fully internalise external environmental and health costs and to move from a piecemeal approach to more comprehensive tax reform. Several OECD countries have started to change their tax structure accordingly. These efforts need to be strengthened and better integrated into air management policies.

Taxes on vehicle use and emissions have been broadly applied in many OECD countries, and are an important source of government revenue. Differentiated taxes targeting sulphur levels and lead content of fuels have contributed significantly to pollution prevention. Pricing and tax incentives should also be enhanced to promote fuel substitution (*e.g.* natural gas, electricity) and encourage energy saving measures. Energy taxes for all types of fuels (but less for aviation fuels and kerosene) have also been introduced and should be promoted where possible (see Chapter 12).

External environmental and health costs can also be integrated through the use of fiscal incentives to support the development and use of low-sulphur, and ultimately sulphur-free, fuels which have numerous environmental advantages (*e.g.* reduced sulphur and PM emissions). The use of advanced diesel NO_x and PM control technologies, as well as improving the performance of other advanced technologies that lower fuel consumption and particulate emissions, should also be encouraged through economic incentives.

Market-based emission allowance and trading schemes among stationary sources have been used primarily in the US to reduce primarily SO_x, but also NO_x, CFCs and VOC emissions. The sulphur emission trading scheme is expected to reduce compliance costs of industry by 25-50% by giving industry maximum flexibility to reduce emissions.

Particular attention should be given to the application of market-based instruments to improve the environmental performance of small- and medium-sized enterprises. They are a growing source of pollution and have not yet been targeted by regulations. Pricing and fiscal incentives may prove to be very effective in this regard.

Voluntary agreements

Voluntary approaches and agreements have been increasingly used with industry to supplement air pollution regulatory programmes. They may take several forms, and range from relatively less stringent voluntary initiatives and arrangements to more stringent, negotiated agreements, with goals and targets, deadlines and reporting requirements. Most of them have been used to promote energy efficiency improvements in industry and electricity generation, fuel efficiency improvements of passenger cars, and a less polluting use of chemicals and consumer products.

Voluntary agreements can be useful to supplement conventional types of regulatory approaches in areas where regulation and enforcement are difficult, or if the sector is too heterogeneous or has too many units (*e.g.* small- and medium-sized enterprises). If carefully designed, negotiated and implemented, the limitations of voluntary agreements can be reduced and their role in air quality management enhanced. Experience is still too recent to come up with a final assessment of their effectiveness, but generally voluntary agreements should be combined with or linked to other instruments, such as permitting, regulations, or pricing and fiscal incentives, to ensure effectiveness. Their credibility and transparency can greatly be improved by public reporting and independent verification of their compliance with binding targets.

Information and other instruments

Improving urban air quality will not only depend on the effectiveness of pollution prevention and control policies, but also on the availability of green areas in cities. The size, type and distribution of green areas are becoming more important in large cities. Green areas, such as parks and tree-lined avenues, improve the urban climate, enhance air circulation, absorb atmospheric pollutants and enable residents to enjoy nearby recreational activities. It has been estimated that urban trees improve air quality by removing several tonnes of pollutants. Therefore, maintaining appropriate green areas in cities should be an important component of overall air quality management that also contributes considerably to the well-being and quality of life of residents.

Regular monitoring and public reporting, as well as access to information on the status and trends in air quality, are very important for raising awareness about these problems, showing progress, and motivating participation of the business sector, governmental bodies and the public. Also, targeted information and education programmes are of increasing importance to demonstrate effectiveness of programmes and good practices.

REFERENCES

EEA (European Environment Agency) (1999), *Europe's Environment: the Second Assessment*, EEA, Copenhagen.

OECD (1995*a*), *Control of Hazardous Air Pollutants in OECD Countries*, OECD, Paris.

OECD (1995*b*), *Motor Vehicle Pollution: Reduction Strategies Beyond 2010*, OECD, Paris.

OECD (1999*a*), *Environmental Data – Compendium 1999*, OECD, Paris.

OECD (1999*b*), "Advanced Air Quality Indicators and Reporting", ENV/EPOC/PPC(99)9/FINAL, OECD, Paris.

UNECE (United Nations Economic Commission for Europe) (1999) "Gothenburg Protocol – to the Convention on Long-Range Transboundary Air Pollution", ECE/ENV/99/8, UNECE, Geneva.

US EPA (United States Environmental Protection Agency) (1999), "National Air Quality and Emissions Trends Report, 1998", US EPA, Washington, DC.

Wiederkehr, P. (2001), "Environmental Outlook to 2020 for Air Quality", background document for the OECD *Environmental Outlook*, OECD, Paris.

Section V

HOUSEHOLDS, SELECTED INDUSTRIES AND WASTE

This section examines the environmental implications of developments in the household sector, the steel, pulp and paper, and chemicals industries, and analyses recent trends and projections for waste generation and management in OECD countries. The households chapter focuses on the direct impact household activities have on the environment through their energy and water use, waste generation, travel patterns and consumption levels. The industry chapters (steel, pulp and paper, and chemicals) first discuss recent trends and future projections for developments in the industries themselves, and then present an analysis of the impact of these developments on the environment. The chapter on waste examines the recent trends and future outlook for the generation of different categories of waste, developments in waste management, and discusses the associated environmental impacts. All the chapters conclude with an indication of the policy options available for addressing the environmental problems identified.

16 Households

KEY SIGNALS

- Consumption by households has increased considerably in OECD countries in the past decades along with economic growth, and is projected to continue to grow to 2020.
- Pressures on the environment from household activities in OECD countries have increased in recent decades and are expected to continue to increase to 2020.
- Future environmental pressures from the household sector will arise from growing energy and water consumption, greater use of motorised transport and air travel, and increases in household waste generation.
- Households in OECD countries are generally aware and concerned about environmental pressures but, for a variety of reasons, that concern has not led to a change in most households' behaviour.
- To reduce the environmental impacts from household purchases and behaviour, existing policies that address the full lifecycle of key consumer goods need to be more widely implemented and new measures introduced. Efforts should be made to raise the environmental awareness of households and encourage them to act in a more environmentally sustainable manner.

16.1. Introduction

Households in OECD countries affect the environment through both their day-to-day decisions on what goods and services to buy and the use they make of them, and their decisions on where to live and work, what kind of dwelling to have, and where to go on vacation. Although the environmental pressures of individual households are sometimes minor compared to environmental impacts from the industrial and public sectors, the combined impact of many households is an important contributor to a number of environmental problems, including air and water pollution, habitat alteration and climate change. In areas such as household energy use, travel and waste generation, material and energy efficiency gains have been outweighed by the absolute increase in the volume of goods and services that are consumed and discarded.

One of the defining characteristics of the 20th century is the development of a "consumer culture" and the progressive multiplication of the products and services offered to, and demanded by, households. Rising per capita income and accompanying changes in lifestyles have led to more individualised buying patterns, a shift towards more processed and packaged products, higher levels of appliance ownership, and a wider use of services. Higher incomes have also increased the number of objects households purchase. For instance, whereas before households used one standardised product for the majority of household cleaning tasks, they now use a range of products, each with its own specific application. The purchase, use and disposal of some semi-durable (clothing, linens, small electric appliances) and even durable goods (motor vehicles, large household appliances, computers) are accelerated by considerations of style, product attributes or obsolescence.

The environmental impacts from household purchase and use of consumer goods depend on the characteristics of the good, including its material and energy content; the resource (energy, water) load associated with its use; the type, the amount and the reusability of product packaging; hazardous pollutant content; and the disposal method.

16.2. Household food consumption

Households in OECD countries increasingly demand a food supply that offers convenience and variety. These trends, along with lifestyle changes and more individualised time allocation within households, have led to a greater consumption of processed and imported food, and to individual portions and packaging. More meals are prepared outside of the home. At the same time, consumer awareness of food quality and safety issues has grown, leading to a greater demand for organic produce.

Households consume more meat, fruit, vegetables, fish and seafood than in the past. Although it is already high, the total calorie intake is increasing in many OECD countries. A few of these have not yet reached saturation levels for per capita consumption of individual commodities and/or aggregate food consumption, but per capita consumption is continuing to grow even in countries that have very high levels of national consumption. Projected food demand for 2020 (compared with 1995/97 levels) includes a 7% increase of per capita meat consumption for OECD regions as a whole, with higher increases for Mexico, Korea and Hungary (OECD, 2000). Per capita consumption of cereals (mostly in refined form) and vegetable oils is also likely to increase (Figure 16.1).

Figure 16.1. **Apparent food consumption in OECD countries, 1974-2020**

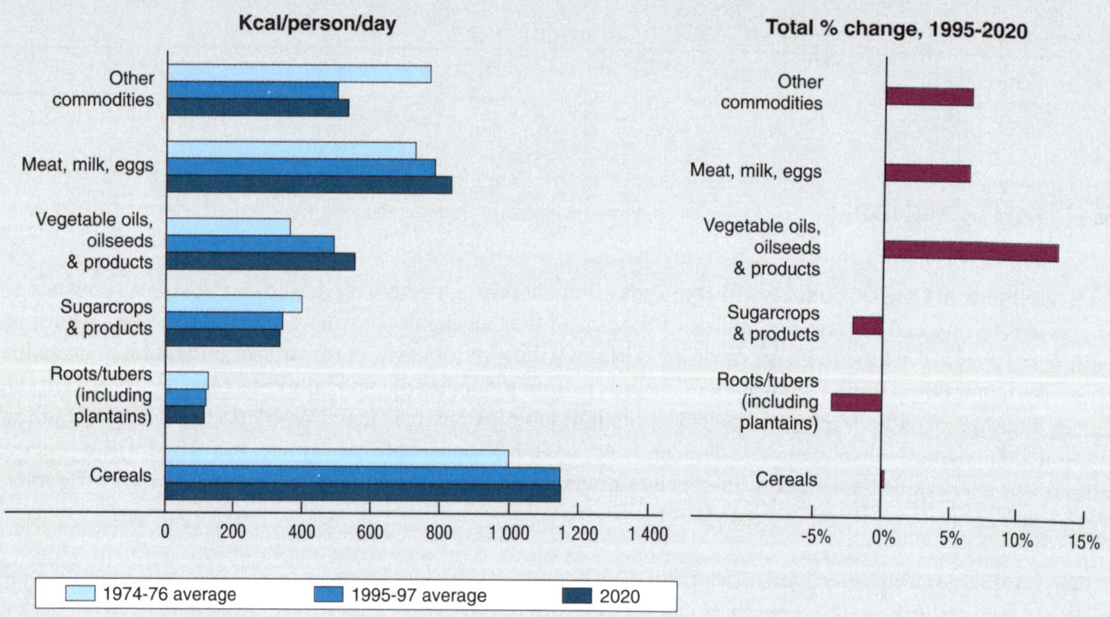

Source: Alexandratos (2000)

Several factors play a role in how patterns of household food consumption will affect the environment, including where and how food is produced, processed, packaged, preserved, distributed and prepared. The most significant environmental impacts occur early in the production chain. However, households in OECD countries influence trends in these areas through their choice of diet and their demand for food-related services. The demand for year-round availability of fresh fruit and vegetables, for example, means more energy for greenhouse production or long-distance transport by road or air. The growing demand for meat has led to intensive pork and poultry production, which are important sources of water pollution. Other impacts result from the food processing and services sectors (*e.g.* energy consumption for the preparation and distribution of chilled and frozen foods).

Households also affect the environment directly by how they purchase, store and prepare their food and how much organic and packaging waste they generate. In one OECD country where trends have been explored, households use almost as much energy in driving their food home as producers do in transporting it to the shops (OECD, 2001*a*). Food waste, although a declining share of household waste streams, appears to be increasing in many countries. Packaging waste from food is also increasing as a share of household waste, and has diversified to include a larger percentage of plastic waste which is generally more difficult to recover (OECD, 2001*b*). Some household food consumption patterns may be leading to a trade-off of impacts between actors in the food chain. For instance, the combination of out-of-home food processing and quick home preparation may result in lower energy use for food preparation at the household level, but potentially greater energy use for household food preservation. Lifecycle analysis of the preservation and preparation of food products is needed to identify the net effect of these trends and determine both where environmental impacts are heaviest and where technological or policy measures are most necessary.

As consumer awareness of food quality and safety, and of potential environmental effects of food production, has grown in the past decades, household demand for organic food products has increased steadily. Organic agriculture has experienced growth rates of over 20% annually in a number of OECD countries (see Chapter 7).

16.3. Household energy use

Two trends are shaping household energy demand in OECD countries. Rising incomes have led to a demand for larger houses. At the same time, family size has decreased, so per capita dwelling area has increased. Bigger dwelling area leads to a greater demand for energy for heating and cooling rooms and for heating water. Increased ownership of energy- and water-consuming appliances are driving a greater demand for energy and water. On the other hand, household heating systems and electrical appliances are more efficient and more households are equipped with energy conserving technologies (*e.g.* double-glazing and insulation). Moreover, in many countries, households now have better access to reliable information (*e.g.* household energy audits) on their energy use.

Higher energy prices, stricter building codes, subsidies for conversion from oil to other energy sources, and programmes to encourage home insulation, have significantly reduced the energy intensity[1] of household space and water heating. Energy intensities for electric appliances have also fallen, although more modestly. New appliances may use less electricity than older ones, but the growing number of household appliances and electronic communication and information devices has outweighed improvements in the efficiencies of almost all their uses. Of particular importance is the rapid growth in the number of dishwashers, clothes dryers, freezers, and air conditioners in many OECD countries (Table 16.1). These trends explain the rising importance of electric appliances and electricity in general in the household total final demand of energy by end use.

Household energy demand is likely to continue to rise in OECD countries to 2020, but there is a large potential for further energy efficiency gains.

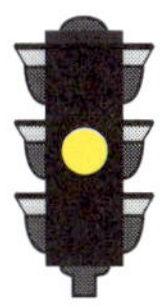

Household energy demand is expected to continue to rise in the short- to medium-term (see Chapter 12). In the Reference Scenario, electricity demand by households is projected to increase substantially in all OECD regions. In the period from 1995 to 2020, the highest relative changes are expected in Central & Eastern Europe (approximately 200%), but significant increases are also expected in the other OECD regions (Figure 16.2).

The environmental impacts of increased household energy consumption will depend on future approaches to residential space and water heating (*e.g.* small-scale co-generation, decentralised energy production in the residential sector, solar power) and the interaction between larger dwelling areas and higher levels of appliance ownership and energy efficiency improvements achieved *via* building standards, technology or product modification, or changes in household behaviour.

1. Energy intensity refers to the amount of electricity needed for a particular end-use, such as space and water heating or running household appliances.

Table 16.1. **Household appliance ownership for selected OECD countries, 1973-1997 (units per 100 households)**

		Refrigerators & combination units	Freezers	Clothes washers	Clothes dryers	Dish-washers	Air conditioners
US	1973	100	34	70	38	25	47
	1997	115	36	79	55	50	86
JAPAN	1973	104	..	101	..	..	15
	1997	120	..	107	23	..	191
UK	1973	73	8	68	4	3	..
	1997	104	42	92	51	21	..
DENMARK	1973	97	42	41	1	6	..
	1997	107	65	72	30	35	..
NORWAY	1973	89	57	72	15	3	..
	1997	141	92	92	35	47	..

Source: IEA (2000).

Figure 16.2. **Electricity demand by households, 1995-2020**

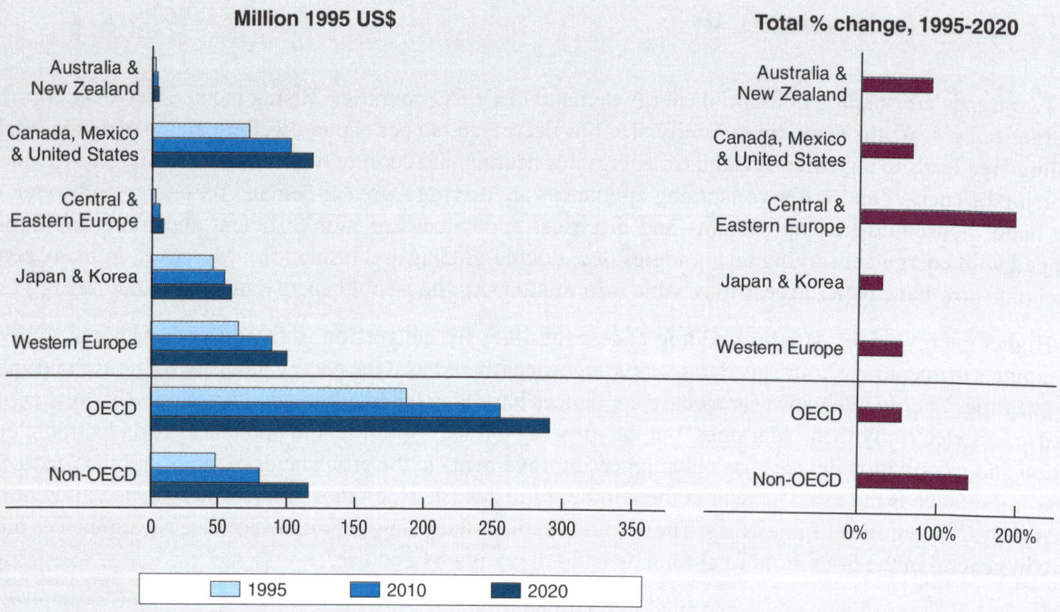

Source: Reference Scenario.

16.4. Household water use

> **Household demand for freshwater has stabilised or declined in a number of OECD countries.**

Household water consumption varies considerably among OECD countries, ranging from 100 to 300 litres per capita per day. Pressures on water supplies stem from the growing number and capacity of water-consuming household plumbing units and appliances (*e.g.* showers, toilets, water heaters, dishwashers, and clothes washers) and other uses of water (*e.g.* lawn watering, swimming pools). The increases in total water use are somewhat offset by increases in the efficiency of water use, such as through reduced municipal pipe leakage, increased water efficiency of household appliances or toilets, etc. As a result, household demand for water has declined or stabilised in a number of OECD countries in recent years (OECD, 1999*a*), including in countries with an already low consumption level (*i.e.* 100 litres/cap/day). OECD countries using water pricing systems that approximate marginal costs (*i.e.* where households pay per unit of water used, rather than by household size or number of occupants) have seen the most significant reductions in household water use. The impact of such systems is strengthened when they are combined with relatively high or increasing water tariff rates (OECD, 1999*a*). These trends suggest a strong potential for households in other OECD countries (and particularly those with high per capita water consumption levels) to reduce their use levels once they have appropriate information, water-efficient technologies and strong incentives.

Households are relatively low consumers of water compared to other sectors (see Chapter 8). However, the pressure on water supplies from household water demand is important, particularly in regions of OECD countries that experience periodic or permanent water shortages. Urban and rural municipalities are also increasingly competing with other users for water where supplies are limited. A growing number of cities in OECD countries are facing rising costs to supply water of drinking quality and sanitation facilities to urban residents. Households also have an impact on water quality through the effects of wastewater releases on oxygen availability in aquatic ecosystems (known as biochemical oxygen demand or BOD), contamination by household products (including phosphate detergents, oil and grease), and run-off of lawn chemicals applied on gardens. OECD countries have improved basic water pollution abatement from households by increasing the number of households connected to basic sewage treatment facilities, and expanding secondary and tertiary treatment. BOD levels due to household sewage release, however, are still projected to increase to 2020 for OECD regions (see Chapter 8).

16.5. Household travel

Household private car use and air travel will increase to 2020, contributing strongly to growing emissions of greenhouse gases.

Individual private car use and air travel are two of the most significant ways in which households contribute to environmental pressures, including global climate change, local air and water pollution, noise exposure, and land alteration for roads and transport infrastructure. Household travel meets the need for three kinds of mobility: commuting and other work-related travel, family and civic excursions (shopping, school, medical visits, various meetings, etc.), and social and recreational trips. Commuting accounts for around 25% of household travel in OECD countries (IEA, 1997). Households use their cars mostly for social or recreational purposes.

Although the demand for all forms of transport has risen (see Chapter 14), car ownership in particular has increased steadily since 1970. Day-to-day car use has grown substantially while road distances travelled by car have nearly doubled. Rising incomes in OECD countries have driven the increases in car ownership and a shift towards heavier and more powerful cars. Variables such as family size and lifestyle, children, the number of workers in the household and geographical factors – where people live, work and shop – are also important, as is access to mass transit. For example, urban sprawl increases the road distances travelled by car (see Chapter 2). To 2020, private vehicle stock and kilometres travelled are projected to increase substantially both in OECD and non-OECD countries (see Chapter 14).

Significant gains may have been made in car fuel efficiency in all OECD countries, but greater car use and preferences for heavier, more powerful, and more comfortable cars have offset these improvements. Moreover, the frequency of trips has increased while the number of people per car has dropped. Among the different types of vehicles, cars collectively emit the greatest amount of the pollutants carbon monoxide (CO), volatile organic components (VOC) and carbon dioxide (CO_2).

Household tourism patterns are a second important source of environmental impacts from household travel. International tourism has grown at an annual average rate of 7.1% since the 1950s, and is projected to grow at an annual average rate of 4.3% to the year 2020 (Figure 16.3). Overall, the World Tourism Organisation projects that by 2020 tourism arrivals worldwide will be triple their 1995 levels (WTO, 1999). One characteristic of this growth is that long-haul tourism travel (travel that takes place between world regions) could grow at a faster rate than intra-regional travel, particularly in Europe and the Americas. Inter-regional (mostly intercontinental) travel involves great distances of air travel where environmental impacts tend to be highest.

Figure 16.3. **International tourist arrivals worldwide**

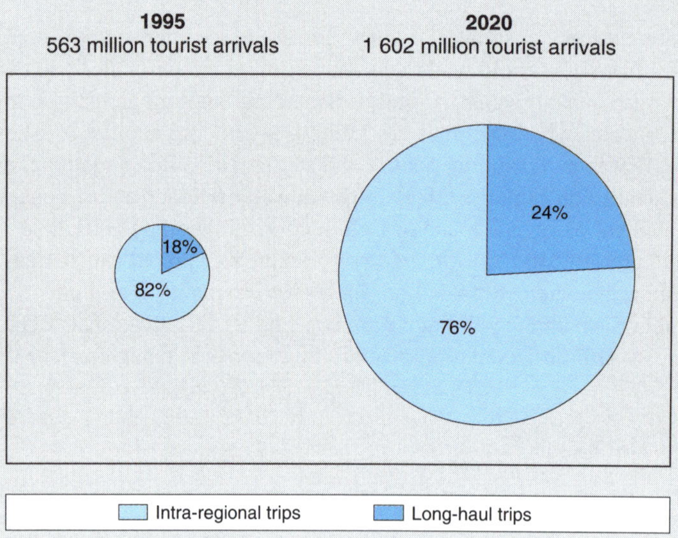

Source: WTO (1999).

Household leisure travel currently represents a relatively important source of transport energy use and CO_2 emissions, roughly of the same magnitude as emissions from other household travel (*e.g.* commuting, shopping). However, aviation-related CO_2 and other emissions have increased rapidly in recent years, largely driven by long-distance leisure travel. Of all tourism-related activities, travel represents the most important source of energy use and greenhouse gas emissions.

16.6. Household waste generation

Rising per capita income and accompanying changes in lifestyles have increased the number of objects households purchase and dispose of. In 1997, households generated on average 67% of municipal waste loads (with a range of 41-96%) (OECD, 1999*b*). Household waste is made up of organic waste (garden and food residue), durable and non-durable goods, inert materials, metal, containers and packaging (glass, paper and paperboard, plastics, composite packaging), motor oils, and textiles. Households also produce very small quantities of hazardous waste, such as aerosol cans, paint, household and automobile batteries, and home and yard chemicals. The composition of household waste has changed significantly. Food wastes as a percentage of municipal waste have declined, although the amount of food waste in the waste stream has increased. This is due to better storage (refrigeration, packaging) and also to a greater consumption of processed food (OECD, 2001*b*). Packaging materials account for a relatively larger share of household garbage. This trend began earlier than most might realise. The term "throw-away society", which many equate with current lifestyles and consumption patterns, was coined in 1955 – even before the advent of plastic packaging (Brower and Leon, 1999).

Most OECD governments have been shifting waste strategies from simple collection and disposal to a "waste hierarchy" approach with an emphasis on preventing waste generation. While household awareness and participation in waste management have improved significantly, the generation of household waste is nevertheless still growing and is projected to increase further to 2020 (see Chapter 20). Municipal waste generation in OECD regions is projected to grow by 43% from 1995 to 2020, for total generation amounting to about 770 million tonnes per year by 2020. In the EU, household waste is projected to increase 22% from 1995 to 2010, with increases in paper and cardboard waste of between 44-62%, and glass waste of between 24-53% (Fischer, 1999). End-of-life (scrapped) vehicles are expected to increase by roughly 30% in the EU region (European Topic Centre on Waste, 2000). With increasing awareness and participation of households in waste-related environmental issues, recycling has improved remarkably over the last two decades. However, it has not been sufficient to counter the overall environmental impacts from greater waste generation.

16.7. Policy options

Most OECD countries have in place, or are developing, measures to modify the environmental impacts from households. These measures are often part of strategies for tackling major environmental problems, such as climate change or local air pollution. They aim to influence household decision-making both directly, by encouraging changes in consumer behaviour, and indirectly, by requiring that more environmentally benign goods become available on the market or by imposing regulatory standards which increase the relative prices of products with greater negative environmental impacts (OECD, 1998).

Some of these policies have resulted in limited changes in behaviour. Nevertheless, since many of the environmental impacts resulting from household decisions are expected to intensify over the next couple of decades, OECD governments will need to reinforce their policies to help households make day-to-day and major investment decisions at a lower cost to the environment. From a household perspective, changing purchase and behaviour patterns may incur a high personal cost for a shared public benefit (*e.g.* better air quality). As a result, governments face several challenges in promoting more sustainable household behaviour.

To tackle these problems more coherently, the environmental objectives of technology, infrastructure and research and development (R&D) policy will need to be expanded, and regulations and standards that define a minimum level of effort (*e.g.* waste recycling targets, minimum energy efficiency performance) more widely applied. These, together with the use of economic instruments and better information-based instruments and initiatives to raise awareness, will be important in sending clearer signals to households on the full costs associated with their decision-making. There is a broader set of policies that also influences household decision-making, and their design should take into consideration their environmental effects. These include landuse and physical planning, labour policies, and education and social policies (*e.g.* childcare).

Technology, infrastructure and R&D policy

The current lack of options to reduce pressures on the environment from the household sector underlines the importance of medium- and long-term government promotion of the development of technologies and infrastructures (energy, transport, waste) that will support sustainable household behaviour. In the area of waste, for example, waste policies need to be better integrated into an economy-wide approach of lifecycle resource management to reduce the material input into the economy and achieve closed materials cycles up to the point where the economic costs involved do not exceed the environmental benefits achieved. At the household level, resource efficiency should translate into a reduced flow of materials into the household to begin with.

Reducing environmental impacts from personal travel will require measures to induce the use of new or improved motor vehicle technologies; reorienting and augmenting investment for public transport systems; and re-examining incentives for private car use (*e.g.* company car policies, flexible working hours) (see Chapter 14). But it will also require going beyond the current public vs. private transport debate which tends to focus on means of transport rather than on transport system objectives. A new and promising approach of multi-modal mobility services offers a wide palette of transport options for "seamless" multi-modal trips that would reduce the need for personal car use. Such systems are already being successfully marketed in a number of OECD countries.

In the area of food consumption, some OECD governments have begun to develop technical assistance programmes to improve waste reduction and management skills in the processing and food service sectors. Measures will also be needed to improve recycling of packaging wastes and composting organic wastes. The low recovery rates for plastics are of particular concern, and call for advances in private sector technology and public policy to make the safe reuse of plastic packaging possible (*e.g.* minimum recycling content for food and cosmetic packaging) (OECD, 2001*b*). An important area for future policy research will be the possible environmental, economic, and social implications of expanded e-commerce, for example in the food sector with home food delivery (see Chapter 6).

Regulatory instruments

Governments can act directly to influence or constrain household decision-making by rationing, or through other restrictions on product ownership. Examples include the enforcement of speed limits to slow and harmonise traffic flows, or restrictions on the use of disposable goods in the service sector (restaurants, hotels, etc.). Some regulations can also directly influence the way in which households operate and use products, such as limitations on hosepipe use to reduce household water consumption, or curbs on personal car use on peak pollution days. The use of direct regulations in the household sector is relatively rare, but industry regulations regarding products and production processes may influence household choices indirectly. For example, bans on the production or sale of certain products necessarily change the composition of household consumption.

Governments have generally taken a more gradual approach to influencing household consumption patterns through imposing or increasing minimum product standards (*e.g.* minimum levels of energy or water efficiency for household products or tightening building regulations). Labelling requirements and minimum energy efficiency standards for electric appliances and equipment, for example, have proven to be an effective environmental policy instrument – delivering tangible results at relatively low cost (IEA, 2000). Despite early successes in North America, more recent attempts to extend and develop new fuel efficiency standards for motor vehicles have fared less well. Efficiency standards could be applied to a wider range of consumer goods.

Economic instruments

Economic instruments – including full-cost pricing, environmental taxes and charges, green tax reform, and the removal of environmentally harmful subsidies – have an important role to play in influencing consumer behaviour. Where the price of energy, road fuels, water or waste does not fully reflect the associated environmental costs, households have an incentive to consume more than they would if they faced the full costs of their consumption patterns.

In the area of transport, many countries have successfully used differential fuel levies to encourage the initial development of a market for lead-free petrol and the subsequent major shift to cars that use lead-free petrol. Some countries are now using differential tax rates to encourage gas-powered cars and low-sulphur diesel. Countries can also use a package of measures, for example imposing a lower tax rate for biofuels and, at the same time, supporting their production and distribution to encourage more rapid introduction and uptake of the environmentally less damaging options. However, while getting the price right is a necessary condition, it is not always a sufficient condition to change household patterns, especially where purchases and use are driven by non-economic factors such as cultural habits (OECD, 1998). The environmental effectiveness of pricing is closely related to price elasticities, the amount by which consumer demand for a given product will change if the price of the product changes. If environmentally damaging products are price inelastic – *i.e.* consumer demand for the product does not change significantly in response to changing prices – price signals may not be sufficient to produce the desired environmental results. In such cases, regulatory instruments, such as product bans, may be more effective than economic instruments.

In the area of household waste production, a new generation of policies uses unit pricing to make the cost of waste management visible to households. Unit pricing ("Pay-As-You-Throw" – PAYT) policies use marginal price structures that penalise higher levels of waste generation. In the US, PAYT has been adopted in over

3 900 communities. Although success rates vary, PAYT communities on average have reduced total waste generated by about 14-27%, while increasing recycling by about 32-59% (Miranda and LaPalme, 1997). Complementary programmes (curbside recycling, yard waste collection, bulky item pick-up, and education campaigns) can increase PAYT effects on source reduction and recycling rates, and discourage illegal dumping.

Information and other instruments

Along with providing information, OECD countries use other instruments in their initiatives to influence household behaviour. These encompass a range of policy instruments, such as awareness raising schemes, education in the broadest sense, labelling and information campaigns, encouraging action by community and consumer groups, support for work by non-governmental organisations on sustainable consumption, and working with industry and others through voluntary agreements. Providing appropriate environmental information combined with adequate price signals can incite households to make environmentally sound choices and change their consumption and waste disposal patterns. Changes in household demand will then lead to concomitant changes in production patterns.

OECD countries use different measures to make consumers aware of how they could adopt more sustainable lifestyles. Campaigns to promote energy efficiency in the home, limit the use of water, reduce waste and increase recycling are widespread. Publicising relevant national indicators of sustainability and consumption could also improve public understanding of the subject, but so far this has not been sufficiently developed. Information and labelling schemes can improve consumer understanding of a specific environmental issue or of a product's environmental effects. There is a wide range of national and regional ecolabelling schemes that intend to inform and influence consumer decisions either by focusing on some key aspects of the environmental impact a product or service may have, or by trying to assess the relative impact on a lifecycle basis. Ecolabelling programmes face a variety of problems, however, as "green" labels and claims proliferate, creating an often overwhelming information environment for the consumer. Partly as a result of this, consumer trust in most sources of information on the environment has declined in recent years.

Some OECD governments are using information instruments in combination with other measures to stimulate environmentally friendly consumer decisions.

In the future, governments will have to better target and communicate information to households. Several governments provide guidance to households on voluntary waste minimisation strategies, including guidance on making careful product purchases (*e.g.* re-use potential, product life, minimum packaging, minimum toxicity, "no-buy" options). Not all this information needs to concern the environment directly; environmental objectives can be achieved even when households are more concerned with financial or health considerations.

Innovative approaches in some OECD countries include combining information with co-ordinated media campaigns, training and motivation of retail store staff, voluntary agreements with manufacturers and retailers to broaden the choice of environmentally friendly products and services, and economic incentives (*e.g.* tax rebates) to stimulate environmentally aware consumer choices. These approaches recognise that information is only useful if there is an environment in which the consumer can put that information to use.

Finally, the education system has a central role to play in promoting environmentally responsible household attitudes and behaviour. Within schools and universities, students can learn not only about the nature of environmental problems, but also about the consequences of their own actions and how changes in their patterns of consumption could reduce those impacts. Furthermore, educational institutions are major consumers of energy, water and other resources in their own right. An increasing number of schools and universities should address the greening of their own organisations. Efforts to integrate the activities of schools into their communities, and to develop links with local businesses, could be encouraged. Professional training, continuing education and non-formal schooling are all vital for educating, empowering and involving consumers in environmental choices and the decision-making process.

REFERENCES

Alexandratos, N. (2001), "World Agriculture, Food Security and the Environment", background document for the OECD *Environment Outlook*, OECD, Paris.

Brower, M. and W. Leon (1999), *The Consumer's Guide to Effective Environmental Choices: Practical Advice from the Union of Concerned Scientists*, Three Rivers Press, NY.

European Topic Centre on Waste (2000), "Scenarios and Projections", [*www.etc-waste.int/activities/0000121.htm*], European Topic Centre on Waste, Copenhagen.

Fischer, C. (1999), "Future Waste Amounts in the EU," European Topic Centre on Waste, Copenhagen.

Geyer-Allely, E. and H.-S. Cheong (2001), "Consumption Patterns", background document for the OECD *Environmental Outlook*, OECD, Paris.

IEA (International Energy Agency) (1997), *Indicators of Energy Use and Efficiency: Understanding the Link between Energy and Human Activity*, IEA/OECD, Paris.

IEA (2000), *Energy Labels and Standards*, IEA/OECD, Paris.

Miranda, M.L. and S. LaPalme (1997), "Unit Pricing of Residential Solid Waste: A Preliminary Analysis of 212 US Communities", [*www.epa.gov/payt/research.htm*].

OECD (1998), *Towards Sustainable Consumption Patterns: A Progress Report on Member Country Initiatives*, OECD, Paris.

OECD (1999*a*), *The Price of Water: Trends in OECD Countries*, OECD, Paris.

OECD (1999*b*), *OECD Environmental Data: Compendium 1999*, OECD, Paris.

OECD (2000), *The Agricultural Outlook*, 2000-2005, OECD, Paris.

OECD (2001*a*), "Sustainable Consumption: Tourism and Travel Patterns – Part I, Tourism Travel Trends and Environmental Impacts", [ENV/EPOC/GEEI(2000)5/FINAL], OECD, Paris.

OECD (2001*b*), *Household Food Consumption*, OECD, Paris.

WTO (World Tourism Organisation) (1999), *Tourism: 2020 Vision*, WTO, Madrid.

17
The Steel Industry

KEY SIGNALS

- The steel sector is very intensive in material and energy use. The main environmental impacts from this sector include air pollution and greenhouse gas emissions, water contamination and waste generation.
- The steel industry in OECD countries has made much technological progress over the last 20 years, fostering increased efficiency in energy and resource use, and reducing pollutant emissions. The use of recycled materials in the industry has also been constantly growing.
- In the Reference Scenario, world steel production is projected to double from 1995 to 2020 while steel production in OECD countries may increase by around 65%. CO_2 and SO_x emissions from the steel sector in OECD countries are projected to increase by about 63% and 51% respectively, while water and fuel use may grow by 79% and 64% each to 2020.
- The most important policy options for tackling the negative environmental effects of the steel industry are the use of economic instruments (*e.g.* subsidy removal, green taxes, emission trading), continued support to technological development and uptake, and voluntary agreements with the industry.

17.1. Introduction

The iron and steel sector is by far the largest and most important metallurgical industry. Steel plays a key role in modern economies because its physical properties and chemical resistance makes it an important material for structural and engineering works, industrial and automotive manufacturing, road and rail infrastructure, construction, and for other diverse uses (*e.g.* electronic components, packaging, medical equipment).

Steel production has significant negative environmental and health effects. The most important environmental effects of the sector are its contribution to climate change through greenhouse gas emissions, its air pollution and particulate emissions, and water and soil contamination. Since the 1970s, the industry has made efforts to reduce greenhouse gas and other emissions, and has introduced processes for water treatment and reuse. Significant reductions have been achieved in energy requirements. The industry has also increased output per tonne of crude steel input: it is using more recycled materials and by-products, reducing the demand for natural iron ore mining. As global steel production is projected to increase to 2020, the need to de-couple the environmental impacts of steel production from continued growth in the sector will persist.

17.2. Developments in the steel sector

Table 17.1. **Key steel sector statistics and projections**

		1980	1995	2020 (projected)	Total change 1995-2020
Gross production (1995 US$ billion)	OECD	..	662	1 088	64%
	World	..	899	1 852	106%
Share of value-added in the economy	OECD	..	0.9%	0.9%	–0.6%
	World	..	1.0%	1.1%	9.7%
Direct employment (million persons)	OECD	2.1	0.9	..	..
	World	3.1	1.9	..	..
Steel consumption (million tonnes)	OECD	284	386	476	23%
	World	562	653	830	27%

Sources: OECD (1990), ILO (1997), IISI (2000), GTAP Database and Reference Scenario.

Demand for steel

Demand for steel closely follows patterns of economic growth and the highest levels of steel consumption are seen in OECD countries. Current steel consumption per person per year is estimated to range from approximately 20 kg in Africa to around 340 kg in Europe, 435 kg in North America, 635 kg in Japan and 830 kg in Korea (IISI, 2000). Demand for steel increases during periods of steady investment in infrastructure, transport, construction, and in the development of automobiles, etc. Demand for steel as an intermediate input to industries varies with industrial structure, economic trends and the demand for specific products. Oil shocks, the instability of newly independent states and economic crises in some Asian economies have affected steel demand in recent years. Today, OECD country steel producers are concentrating on key sector markets (*e.g.* construction and automotive) as their main drivers of growth.

The share of OECD countries in world steel consumption increased from 49% in 1980 to 58% in 1998, a share which is likely to be maintained to 2005 (OECD, 1990; IISI, 2000). The Reference Scenario confirms this trend and projects that demand for steel in OECD regions will grow by around 68% from 1995 to 2020 at an annual rate of 2.1%; the projected change in world demand is 105% for the same period.

Production of steel

Steel is manufactured using either integrated or non-integrated processes. Integrated plants produce steel by using iron ore, scrap, coal, and limestone as raw materials processed in coke ovens and basic oxygen furnaces (BOF). Non-integrated plants make steel by refining scrap[1] in electric arc furnaces (EAF), the so-called mini-mills. The non-integrated process is the result of major technological changes that have occurred in the sector, bringing significant improvements in productive efficiency and reduced environmental impacts. Scrap, as raw material, accounts for 30% of iron input in basic oxygen furnace plants and more than 90% in electric plants. The old and inefficient open-hearth furnaces have been replaced for the most part by processes based on oxygen conversion, continuous casting and electric arc furnace technology. The share of electric arc furnaces in liquid steel production is currently 33% and is expected to continue to increase, reaching 40% of total production by 2010 (IISI, 2000).

World crude steel production increased from 595 million tonnes (Mt) in 1970 to 777 Mt in 1998. After a boom in steel production from 1970 to 1989 the growth has slowed, averaging only 0.1% per year between 1990 and 1998 (IISI, 2000). At the same time, there has been a significant shift in production towards a number of non-OECD countries, such as China and India, resulting in rapid growth in steel production in these regions. Some Central & Eastern European countries have also entered the global market. The share of OECD countries in world steel production has declined from 54% in 1995 to 48% in 1998. The greatest relative increase in steel production to 2020 in OECD regions is projected for Central & Eastern Europe (Figure 17.1). Because of their production capacity, large markets, and high competitiveness, the largest steel producing countries (*e.g.* China, the US, Japan, [2]Russia, Germany and Korea) are likely to maintain their leading positions. Production in OECD countries is projected to increase by around 64% to 2020, providing that sufficient investments are made in the coming years to build-up the necessary capacity. Worldwide production of steel may more than double.

Trade in steel products

Increased steel consumption in the Asia-Pacific region, and the emergence of some Central & Eastern European countries as important steel exporters, resulted in a sharp expansion of world steel trade during the 1990s. This increase in trade has been accompanied by considerable trade flow volatility as a result of changes in steel production and exports from the former Soviet Union and economic instability in some of the main steel producing Asian economies. The growing share of crude steel production in non-OECD regions has been a direct result of

1. Iron scrap consists of products that have completed their useful life (*e.g.* cars, appliances) and by-products from the processing of metals.

Figure 17.1. **Steel production, 1995-2020**

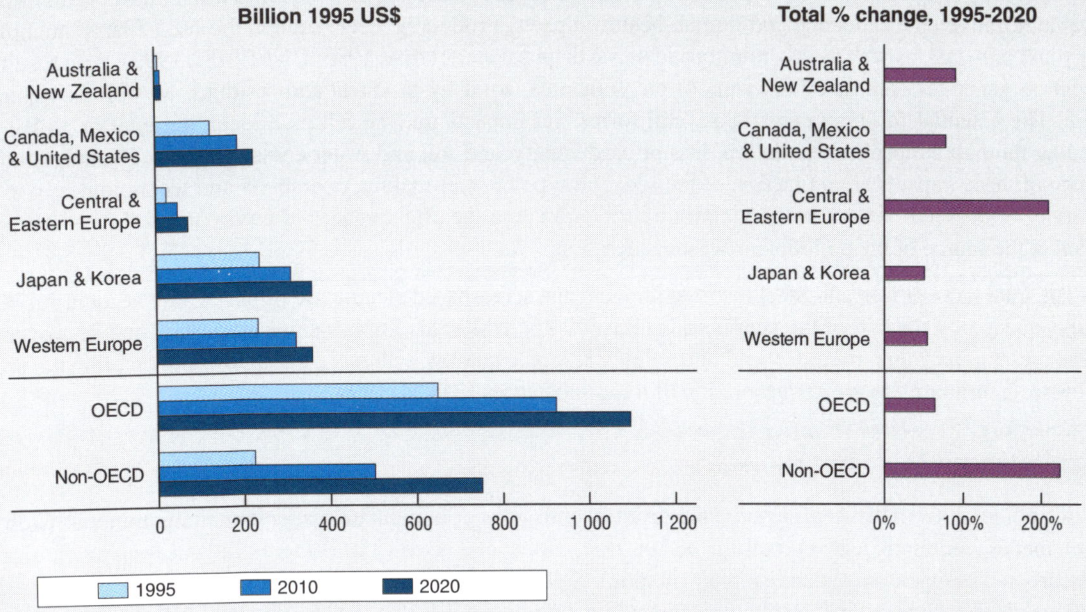

Source: Reference Scenario.

changes in some developing regions, especially in Asia and Latin America, where growing demand is tied to economic reforms and modernisation. Growth in steel consumption in these regions has been met primarily by rapid expansion of domestic production, leading to significant export declines for producers in OECD regions.

These changes in world steel markets are likely to result in more OECD countries shifting to become net importers of steel, a trend already seen in the US and likely in Western Europe (IISI, 2000; Plummer, 1999). In OECD regions, the steel industry is likely to become more oriented towards internal markets and selective high value-added production. In non-OECD regions, the expansion of production capacity and increasing competitiveness could foster a substantial increase in exports. It is expected that Asian economies may see a sharp increase in their steel trade deficit, except China which may reach a trade surplus after 2015. Central & Eastern European and former Soviet Union economies may exhibit large steel trade surpluses by 2020.

Institutional changes

In the last 20 years, the main factors shaping trends in the steel industry have been economic growth, industrial concentration, privatisation, technological development, increasing international trade, and environmental concerns. Although previously considered a strategic sector for some national economies, in recent decades many governments have reduced their involvement in the steel industry through privatisation of state owned steel companies and reductions in support to the industry. Impacts on employment in the steel sector have sometimes been severe. At the same time, deregulation and privatisation of the industry have fostered structural changes, and international trade has been increasingly liberalised through the lowering of tariff and non-tariff barriers, increased international competition, and advancing globalisation. Increased privatisation and the opening of markets have provided new trade and investment opportunities that the steel industry has met through further consolidation of companies. Mergers and acquisitions, rapid integration of innovative techniques, and more flexible and environmentally friendly production are likely to continue over the next two decades.

17.3. Environmental effects from the steel industry

Depending on the amount and types of releases that occur at steel plants, producing iron and steel can have considerable negative environmental and human health impacts. Producing steel requires the use of large quantities of energy and minerals, as well as vast mining and waste disposal areas (IEA, 1995; UNEP/IISI, 1997). As a result, steel production generates significant amounts of air pollutants, solid by-products and residues, as well as wastewater sludge. The potential for harm can take several forms and impacts may be felt on a local, regional or global scale, including through atmospheric pollution, loss of biodiversity, and soil and water contamination. The magnitude and severity of these impacts are a function of the size and type of steel-making operations and technologies in use, the sensitivity of the environment in the vicinity of operation, and the effectiveness of environmental measures implemented at the source of the pollution or waste generation.

The International Iron and Steel Institute projects that scrap-based electric arc furnaces may account for 40% of steel production by the end of the next decade (IISI, 2000). While this implies some environmental improvements, increasing scrap recycling may also lead to declining resource quality, with more tramp material entering the production cycle, contributing to greater generation of toxic substances (RIVM, 1999).

Air pollution and climate change

Air pollution from the steel sector is the result of emissions of particulate matter containing minerals (iron, iron oxide), metals (cadmium, lead, chromium, nickel, zinc, copper and arsenic) and other pollutants (polycyclic aromatic hydrocarbons, nitrogen oxides and sulphur dioxide). Steel plants with electric arc furnaces release dust containing heavy metals and some organic compounds, including dioxins (UNEP/IISI, 1997). The steel industry is also a source of carbon dioxide (CO_2) emissions, contributing 3% of CO_2 emissions from fossil fuel combustion in OECD countries in 1997 (IEA, 1999). CO_2 is generated during iron and steel making operations, either as a result of the reaction of carbon (coke) with iron oxide in the blast furnace, or from a power plant producing electricity used in the production of steel. Integrated plants can emit between 1.6 and 2.4 tonnes of CO_2 per tonne of steel produced, while electric arc furnace plants, which only use iron scrap as raw material are responsible for the emission of about 0.7 tonnes of CO_2 per tonne of steel produced (IISI/UNEP, 1997; Carson, 1999). The introduction of efficient processes, pollution control systems and practices has reduced emissions significantly in recent years. High proportions of many of the waste or pollution releases from steel production are now collected and either recycled internally as feedstock, off-site as saleable materials, or shipped to waste treatment plants. At present, state-of-the-art techniques (*e.g.* fabric filters, scrubbers, biological water treatment plants) may enable the collection of over 90% of some of the air pollutants (IISI, 2000).

CO_2 and SO_x emissions from the steel industry are expected to increase substantially to 2020, but introduction of new technologies could offset this increase.

The projected production growth in the steel sector to 2020 will have environmental implications. Greater material and energy requirements are likely to lead to additional impacts on the environment, whether they result from increased mining operations and energy use, or through greater pollution releases in the production process. The risks of environmental degradation are even higher when the potential effects on climate change of increases in steel production are considered. In the next two decades, CO_2 emissions are likely to increase as a result of rising production levels. However, the introduction of major new production technologies will reduce CO_2 emissions per unit of output, and if the spread of these technologies is accelerated, they could lead to a stabilisation or even reduction of CO_2 emissions. From 1995 to 2020, the Reference Scenario projects that CO_2 emissions related to the steel industry in OECD regions may grow by 63% (and by 141% worldwide), and SO_x emissions by 51% (and by 138% worldwide).

Water contamination

Water pollutants from the steel industry include releases of suspended solids and several toxic substances (heavy metals, phenol, and cyanide). Integrated mills on average emit 3 m^3 of wastewater containing 1.6 kg of suspended solids, 150 g of oil, 110 g of amoniacal nitrogen and 8 g of phenol, meths and cyanides per tonne of steel produced. Suspended solids, mainly made up of iron oxide, represent the largest portion of water pollutants, and excessive discharges reduce the oxygen supply in water and are detrimental to aquatic life. Water emissions from furnace, rolling, and coating steel operations can also contain trace elements of heavy metals and persistent organic compounds. The steel industry is a major user of water but recycling facilities allow reuse of up to 97% of water in some modern steel plants. The main options for preventing and controlling pollution of water by the above mentioned substances include the use of collection and treatment facilities for metal-bearing wastes, as well as substituting to less harmful products for toxic components of oil and grease used. The steel sector currently accounts for around 28% of total industrial water use worldwide. Under the Reference Scenario, this share is expected to increase slightly, to 29% in 2020, as a result of an 87% net increase in the volume of water used by the industry. Water use by the steel industry in OECD regions is expected to increase by 79% to 2020.

Waste

Steel making also gives rise to large amounts of solid by-products, waste and residues. Coal residues amount to approximately 1 kg per tonne of coke produced (around 600 kg of coal and 1 500 kg of iron ore are required to produce 1 tonne of crude steel in integrated plants). Integrated steel plants produce about 585 kg of solid by-products and waste per tonne of crude steel, whereas electric arc furnace plants generate about 200 kg/tonne. However, slag, which represents 70-80% of by-products and waste per tonne of crude steel produced, is almost completely recycled. It may be used in public works and road construction, or can be recycled back into the furnace. Over 90% of all solid waste and by-products are used either internally as material input or fuels, or sold to other users. The steel industry has achieved significant reduction of by-products and solid waste, notably at blast furnace operations, and from the 1950s to 1995 by-products from iron making have been reduced by 60% (UNEP/IISI, 1997).

Resource and energy efficiency

The steel industry is energy and resource intensive, with energy consumption ranging from 10-15% of world industrial energy requirements (IISI, 2000). The amount of raw material required per tonne of crude steel produced by integrated plants is approximately 2.24 tonnes, and in electric arc mills only 1.03 tonnes. (UNEP/IISI, 1997). Since energy represents about 15% of total production costs, the industry has had an incentive to implement more energy efficient processes over the past three decades. The use of alternative sources of energy (*e.g.* slag and waste used as fuels) and more efficient processes have helped to increase combustion efficiency, while reducing energy losses and overall energy demand (Carson, 1999). Energy input per unit of manufacturing value-added declined by 1.9% per year on average for the steel industry in 10 OECD countries between 1971 and 1991, contributing to a reduction of nearly 20% in energy consumption in the steel sector since the 1970s, and the gross energy recovery ratio improved consistently to nearly 30% of gross energy consumption (OECD, 1999*a*).

The steel industry has made progress in increasing its efficiency of resource and energy use, but still uses large amounts of energy, water, and iron ore.

The high recycling capacity of steel has played a key role in improving the environmental performance of the sector with respect to both energy and material intensity. Increasing electric arc furnace production translates into a growing share of scrap in total material input which, in turn, contributes to a reduction in total energy input. Steel produced from scrap requires 60% less primary energy than that produced from ore. Steel is totally recyclable, without downgrading the quality of the metal, and approximately 350 Mt of steel scrap is recycled each year (UNEP/IISI, 1997). In terms of resource productivity, the use of more efficient technologies and production practices, and the adoption of environmentally oriented equipment and better plant design, has continuously lowered material use over the past 25 years (Angulo, 1995).

Human health impacts

Occupational and public concerns over potential effects on human health of steel production include effects relating to the emission of particulate matter, carcinogenic substances, and heavy metals. Between 60-90% of particulate matter emitted by iron and steel facilities are less than 10 micrometre (μm) in size, and are therefore particularly harmful to respiratory tracts. Short-term exposure to particulate matter can also lead to eye, nose, and throat irritation and aggravate cardiovascular diseases. Long-term exposure may lead to permanent lung damage, carcinogenesis, and premature death (UNEP/IISI, 1997).

In recent years, the steel industry has made improvements in the area of occupational health and safety (IISI, 2000). Nonetheless, continuous exposure to heavy metals (*e.g.* lead, mercury, zinc, chromium, manganese) may lead to short-term and chronic effects on the health of steelworkers. Depending on the level and rate of exposure, impacts on human health may include damage to the central nervous system, the brain, kidneys and the reproductive system. Steelworkers may also suffer from higher blood pressure, weakness in joints, memory loss and anaemia, as well as "metal fume fever", characterised by symptoms such as fever, chills, nausea, breathing difficulties and extreme fatigue.

Steel production has negative impacts on human health, but these effects have been reduced in recent years.

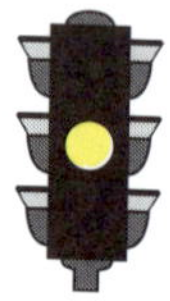

17.4. Policy options and their potential effects

Although OECD countries account for the majority of steel production and consumption, steel production has been growing fastest in non-OECD countries in recent decades. Reference Scenario projections indicate that the non-OECD regions overall will have a higher growth rate for the steel sector to 2020; with China projected to have the highest absolute increases in steel production.

In recent decades, the steel industry in OECD countries has achieved important reductions in its environmental impacts. However, large challenges for the sector remain in terms of further reducing and controlling air emissions; improving solid waste management, recycling and wastewater treatment; reducing energy and resource intensity; and developing cleaner technologies.

Technological development and diffusion

Technological change has been a key factor in environmental improvements in the steel sector. Policies that provide incentives for technical innovations in steel production can lead to less energy use and material consumption, increased valorisation and recycling of steel by-products and iron scrap, quality control of iron scrap, and further reduction of air and water emissions. Available policy tools include regulations and economic instruments that encourage research into and uptake of new steel production processes, plants, and equipment that help to minimise environmental and health impacts of steel production. Technological benchmarking and some regulations can lead to a wide diffusion of the best available technologies and practices to increase environmental performance in the steel industry. Policies can also be employed that encourage the development of new steel materials and increase the recycling of steel products.

Regulatory instruments

Regulatory options for reducing the environmental impacts of steel production include adopting an integrated approach to operating steel plants, in particular with more attention paid to pollution prevention options, and taking into account the latest environmental and technological standards as well as quality objectives (OECD, 1999*a*). Likewise, strengthening the regulatory control over hazardous compounds, such as chlorinated oils and emulsions, could help to reduce emissions in the long run. The regulatory framework has been effective in reducing environmental releases. However, better co-ordination and harmonisation of regulatory measures addressing the impacts of the industry's activities could lead to higher standards of environmental protection.

Economic instruments

The use of market-based instruments to conserve and improve the quality of the environment should be extended in the steel sector. This includes the use of tradable emission permit schemes or taxes to reduce CO_2 or other air pollution emissions. The use of taxes and removal of implicit or explicit subsidies, for example concerning energy use, would also help to decrease environmental impacts. For OECD countries, model simulations indicate that eliminating all subsidies to the steel sector and introducing an *ad valorem* tax on steel use that increases by 2 percentage points per year would reduce CO_2 and SO_x emissions by the sector in OECD regions by 9% each in 2020 compared with the Reference Scenario, and water use by 13% (see Table 17.2).

Table 17.2. **Subsidy and tax policy simulations: effects on the steel industry in OECD regions and its environmental impacts**

(% change from Reference Scenario in 2020)

	Policy simulations	
	Steel industry subsidy removal and tax on all steel use	Energy subsidy removal and tax on all energy use
Gross steel production	–11%	–1%
CO_2 emissions	–9%	–35%
SO_x emissions	–9%	–54%
Water use	–5%	–2%

Note: The energy subsidy and tax policy simulation was applied to all sectors, not just the steel industry. See Chapter 12 for more details.
Source: Reference Scenario and Policy Simulations.

A simulation of the removal of all energy subsidies in OECD regions, combined with the application of a tax on all energy use (see Chapter 12 for further details), was found to have less of an effect on OECD steel production and water use, but more significant reductions in CO_2 (35%) and SO_x (54%) emissions from the sector.

Voluntary agreements

Voluntary approaches are increasingly adopted by steel producing companies in OECD countries that are willing to improve their environmental performance. Voluntary commitments can be used to help reduce CO_2 and air pollution emissions from the steel industry, for example through setting voluntary industry targets for energy efficiency improvements of recycling programmes. Voluntary agreements are a flexible approach that will generally be most successful in reducing the environmental impacts of steel production when they are combined in a policy mix along with economic instruments and the adoption of environmental standards in steel plant management. Voluntary approaches could be used to allocate some of the actions needed to achieve internationally agreed targets on reductions in emissions from the steel sector. They can also be used in the development of take-back programmes for steel products, such as household appliances and computers, and the promotion of increased household participation in steel scrap collection.

Information and other instruments

Life Cycle Assessments (LCAs) of steel products should be conducted on a more regular basis, and the results better disseminated. Product labelling based on standards derived from LCAs of steel products could contribute to increased consumer awareness. Ensuring the collection of regular, comparable information on the efficiency of resource and energy use, and the waste reduction, reuse, and recycling of individual steel producing firms could help companies to upgrade their environmental performance. This would also contribute to integrating environmental management systems into the corporate structures, and involving the public and stakeholders in the policy-making process.

REFERENCES

Angulo, F. (1995), "Implications of the Globalisation Process in the Metallurgical Industry", First Global Consultation on Environmental Management/Cleaner Technologies in the Metallurgical Industry, UNIDO, Vienna, 16-18 October.

Carson, C. G. (1999), "Steel Industry", Industry Sector Reports, BIAC/OECD/IEA Workshop on Climate Change, 8 March: 34-37, OECD, Paris.

IEA (International Energy Agency) (1995), *World Energy Outlook*, IEA/OECD, Paris.

IEA (1999), CO_2 *Emissions from Fuel Combustion*, IEA/OECD, Paris.

IISI (International Iron and Steel Institute) (2000), *Worldsteel*, The global forum for the world steel industry and IISI/ISSF, various statistics and information, *www.worldsteel.org*.

ILO (International Labour Organization) (1997), "Meeting at ILO Examines Future of Steel Industry Workforce", Press Release (ILO 97/24), 23 October 1997, *www.ilo.org/public/english/bureau/inf/pr/1997/24.htm*.

OECD (1990), *The Steel Market in 1989 and Outlook for 1990*, OECD, Paris.

OECD (1999*a*), "Environmental Requirements for Industrial Permitting: Vol. 2", OECD Workshop on the Use of Best Available Technologies and Environmental Quality Objectives, 9-11 May 1996, Paris.

OECD (1999*b*), *Voluntary Approaches for Environmental Policy: An Assessment*, OECD, Paris.

Plummer, C. (1999), "The Competitive Impact of Iron Ore and Steel Industry Consolidation", Gorham Intertech Consulting Conference – Iron Ore for Alternative Iron Units, Charlotte, NC, 26 October.

Rajotte, A. (2001), "Iron and Steel Industry", background document for the OECD *Environmental Outlook*, OECD, Paris.

RIVM (Dutch National Institute of Public Health and the Environment) (1999), "Long-Term Perspectives on World Metal Use – A Model-Based Approach", RIVM report 461502 021, RIVM, Bilthoven.

TRI (Toxic Releases Inventory) (1999), *Toxic Releases Inventory Data for Primary Metals*, Sic Code 33, Government Printing Office, Washington DC.

UNEP/IISI (United Nations Environment Programme/International Iron and Steel Institute) (1997) "Steel Industry and the Environment – Technical and Management Issues", Technical Report No. 38, UNEP/IISI, Brussels.

18

The Pulp and Paper Industry

KEY SIGNALS

- Global production in the pulp, paper and publishing sector is expected to increase by 77% from 1995 to 2020, with production in OECD regions projected to increase by 60% and in non-OECD regions by 170%.
- The pulp and paper industry is highly intensive in its use of capital, energy and natural resources. It is estimated that the growth in production by the sector to 2020 would require some 71% more water, 53% more energy, and substantially more wood or recycled paper.
- Environmental impacts from this sector include contributions to deforestation, water contamination, soil and air pollution, and loss of biodiversity. Carbon dioxide (CO_2) emissions from the sector are expected to increase by 62% in OECD regions between 1995 and 2020, and sulphur oxide (SO_x) emissions by 23%. Worldwide, CO_2 and SO_x emissions from the sector are expected to roughly double.
- The most important policy options are the diffusion of best available technology, integrated regulatory approaches, the use of taxes, and voluntary agreements.

18.1. Introduction

The pulp and paper industry[1] produces a wide range of goods that are used for packaging, communications, printing, and for sanitary and household purposes. Paper is made from wood and non-wood fibres, and requires huge amounts of water and energy in its production. The sector is very capital intensive and the most rapidly growing among the wood-based industries. Its development is highly cyclical, and strongly affected by economic instabilities in the world market. While the downstream manufacturing of paper products is often still characterised by small units and conventional technology, pulp and paper manufacturing is dominated by large and globalised companies.

Pulp and paper production is associated with increased deforestation, water contamination, air pollution, and the release of greenhouse gases. As a result, the industry contributes to negative impacts on biodiversity, soil fertility and wildlife, and can have negative consequences for human health. In OECD countries, the industry has introduced new technologies over the last decade which have resulted in lower environmental impacts and greater efficiency in resource use. Nonetheless, the continuing expansion of paper and board consumption to 2020 is expected to increase pressures on the environment from the sector. Growing demand for paper products is expected, with increased requirements for wood pulp, non-wood fibres, and recovered paper.

18.2. Developments in the pulp and paper industry

Demand for pulp and paper products

Over the last 30 years, pulp and paper production increased threefold. Developments in industrial goods, information technologies, household consumption and personal care have fed demand for all kinds of paper

1. The Reference Scenario and the modelling simulations cover publishing in addition to pulp and paper production, which is the main focus of this chapter.

products. Between 1980 and the mid-1990s, total demand for paper products increased by 70% from 156 million tonnes (Mt) annually, to 266 Mt. In this period, paper and board for packaging and other purposes generated the highest demand, followed by printing and writing paper. Demand for newsprint paper increased only slightly over the last decade. OECD countries account for around 76% of global paper consumption measured in tonnes, but the demand in OECD countries is oriented towards strong, glossy and sophisticated paper products, so the value share for OECD countries in total paper consumption is somewhat higher.

World demand for pulp and paper is projected to increase by 77% from 1995 to 2020 under the Reference Scenario, with an annual average growth rate of 2.3% (Table 18.1). Although the demand for products from the sector is projected to grow particularly fast in the regions of China and South East Asia, OECD regions are expected to maintain the dominant share of paper product consumption in the world market.

Table 18.1. **Key statistics and projections for the pulp, paper and publishing sector**

		1980	1995	2020 (projected)	Total change 1995-2020
Gross production (1995 US$ billion)	OECD	..	1 029	1 646	60%
	World	..	1 213	2 150	77%
Share of value-added (%)	OECD	..	1.9%	1.9%	–1.0%
	World	..	1.8%	1.8%	–2.1%
Total paper consumption (Million tonnes)	OECD	123 612	203 367	288 700	42%
	World	155 744	266 440	391 004	47%

Sources: FAO (1999), GTAP database and Reference Scenario.

Pulp and paper production

One of the main drivers of production changes in the pulp and paper industry has been the increasing demand for stronger, more versatile, and brighter paper products. This has had a major long-term influence on product diversification and technological development. Since the 1960s, new product specifications have led to significant growth in production of better quality, high-priced pulp, leading to higher operation costs and in, some cases, to increased environmental impacts. On the other hand, increasing production costs have provided strong incentives to develop more cost-efficient and less environmentally disruptive pulping and bleaching techniques.

Total production of paper and board was 294 Mt in 1998. Production of wood pulp reached 175 Mt (of which 11% was from recycled or non-wood fibres) and around 40% of all paper was manufactured from recycled paper and board (FAO, 2000). OECD countries account for about 77% of world pulp and paper production. Among these, the main producers – representing more than 80% of production – are the US, Japan, Canada, Germany, Finland, Sweden, Italy, and Korea. Figure 18.1 shows how production in the pulp and paper industry is structured.

Figure 18.1. **Production flow in the pulp and paper industries, 1998**

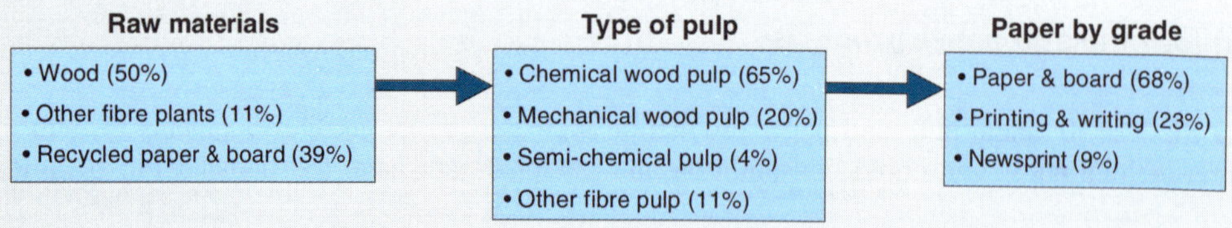

Source: FAO (1999).

The Reference Scenario projects a global increase in pulp and paper production, publishing included, of 77% from 1995 to 2020. Figure 18.2 shows that for this period pulp and paper production may grow at higher rates in non-OECD regions than in OECD regions, which will however maintain the largest share in production volume (around 75% of world production). In OECD regions, the highest growing production rates are projected for the regions of Central & Eastern Europe, Australia & New Zealand, and Canada, Mexico & the US.

Figure 18.2. **Production in the pulp, paper and publishing sector, 1995-2020**

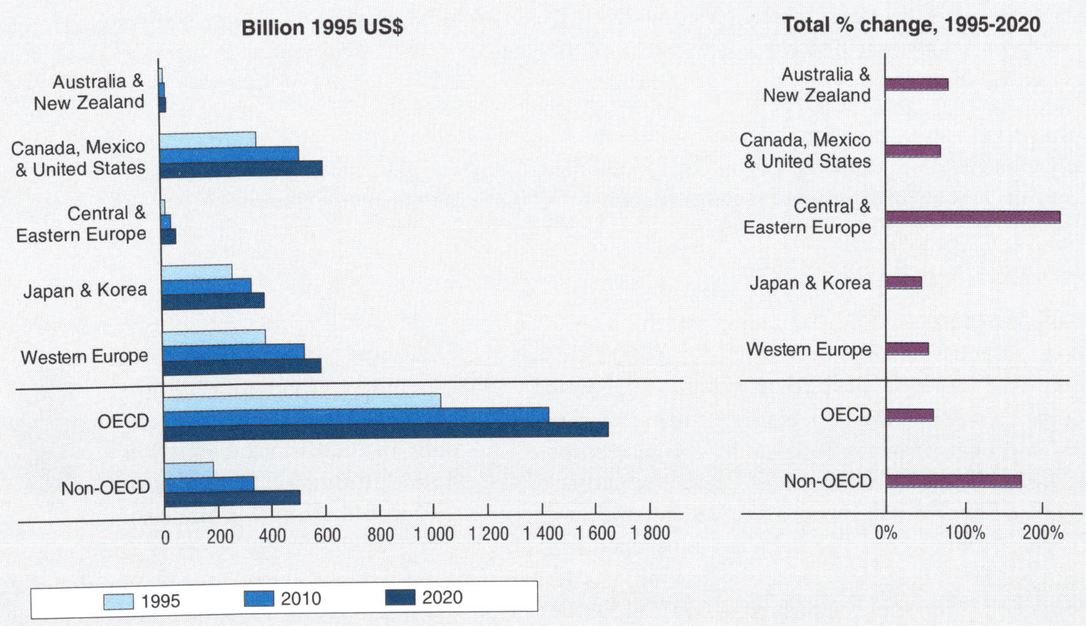

Source: Reference Scenario.

Production in this sector is mainly destined for domestic markets. World trade of pulp, paper and printing accounts for about 30% of total production and is dominated by OECD countries (78% of import and 88% of export volume in 1998). While developing countries currently account for around 12% of world export value, the Reference Scenario projects that this share will increase to 18% in 2020.

Institutional framework

The pulp and paper industry requires a large amount of capital investment in production plants. In response to the high capital requirements, combined with the volatility in paper markets, the industry has extended the economies of scale, introduced more cost-effective technologies, and shut down inefficient mills. Pulp and paper companies have been part of a continued concentration process, particularly over the last two decades. By 1990, over 20% of the paper mills that existed in 1980 in the main OECD pulp and paper producing countries had closed. The mills affected were generally small, with an average production capacity of 50 000 tonnes per year. Bigger and more efficient papermaking facilities now often have an annual capacity of more than 450 000 tonnes, while the largest production lines can produce 1 Mt per year, compared to the maximum of 100 000 tonnes in the 1960s. The current industry profile shows a small number of global companies participating intensively in the development of world markets.

The pulp and paper industry is becoming global and highly competitive, with an increasing share in world markets and rapid growth of paper consumption in developing regions. Better investment policies and changes in papermaking technologies that allow a greater use of short fibres have facilitated the entry of new competitors and the expansion of paper consumption in developing markets. The need to expand production capacity to meet global demand has provided incentives to acquire existing installations in developing countries as a lower cost option than investing in new capacity, thereby fuelling consolidation trends.

18.3. Environmental effects of pulp and paper production and use

At different points in its lifecycle – from forest harvesting, to pulp and paper processing and waste generation – the pulp and paper industry has significant environmental impacts. These impacts have been extended by increased manufacturing of glossier and stronger paper products. The industry has to some extent offset effects and impacts by using environmentally friendly technologies and more efficient processes with lower requirements of wood, water and energy per unit of product.

However, despite these improvements, the continuing expansion of paper and board consumption up to 2020 is likely to increase negative impacts on the environment. There is likely to be a growing demand for greater production and consequently for more wood pulp, non-wood fibres, and recovered paper.

Air pollution and climate change

CO_2 and SO_x emissions from the pulp and paper sector are expected to increase substantially to 2020, but introduction of new production technologies could offset this increase.

Pulp and paper facilities emit nitrogen oxides (NO_x), carbon dioxide (CO_2), sulphur dioxide (SO_2) and fly ash. Chlorine and chlorine dioxide may also be emitted. Although considered less troublesome than water discharges, air contaminants released at the chemical pulping stages, notably sulphur compounds, present a particular problem, and some of the chemicals released to the air are carcinogens (EPA, 1999). The pulp and paper industry is the third greatest industrial greenhouse gas emitter, after the chemical and steel industries and contributed 1% of CO_2 emissions from fuel combustion is OECD countries in 1997 (EIA, 1999). In 1995, emissions from the pulp and paper industry amounted to 175.4 Mt of CO_2 and 0.6 Mt of SO_x. Under the Reference Scenario CO_2 emissions from the industry in OECD countries are expected to increase by 62% to 2020, while SO_x emissions rise by 23%. Both CO_2 and SO_x emissions from the industry worldwide are expected to increase by roughly 100% between 1995 and 2020. As with other industry sectors, the introduction of new production technologies and increased efforts towards energy efficiency improvements could reduce the emission growth rates in OECD countries.

Freshwater use and water pollution

Pulp production requires large amounts of water and chemicals, giving rise to high volumes of polluted effluents. In 1995, the pulp and paper sector was responsible for around 11% of the total volume of water used in industrial activities in OECD countries. Reference Scenario projections indicate that worldwide water use by the sector is likely to grow from 11 to 18 billion m^3 between 1995 and 2020. Although OECD countries are using 97% of the water currently used for pulp and paper production, non-OECD countries are increasing their share, and may increase their water use by 154% to 2020. The discharges resulting from pulp and paper manufacturing include chemical compounds formed during processing, and natural substances, such as wood fibre fragments and other wood components, that are washed out during manufacturing. These effluents lower oxygen levels in water bodies, slowly disseminate biodegradable residues, and introduce known and unknown compounds into surface and underground waters.[2]

2. The key variables for assessing the quality of mill effluents are:
   - Biochemical Oxygen Demand (BOD): the oxygen required for oxydation of organic matter present in effluent;
   - Total Suspended Solids (TSS): the fragments and particles of wood in the waste water from mills;
   - Dioxins and Furans: a family of chemical compounds; and
   - Absorbable Organic Halide (AOX): chlorinated compounds.

The most important and direct environmental impacts of pulp and paper production arise at the pulping and bleaching stages.[3] Mechanical pulping generates few pollutants, while chemical pulping and bleaching mills account for the greatest environmental impacts. Severe eutrophication and oxygen depletion of water bodies result from chemicals and residues in pulp mill effluents. Bleaching plants also produce chlorinated compounds, and only half of these pollutants are readily biodegradable. Chemicals used in pulping and bleaching operations also contribute to environmental acidification through sulphur releases and react with organic matter to produce organochlorines.

Since 1980, water pollution from the pulp and paper sector in most OECD paper producing countries has been reduced significantly. The emission of chlorinated organic compounds has been reduced by around 90%, oxygen consuming organic compounds by over 85%, and sulphur dioxide compounds by more than 90% (CEPI, 1998; CPPA, 1999). Since 1988, the use of dioxins and furans has dropped by 99% in the Canadian and European pulp and paper industries.

Waste

The use of recovered paper in the pulp and paper industry is increasing rapidly.

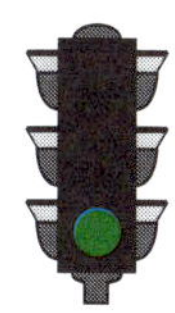

The use of recovered paper in papermaking is growing worldwide, but landfilling is still the dominant disposal route for discarded paper products in most countries. In 1980, global generation of wastepaper amounted to 50 Mt. This increased by more than 120% to reach 112 Mt by 1998. With the prospect of continued growth in the consumption of paper products, more wastepaper will have to be recycled or diposed. When wastepaper is not recycled or incinerated, its decomposition contributes to the generation of methane emissions, a potent greenhouse gas. Around 80% of wastepaper is generated in OECD countries where technical development, coupled with social awareness, have encouraged significant increases in wastepaper recycling. In OECD countries, the share of recycled paper, imports included, in fibres used for paper and board production has thus increased from 33% in 1980 to 45% in 1998 (FAO, 2000).

Forests

Pulp and paper production contributes to world forest depletion *via* the use of virgin wood fibre, which makes up around 45-54% of the raw materials used in the industry. While in OECD countries sustainable management practices try to balance roundwood requirements with forest regeneration rates, this is not always the case in non-OECD countries. The pulp and paper industry utilises large amounts of virgin fibre from woodwaste and sawnwood derived from other production processes that use wood (*e.g.* housing, furniture and packaging production). Despite the growing use of recycled materials in paper production, the Reference Scenario projects an increase of around 73% in world demand for roundwood from 1995 to 2020 (see Chapter 10).

Resource efficiency

The pulp and paper industry has reduced its use of energy and water in the production process. Through the use of new technologies, less water is required in the production process, and recycling systems have been introduced to close the loop and clean the water before it is discharged. As a result water use in the paper industry per unit of output has dropped by 50-80% from 1975 to 1997, although total water use is still increasing (CEPI, 1998; CPPA, 1999). Most of the pulp is already converted into paper products at the manufacturing site (integrated mills), with only about 11% of the production being shipped as market pulp. In integrated mills, energy recovery from the

3. Pulp is bleached to increase the brightness of the final paper product and, for pulps produced through chemical pulping, to help remove the lignin that binds cellulose fibres. The industry has been moving from using elemental chlorine (a gas that produces the highly toxic dioxins) to elemental chlorine-free (ECF) and totally chlorine-free (TCF) technologies, which are less environmentally damaging processes.

wood by-products and black liquor (spent cooking chemicals and wood-based organic matter) is enough to supply the paper mill with an increasing amount of energy (MEB, 1993).

The use of recovered paper has also contributed to significant energy savings, since processing chemical pulp from recovered paper requires about 1.3 MWh/tonne, compared to the 4.2 MWh/tonne required for chemical pulp produced from wood and the 5 MWh/tonne needed for producing printing and writing paper from wood fibres (UNIDO, 1993). Since 1990, total energy consumption per tonne of production in the Canadian and European pulp and paper industries has decreased by 10.5%, and the average amount of recycling content in paper production has increased by 22% from 1990 to 1998 (CPPA, 1999; CEPI, 1998). Moreover, most of the energy needed to make pulp from virgin fibre is produced on site, primarily from biomass-based energy sources that do not contribute to greenhouse gas emissions. Nonetheless, the Reference Scenario projects that to 2020 production growth by the sector may require 53% more energy globally, and 38% more energy in the pulp and paper sector in OECD regions.

18.4. Policy options and their potential effects

Owing to a combination of new regulations and industrial innovations, the intensity of the negative environmental effects from the pulp and paper industry per unit of product has been decreasing in OECD countries, although the pressure of net increases in future production may offset these positive results. Regulations now strictly control emissions to different environmental media, and the new mill technologies used in pulping, bleaching and recycling limit much of the pollution. However, these trends could be accelerated through the further use of policy instruments – such as economic instruments, voluntary agreements and information provision – to improve the positive environmental effects. A general objective of the policy instruments used should be to provide incentives for the adoption of best available techniques and practices, with a focus on pollution prevention and the effective co-ordination of policies. From forestry use to paper manufacturing and waste generation, sustainable practices will need to be reinforced and co-ordinated across the related sectors. Important objectives for more environmentally responsible pulp and paper manufacturing may include harvesting wood fibre from sustainably managed forests, improving resource and energy efficiency in the production process, and increasing the share of non-wood materials and recovered paper in production.

Technological development and diffusion

Finding solutions for the environmental problems to which the pulp and paper industry contributes requires the development of new environmentally friendly technologies and the better up-take of existing ones. The promotion of technological innovation in the pulp and paper sector should integrate explicit environmental objectives. Policies should take into account – and provide incentives for – the development of cluster-oriented research and clean technologies, involving foresters, equipment and chemical suppliers, research institutes, and pulp and paper companies as key actors for promoting continuous innovation in environmentally sustainable techniques and systems.

New technologies can encourage innovations in energy and resource use to render them more efficient and also enable the cost-effective substitution from fossil fuels to renewable forms of energy (*e.g.* alcohol fuel, spent liquor, bark, sawdust). Moreover, there are possibilities to further reduce environmental impacts associated with pulping and bleaching operations, for example by introducing totally effluent-free mills and closed loop processes. The growth of chlorine-free processes and the introduction of primary and secondary water treatment, as well as other less polluting processes, can continue to reduce the total effluent load from the industry (AET, 1999).

Regulatory instruments

Regulations are used to control emissions to different environmental media. Water discharges are controlled, limiting suspended solids and biochemical oxygen demand, as well as the loading of organochlorines. Air emission standards generally limit the output of gases and particles. Regulations in OECD countries continue to be tightened, particularly with the implementation of the EPA Cluster Rule in the US to restrict toxic releases from pulp and paper mills, and the Integrated Pollution Prevention and Control Directive in the European Union. The relationship between environmental controls and results will need to be assessed in order to ensure that environmental quality is improving in pulp and paper production and measures are cost-efficient. The monitoring of environmental effects should play an increasing role in shaping future regulatory requirements in the pulp and paper sector. The statutory framework must take into account the need for integrated approaches in order to avoid dispersion in separated regulatory areas such as water, air, waste, toxic substances, occupational exposure, etc. Within the pulp and paper industry, for example, mills with no effluent treatment systems should be targeted for regulatory action, so that the load of conventional and toxic pollutants discharged into water bodies is reduced. Regulations to prevent polluting discharges and dioxin releases to air by pulp and paper mills need to be extended.

The development of new legal instruments (*e.g.* Extended Producer Responsibility – EPR) could integrate the producer responsibility concept as one of the main elements of waste paper management. The producer can be given partial or full responsibility for the management of the product once its lifecycle has been completed, and sharing responsibility can contribute to closing the fibre loop by increasing paper collection for recovery. Regulations aimed at imposing an industry-wide cap on CO_2 and methane emissions, recycling paper quotas and take-back programmes should also be considered.

Economic instruments

The use of market-based incentives to address the environmental effects of the pulp and paper industry should be further evaluated and extended, including the application of charges and levies on industrial emissions. This could include applying green taxes at different stages of the paper cycle, such as a carbon tax to address fossil fuel emissions, and a landfill tax to cover the environmental impacts of paper product waste disposal. Optimising both market and policy signals can be a powerful strategy to foster investment and rapid changes in the industry that will favour environmental protection, such as environmentally bleached and recycled products, and the sourcing of wood fibres from sustainably managed forests. Incentives can be used to improve production yields and increase the portion of recovered paper and non-wood raw material in pulp and paper manufacturing. Taxes can be used to reduce emissions and increase energy and resource efficiency.

For OECD countries, model simulations project that eliminating all subsidies to the pulp, paper and publishing industry and introducing an *ad valorem* tax on the use of products from the industry that increases by 2 percentage points per year would reduce environmental impacts from the industry significantly (see Chapter 12 for further details). CO_2 emissions by the sector in OECD regions would be roughly 15% lower in 2020 than under the Reference Scenario, while SO_x emissions would be 13% lower and water use 16% less (see Table 18.2). A policy simulation of the removal of all energy subsidies in OECD regions, combined with the application of a tax on all energy use, was found to have an almost negligible effect on OECD pulp and paper production and on water use by the sector, a comparable effect on CO_2 emissions (15% lower), but a more significant reduction in SO_x emissions (37% reduction).

Table 18.2. **Subsidy and tax policy simulations: effects on the pulp and paper industry in OECD regions and its environmental impacts**

(% change from Reference Scenario in 2020)

	Policy simulations	
	Pulp, paper and publishing industry subsidy removal and tax on all paper product use	Energy subsidy removal and tax on all energy use
Gross pulp, paper and publishing production	–13.8%	–0.2%
CO_2 emissions	–15%	–15%
SO_x emissions	–13%	–37%
Water use	–16%	0%

Note: The energy subsidy and tax policy simulation was applied to all sectors, not just the pulp and paper industry. See Chapter 12 for more details.
Sources: Reference Scenario and Policy Simulations.

Voluntary agreements and information

In recent years the increasing use of voluntary programmes and stakeholder initiatives in the pulp and paper industry has emerged as a complement to command and control regulations. Programmes that aim at total quality management and environmental protection in productive processes, such as the ISO 14000 standard, Forest Stewardship Certification and Environmental Management Systems, have brought about a shift towards environmental protection and sustainable practices. There have also been several initiatives across OECD countries, namely the adoption of mandatory and voluntary recycling quotas in paper manufacturing, take-back and recycling programmes required by law and/or voluntary industry schemes, and more comprehensive municipal waste management. These approaches should be carefully designed, focusing on lifecycle impacts, taking into account the economics of supply and demand of fibre, and respecting different environmental concerns, such as climate change and energy recovery.

Voluntary agreements by the pulp and paper industry could contribute to reductions in carbon dioxide emissions from the sector. Such an approach could both reduce contributions to climate change and contribute to increased energy efficiency by improving technologies, practices and management at every stage of the paper production cycle. Voluntary schemes can also be used to separate paper and paperboard at the waste source, *e.g.* in households or businesses, for recycling. Green procurement policies by governments can also be used to favour the use of more environmentally friendly paper products.

The pulp and paper industry should extend the use of ecolabels that are based on the lifecycle approach, certification of sustainable and non-polluting practices, and that cover the range of environmental requirements. Non-governmental organisations, working with consumers, can encourage pulp and paper importers to reduce the sourcing of pulp and paper products from non-certified forests and from polluting mills, and upgrading and promoting the use of the best available technologies and practices throughout the pulp and paper industry could bring many environmental benefits, increasing production while reducing emissions and the use of energy and natural resources.

REFERENCES

AET (Alliance for Environment Technology) (1999), "Sending Ripples Around the World", AET, November 1999, *www.aet.org/facts/ripple.Ad.html.*

CPPA (Canadian Pulp and Paper Association) (1999), *Reference Tables*, CPPA, Montreal.

CEPI (Confederation of European Paper Industries) (1998), *Environmental Report* 1998, CEPI, Brussels.

EPA (US Environmental Protection Agency) (1999), *Toxic Release Inventory: Release Inventory Data for Pulp and Paper* (SIC code 26), United States Environmental Protection Agency, Washington DC.

FAO (Food and Agriculture Organization of the United Nations) (1999), *Global Forest Products, Consumption, Production, Trade and Prices: Global Forest Products Model Projections to* 2020, FAO, Rome.

FAO (2000), "FAOSTAT November", *http://apps.fao.org/default.htm.*

IEA (International Energy Agency) (1999), CO_2 *Emissions from Fuel Combustion*, IEA/OECD, Paris.

IIED (International Institute on Environment and Development) (1996), *The Sustainable Paper Cycle*, IIED, London.

MEB (Management Institute for Environment and Business) (1993), *Competitive Implications of Environmental Regulations in the Pulp and Paper Industry* (Draft), MEB, United States.

Rajotte, A. (2000), "Pulp and Paper Industry", background document for the OECD *Environmental Outlook*, OECD, Paris.

UNIDO (United Nations Industrial Development Organisation) (1993), "Pulp and Paper: Case Study No. 1", Conference on Ecologically Sustainable Industrial Development, Copenhagen, 14-18 October.

WBCSD (World Business Council for Sustainable Development) (1996), *A Changing Future for Paper*, WBCSD, Geneva, *www.wbcsd.ch*.

19

The Chemicals Industry

KEY SIGNALS

- In OECD countries the production, consumption and trade of chemicals and chemical products has grown steadily, and is projected to continue to grow through 2020 roughly at the same rate as GDP (with faster growth in developing countries). However, production in OECD countries is expected to shift from high volume basic chemicals to speciality chemicals and products derived from the life sciences.
- The chemicals industry in OECD countries has made significant progress in reducing the release of pollutants to the environment during the manufacturing process.
- Major concerns exist about the possible impact on the environment and human health of substances produced by the chemicals industry, which are found in virtually every man-made product. Many are being detected in the environment, where particular problems can be caused by persistent, bioaccumulative and toxic chemicals. Concern is growing, for example, about chemicals which cause endocrine disruption and which persist in the environment.
- Priority should be given to filling the immense knowledge gap about chemicals on the market. A scientific, rules-based approach requires information on the effects of chemicals and chemical exposure as the basis for risk management decisions. A variety of instruments – including economic incentives, voluntary approaches, and regulations – can be used to encourage the development of better chemicals information. Where such information is not available, more and more countries will take a precautionary approach.

19.1. Introduction

Chemicals produced by the chemicals industry are used to make virtually every man-made product. The global chemicals industry today produces many thousands of substances. Chemical companies sell these to other industries, companies or consumers. These substances can be mixed by the chemicals industry and sold as preparations – mixtures of two or more substances which do not react with each other – or be used by customers in the chemicals industry to create preparations. It is estimated that there are between one and two million preparations on the market today, as well as countless manufactured articles made from chemicals. The chemicals industry makes up an important part of the world economy. With an estimated US$1 500 billion in sales in 1998 (CMA, 1999), the chemicals industry accounts for 9% of international trade (WEC, 1995).

There is a lack of adequate safety information about the great majority of chemicals on the market.

Over the years, concerns have been raised about the impact the chemicals industry has on the environment. This industry, like other industries, releases pollutants to the environment that can negatively impact on air, water and soil quality. Due to the nature of the products used and produced by the industry, the potential risk of occupational health hazards, or accidents involving hazardous substances, must be considered. In addition, because of their persistence, bioaccummulative properties and/or toxicological effects, certain chemicals can pose a risk to workers in other industries, consumers, the general population and/or the environment. The primary problem today is the lack of knowledge about the properties, effects and even exposure patterns of the great majority of chemicals (and, by extension, of preparations and consumer products made with them) on the market today. Whereas known hazardous chemicals are being managed to a large extent, there may be many unknown hazardous chemicals whose potential risks are neither being evaluated nor managed because the necessary information is not available.

19.2. Developments in the chemicals industry

Table 19.1. **Key chemical industry statistics and projections**

		1998 (or latest available year)	2020 projected	Total change 1995-2020
Gross production (1995 US$ billion)	OECD	2 103	3 388	62%
	World	2 655	4 916	85%
Share of value added in the economy (%)	OECD	3.5%	3.4%	–0.1%
	World	3.4%	3.4%	0.0%
Exports (1995 US$ billion)	OECD	472	787	67%
	World	570	1 062	86%

Note: Gross production figures are 1995 GTAP figures and include rubber and plastics manufacturing and can therefore differ from other statistics on the chemicals industry.
Sources: GTAP database and Reference Scenario.

The chemicals industry is very diverse, comprising basic or commodity chemicals; speciality chemicals derived from basic chemicals (adhesives and sealants, catalysts, coatings, chemicals for electronics, plastic additives, etc.), or from the life sciences (pharmaceuticals, pesticides and products of modern biotechnology); and consumer care products (soap, detergents, bleaches, hair and skin care products, fragrances, etc.). Given the complexity of the processes used and the constant need for innovation, the chemicals industry is research intensive, with companies allotting 4-6% of their annual sales to research and development (R&D), depending on the sector (CMA, 1999). Some product sectors are undergoing major restructuring, such as agrochemicals, pharmaceuticals and the food industry, where biotechnology is beginning to play an important role.

The chemicals industry is a major employer, with over 10 million people employed worldwide (CMA, 1999). Yet, as the industry has become more productive and production processes have become highly automated, world employment levels in the industry have fallen 7.5% over the last ten years.

Demand for and production of chemicals and chemical products

Today, the per capita consumption of chemicals in OECD countries is far greater than in non-OECD countries. However, the Reference Scenario projects that over the next 20 years, as GDP will grow and the use of chemicals will increase in developing countries, domestic demand in those countries is projected to rise more rapidly than in industrialised regions. By 2020, the developing world may account for 33% of world chemical demand, compared with 23% in 1995.

Almost every country has a chemicals industry, but the bulk of production is accounted for by a small number of industrialised countries, with currently approximately 80% of the world's total output being produced by only 16 countries (in decreasing order): US, Japan, Germany, China, France, UK, Italy, Korea, Brazil, Belgium/ Luxembourg, Spain, the Netherlands, Taiwan, Switzerland and Russia (CMA, 1999).

While the overall volume of the chemicals industry has been growing, the sub-sectors that make up this volume have grown at different rates. Production in all sectors is highest in OECD countries, where growth is now fastest in the speciality chemicals and life science sectors. During the period 1980 to 1997, pharmaceuticals was the major growth sub-sector in the chemicals industry in OECD regions. Basic chemicals production increased most strongly in Japan. Data on sectoral contributions are not available for developing economies, but it can be assumed that over the same period the basic chemicals sector was one of the growth leaders in China and other countries of the Far East, and in the Middle East (CIA, 1999).

Although it is projected that OECD countries will still produce the majority of world chemicals in 2020, with 69% of total world production, this will represent a 10% decrease from their 1995 share (Figure 19.1).

Figure 19.1. **Chemicals production, 1995-2020**

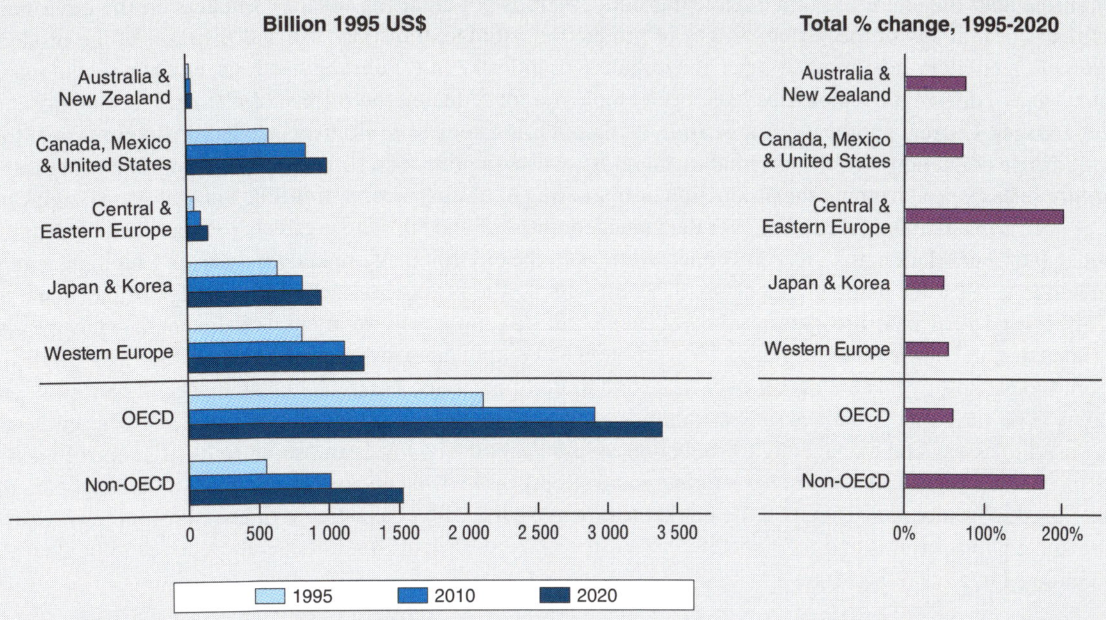

Note: These figures include manufacturing of plastics and rubber.
Source: Reference Scenario.

Changes in sectoral contributions are expected to follow a similar pattern up to 2010 as occurred between 1980 and 1997: pharmaceuticals will be leading the growth, followed by speciality chemicals, agricultural chemicals, textile fibres and basic chemicals. In the US it is expected that by the year 2010 the total sales in the life sciences sector will approach those of basic chemicals, and, by the year 2020, easily exceed the annual basic chemicals revenues. By then, speciality chemicals revenues are expected to be close to those for basic chemicals, and the production of high volume basic chemicals, which is a mature market today, is expected to shift to non-OECD countries (Swift, 1999).

Trade and structure of the chemicals industry

Trade in chemicals is currently dominated by OECD countries, which register trade surpluses with virtually all the other regions of the world. Compared with the mature markets of OECD regions, globalisation in recent years has resulted in high growth rates in exports and imports by non-OECD countries.

It is anticipated that the growth of the global chemicals industry, together with continuing globalisation and greater competitiveness, will intensify recent trends of companies forming alliances in order to achieve economies of scale. To this will be added the impetus of steadily mounting costs. The cost of R&D, bringing new products to market, managing the safe production and distribution and use of chemical products throughout their life-cycle and meeting pressures of environmental health and safety requirements will most likely rise. The trend towards fewer and larger multinational producers is expected to continue, with companies based in OECD regions becoming more and more knowledge-based, concentrating on speciality chemicals and life sciences rather than asset-based basic chemicals (CIA, 1999).

19.3. Environmental effects of the chemicals industry and its products

Throughout the entire life of a chemical product there is potential for negative impacts on the environment. These can occur at any of the various stages of production, storage, transport, use and disposal of the product. In response to regulatory requirements or on the initiative of industry (*e.g.* voluntary actions, environmental management systems), different technologies have been employed for reducing the release of chemical pollutants at each of these stages. Current approaches range from using pollution control equipment (closed or floating roof storage tanks, particle collection systems, thermal incinerators, waste treatment facilities, etc.) to designing processes that minimise releases, and banning the production or marketing of a substance, or limiting some of its uses. Nevertheless, concerns persist. They are either over the management of chemicals whose effects (endocrine disruption, persistence, bioaccumulation, toxicity) and concentrations in the environment are known, or over which management approach to take for the many more chemicals about which little is known. Lack of knowledge of the characteristics, effects and even exposure patterns of these chemicals is a primary environmental challenge faced by the chemicals industry and its regulators today. This problem is compounded by speciality chemicals and products of modern biotechnology that are taking over more and more of the market and may present special concerns related to safety. They are often produced in low volumes and therefore do not fall within the classical programmes now being used to assess and manage high production volume chemicals. Furthermore, introducing control measures even for existing chemicals whose characteristics and effects are known can pose problems. For the most part, these widely used substances have been on the market for many years, with considerable financial resources invested in plants and equipment; this and their specific uses often make it difficult for suitable alternatives to be developed and marketed in a short time frame.

Release of hazardous substances

Certain chemicals released to air, water or soil by the chemicals industry in its production processes or by its products can have direct or indirect impacts on human health and the environment. Exposure to certain hazardous substances – such as polychlorinated biphenyles (PCBs), dichlorodiphenyltrichloroethane (DDT), heavy metals, and endocrine disruptors – can lead to a direct toxic effect on human health and/or the environment. Other substances – such as chlorofluorocarbons (CFCs), carbon dioxide (CO_2), nitrogen oxides (NO_x), and sulphur oxides (SO_x) – are cause for concern especially after they react with other substances. For instance, CFCs are non-flammable and non-toxic, but if they react with other substances in the stratosphere they can destroy some of the ozone layer which, in turn, can lead to diminished protection from UV radiation and therefore to a greater incidence of skin cancer. The same type of indirect effect applies to climate changing substances, such as CO_2 and other greenhouse gases, and to substances which promote the formation of smog (NO_x and volatile organic compounds (VOC)).

While the chemicals industry is, to some degree, responsible for emissions of greenhouse gases, substances that promote the formation of tropospheric ozone and CFCs, most attention has focused on its releases of hazardous pollutants. Unlike climate changing substances, which have been largely identified, the nature and magnitude of hazardous substances released during chemical production processes and use are far from known. Nonetheless, using the data that are available, governments have developed lists of substances they consider to pose risks to human health and/or the environment and that require national control. At present, a very limited number of hazardous substances are included in internationally agreed lists. A major concern persists regarding those substances that may be hazardous, but which have not yet been characterised as such due to a lack of toxicity data, and are therefore not listed anywhere.

The emission of some pollutants to the environment from the chemicals industry is decreasing.

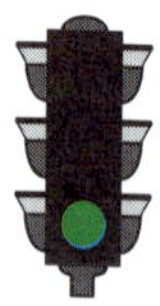

Although there are no consolidated data on emissions of known hazardous substances across OECD countries, it is probable that, overall, such releases from the chemicals industry in these countries are declining. For example, releases by the US chemicals industry of the "core" hazardous chemicals listed on the 1988 Toxic Release Inventory (TRI) were showing a steady downward trend by 1998, with reductions in releases to air and surface water accounting for the bulk of the improvements (US EPA, 2000). US chemicals industry releases of substances covered by US Standard Industrial Classification codes 20-39 also decreased relative to total releases reported from other manufacturing industries. In Japan, chemicals industry emissions of 12 priority air pollutants decreased by 35% between 1995 and 1998. In the UK, chemicals industry emissions of 27 substances, all of which are of particular concern if discharged to water or to sewers, were 96% lower in 1998 than in 1990 (CIA, 2000). In Canada, releases of emissions of 15 priority substances were reduced by 73% from 1992 to 1998 (Caswell, 2000). The situation on releases in non-OECD countries is unclear since no past trends data are available. Nevertheless, it should always be kept in mind that the chemicals which are listed and monitored represent only a limited percentage of those emitted by the chemicals and other industries.

Effects on air quality and climate change

Figure 19.2 shows that, compared to other industries, the chemicals industry in OECD countries is a major emitter of CO_2 (one-quarter of all emissions from fuel combustion from industry sectors). Chemicals industry emissions nevertheless only make up 4% of the emissions from fuel combustion in all sectors. Data reported by the European and US chemicals industries indicate that while production is rising, CO_2 emissions have been stabilising since 1985 (due to a greater use of oil in place of coal as an energy source) (OECD, 2001).[1] The Reference Scenario projects that CO_2 emissions from the chemicals industry in OECD countries will increase by 66% from 1995 to 2020 (following total industry trends), while the increase in non-OECD countries will be 165% during the same

Figure 19.2. **CO_2 emissions from fuel combustion in OECD countries, 1997**

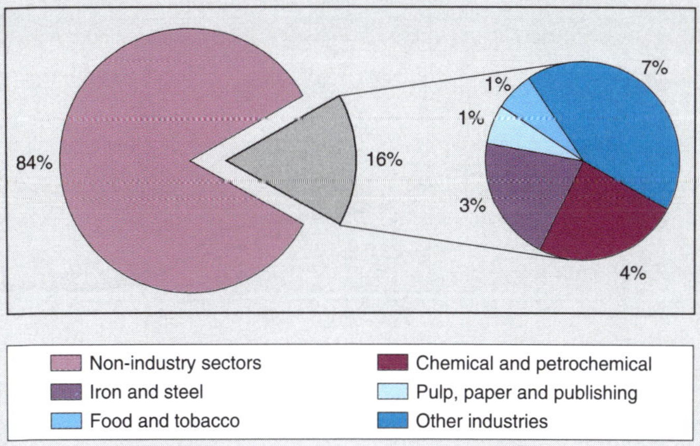

Note: The "chemical and petrochemical" category shows CO_2 emissions from both the chemicals manufacturing industry and the petrochemical industry.
Source: IEA (1999).

1. These data on CO_2 emissions from the chemicals industry are from manufacturing. They therefore take no account of contributions to CO_2 and other greenhouse gas emissions from the industry's products. If a full picture of the chemicals industry contribution from its products to GHG emissions over their complete life-cycle from "cradle to grave" were required, these "downstream" emissions should be added to the reported manufacturing emissions. In addition, the discussion of the contributions of the chemicals industry to climate change is restricted to CO_2 emissions since data on emissions of other GHGs are not readily available.

period. However, if greater energy efficiency gains are achieved in the chemicals industry, CO_2 emissions may increase at slower rates or continue to stabilise in OECD countries.

Little global data are available on the total contribution by the chemicals industry to the release of substances which promote the formation of tropospheric ozone (VOCs, NO_x) and acid rain (SO_x). However, indications suggest that these emissions from the chemicals industry are generally decreasing due to technological changes that are influencing energy use and the operation of chemical plants. The following examples support this assumption. The US Toxic Release Inventory (TRI) data indicate that the US chemicals industry is a relatively small contributor to total US emissions of these pollutants, and that they have generally been declining since 1970 (CMA, 1998). In Japan, SO_x emissions from the industry fell between 1980 and 1987, but have increased slightly since then, and although NO_x emissions rose slightly between 1990 and 1998, emissions per production unit have remained stable (JRCC, 1999). And in the UK, VOC emissions from the chemicals industry have fallen by 46% (CIA, 2000).

Tremendeus progress has been made in OECD countries in phasing out the production and consumption of ozone depleting substances.

Since the adoption of the Montreal Protocol in 1987, tremendous progress has been made in phasing out the production and consumption of chemicals that deplete the ozone layer, as demonstrated by Figure 19.3. Production of most other ozone depleting substances show a similar downward trend, although production of hydro-chlorofluorocarbons (HCFCs) is rising. HCFCs may have only 2-5% of the ozone depleting potential of CFCs, but they are likely to remain in the stratosphere for a long time since, under current agreements, they will not be phased out for at least another 20 years.

Figure 19.3. **Production of CFCs for selected countries and regions, 1986-1997**

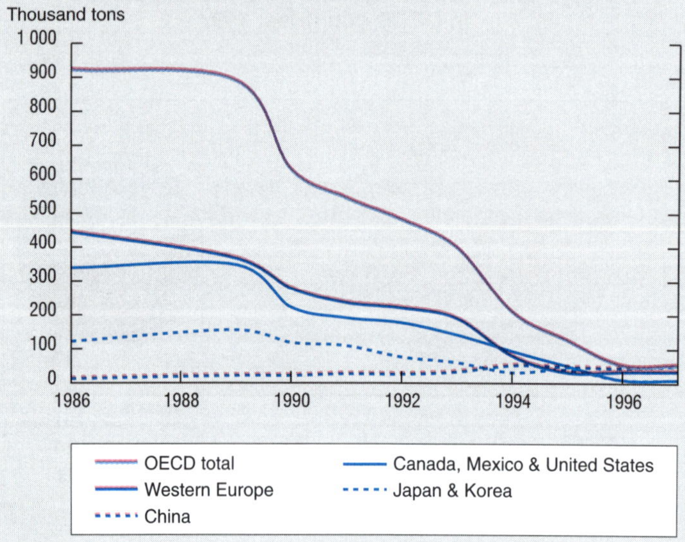

Source: UNEP (1999).

Effects on water use and quality

Compared to all other manufacturing industries, the chemicals industry in OECD countries was the largest consumer of water in 1995, using 43% of all the water consumed by industries and greatly surpassing the iron and steel sector (26%) and the pulp and paper sector (11%). Worldwide, however, agriculture is by far the largest user of water, followed by manufacturing industries (see Chapter 8). Although the figures for the chemicals industry's water consumption are not yet a cause for concern, government policies are nonetheless aiming at water conservation and increased efficiency in this area.

Natural resource and energy use

The chemicals industry also impacts on the environment by depleting non-renewable resources, or by putting pressure on renewable resources, and by creating pollution during their use (combustion). The chemicals industry uses a combination of natural gas, coal and coke, minerals, water, fuel oil, liquified petroleum gas and electricity as sources of energy and raw materials. The chemicals industry is energy intensive and accounted for some 7% of total world energy use in 1998. The share of energy use of the chemicals industry in non-OECD regions is increasing, where it grew from 20% of world chemicals industry energy use in 1971 to 43% in 1998, as compared to a reduction from 80% in 1971 to 57% in 1998 for OECD countries (IEA, 2000).

Significant energy efficiency gains in the chemicals industry in OECD countries can be illustrated by the example of the EU where, between 1985 and 1996, production grew 34% while energy consumption levelled off (CEFIC, 1998). Similarly, in the US, energy consumption per unit of production declined by 43% between 1974 and 1998 (CMA, 1998 and 1999). Since the demand for chemicals is projected to increase to 2020, and energy use by the chemicals industry is also expected to increase, special efforts will be needed to reach even greater energy efficiency improvements.

Waste

The chemicals industry contributes to the generation of waste in several ways. First, hazardous substances produced as a by-product of manufacturing may be disposed of on land, recycled or treated. Hazardous chemicals produced by the chemicals industry and contained in products which work their way through the supply chain will eventually be disposed of after final use of the product. In addition, the chemicals industry produces non-hazardous waste.

Some countries, such as Finland and Ireland, distinguish total hazardous waste by industrial sectors. In Finland, of the total hazardous waste generated by all of industry in 1992, the oil and chemicals industry accounted for 46%. In Ireland, hazardous wastes from organic chemical processes amounted to 84% of total hazardous waste in 1995, and the chemicals industry also generated 0.16% of non-hazardous waste (Ireland EPA, 1996). In Japan, the volume of waste produced by the chemicals industry increased slightly between 1996 and 1998. However, approximately 71% of waste is disposed of by on site recycling and volume reduction, and the volume of final off-site disposal has been reduced from around 15% in the early 1990s to 9% in 1998 (JRCC, 1999).

19.4. Policy options and their potential effects

Overall, the chemicals industry in OECD countries has made significant progress in reducing releases of pollutants to the environment that result from manufacturing processes. Emissions of tropospheric ozone precursors are falling, and the industry has made tremendous progress is phasing out the production of CFCs. Current information suggests that releases of other known hazardous substances from chemical production and processing are probably also declining in OECD countries. Although the chemicals industry is a major energy user, important efficiency gains have been made, and CO_2 emissions are stabilising while production is increasing. CO_2 emissions from the global chemicals industry will probably increase in the future, primarily because of increasing chemicals production in non-OECD countries which use less energy efficient technology and are more reliant on coal as a fuel.

Progress in OECD countries has been due to government policies which gather and make information available (*e.g.* pollutant release inventories) or reduce emissions of pollutants through regulations, voluntary agreements with industry, and economic instruments like charges, tradable permits and taxes. The chemicals industry spends a significant amount of resources on environmental protection as compared to other industries. Pressure from the public, media and environmental NGOs continues to be important in focusing attention on the environmental, health and safety performance of the industry.

Nevertheless, significant concerns remain regarding the health and environmental impacts from products made with chemicals (*e.g.* consumer products). The successes in managing the potential risks of these products have been limited to specific groups of regulated chemicals (new industrial chemicals, pesticides) or uses of specific chemicals which have proven to be unsafe (asbestos, formaldehyde, PCBs, etc.). The lack of knowledge about most chemicals on the market and, consequently, the uncertainty about whether the public and the environment are being adequately protected, is the major challenge to policy makers. Policies will need to be developed involving all stakeholders to quickly and efficiently determine, collect and make best use of the minimum necessary information on these chemicals so that appropriate management decisions can be made. Better ways need to be found to evaluate and manage the risks resulting from the release of chemicals from products. As a complement to an integrated product policy, life cycle management of chemicals (and products containing them) should be promoted, as should management schemes with increased producer responsibility linked with public information policies to ensure that products are developed, used and disposed of safely.

With increased trade in chemical products, and the growing recognition that pollutants travel across national borders, the last decade has seen an increase in international efforts by governments to manage chemicals. In the international context, inter-governmental organisations are facilitating co-operation among countries, reducing duplication, and establishing conventions and other international agreements to take certain actions related to chemicals of concern. As a follow-up to these conventions, attention should be paid to monitoring compliance and taking necessary enforcement action. It will also be important to ensure that environmental problems, which have been addressed to a certain extent in OECD countries, are not reproduced as the chemicals industry develops outside OECD regions, and that products imported from non-OECD countries can meet OECD safety standards. Capacity building in non-OECD countries by governments, industry and NGOs is an important international priority.

The OECD Chemicals Committee plays a leading role in developing chemicals control policies which are harmonised across OECD countries. Not only is chemical safety promoted through these activities, but also non-tariff trade barriers are minimised and duplicative testing efforts by the industry are avoided.

Technological development and diffusion

Companies must examine their entire production process to find more cost-effective ways to reduce the polluting substance that reaches the end of the pipe; and governments, working with industry, should develop policies that aim to prevent – rather than control – pollution, and that promote source reduction, recycling, treatment and disposal. Industry should consider both housekeeping changes, like process controls and in-process recycling, and more far-reaching "sustainable chemistry" approaches. Sustainable chemistry covers the design, manufacture and use of efficient and effective, more environmentally benign, chemical products and processes. It strives to conserve energy and non-renewable resources, prevent pollution, minimise risk and waste at all stages of a product's life cycle and to develop products that are durable and can be re-used and recycled. Policies that foster research and can assist companies to develop and market these new chemicals should be promoted.

As production shifts toward developing countries, policies need to be forged that encourage transferring to non-OECD countries non-confidential technological approaches that are used in OECD countries to minimise releases of pollutants. Further, chemical companies from OECD regions operating in developing countries should help facilitate the diffusion of information on non-proprietary technology in those countries, particularly in the context of industry voluntary Responsible Care programmes.

Legal and regulatory instruments

A number of international conventions exist or are in the process of being finalised. The Montreal Protocol on the phasing out of CFCs is well implemented. More efforts should be made to implement the ILO conventions related to chemicals and major accidents. The implementation of new conventions concerning Prior Informed Consent (PIC) and Persistent Organic Pollutants (POPs) will require immediate attention.

All countries should be encouraged to establish pollutant release and transfer registers (*i.e.* inventories of pollutants released from sources) as they help drive reductions in hazardous substances released to air, water, soil and waste, and can be used for target-setting and monitoring by governments as well as industry and the public.

The large number of existing industrial chemicals on the market – and the general lack of information on them – poses basic policy questions related to priority-setting and interim management solutions. The chemicals industry should be encouraged to provide information on the uses of its products, not only so that they can be managed better, but also to help set priorities for assessment. The chemical-by-chemical approach to testing, assessment and management needs to be replaced, or at least supplemented, by a framework that makes the consideration of groups of chemicals related by their structure, use or other parameters, possible.

Current programmes to fill the information gap on existing chemicals deal primarily with high production volume chemicals. Stronger policy instruments should be used to ensure that the necessary data are produced by industry and that the chemicals are assessed. Faced with the lack of data and the slowness of these programmes, an increasing number of countries will use a precautionary approach as an alternative or complement to data-based decision making for the management of chemicals. It might also be possible to subject existing chemicals, for which sufficient data are not provided by a certain date, to procedures resembling those applied for introducing new substances to the market. Furthermore, procedures could be developed to give industry greater responsibility not only in generating necessary data on all chemicals on the market (and not just on high production volume chemicals), but also in preparing assessment reports and providing them to governments. Governments could provide relevant guidance on data requirements and on methodologies for assessing data and preparing reports, as well as conduct periodic audits to assure that companies are properly following their guidance. In parallel, information should be provided to the public which remains an important stakeholder in this process.

In addition, integrated policies for product design and production processes should be promoted in the framework of existing and planned legislation, which needs to consider the impacts a product may have on human health and the environment throughout its entire life cycle. Stakeholder involvement in policy development and clear designation of responsibility throughout the chain of production, use and disposal of chemicals are basic tenets of such an approach, as is the need to include small and medium-sized enterprises.

Economic instruments

Economic incentives, such as tax deductions for research and development associated with sustainable chemistry, or reduced fees (as well as expedited regulatory reviews) for new chemicals that are "greener" than the substances they replace, could help promote the development of alternative chemicals which are more environmentally friendly. Longer protection time for patent and proprietary rights for greener chemicals would also encourage the chemicals industry's research in this direction; although this might limit the diffusion of such technology.

Conversely, economic disincentives could be used to discourage industry from marketing chemicals with unacceptable hazards. The taxation of chemicals considered by governments to be of concern would provide an incentive for a company to decrease production of the taxed chemicals and shift to making alternatives that are not taxed. The resulting difference in price would encourage consumers to select cheaper alternatives that are more environmentally friendly. The example provided by the phase-out of leaded petrol is illustrative: in 20 OECD countries, a tax differential in favour of unleaded petrol was introduced at the same time as a series of other policy measures to encourage substitution of leaded with unleaded petrol (OECD, 1997). While ideally the chemical taxes should be differentiated according to their environmental effects, a set tax or charge might also be levied on companies that produce chemicals for which sufficient data are not available in order to encourage further information gathering and research into these effects.

A chemicals policy simulation was undertaken to indicate the effects that might be expected from continued globalisation and trade liberalisation, and from attempts in OECD regions to internalise the social and environmental effects of toxic chemicals in the environment over the coming decades. To represent the effects of further trade liberalisation, the simulation included the removal of all subsidies to the production and use of chemicals (including agrochemicals) in OECD regions, and the elimination of tariffs on imports of chemicals to OECD regions. Ideally, a tax or charge on the use of a limited number of toxic chemicals that reflects their social and environmental costs would be levied. However, as the model used in the policy simulations does not distinguish between different types of chemicals, a tax was applied across all chemicals instead. Thus, the simulation included the application of an *ad valorem* tax which increases by 2 percentage points per annum (*i.e.* reaching 50% of the pre-tax price of chemicals by 2020) on the use of all chemicals.

The results of the policy simulation show significant effects, with chemicals production in OECD regions estimated to be 20% lower in 2020 compared with the Reference Scenario, and about 12% lower worldwide (see Table 19.2). The environmental effects of the subsidy removal and tax implementation are even more substantial. Compared to the Reference Scenario, CO_2 emissions from the chemicals sector would be 24% lower in OECD regions in 2020. While in non-OECD countries they would increase by just over 6% due to a "leakage" of chemicals production to these regions, this would not offset the reductions in OECD regions. Thus, compared with the Reference Scenario, the chemicals industry would experience an overall reduction in CO_2 emissions of 6% worldwide in 2020 compared with the Reference Scenario as a result of the policies examined. In addition, SO_x emissions from the chemicals industry would be reduced by 24% in OECD regions in 2020 compared with the Reference Scenario, while they would increase by 3% worldwide.[2] Water use by the industry would also decrease significantly in OECD regions. Finally, as discussed in Chapter 7, an *ad valorem* tax on chemical inputs would impact on the use of fertilisers in OECD agricultural production, with significant reductions in nitrogen loading to waterways from farm chemical run-off.

Another policy simulation that was undertaken (see Chapter 12) involved the elimination of all subsidies to energy production and use in OECD countries, combined with an *ad valorem* tax on fossil fuel use which increases each year by 2 percentage points, 1.6 percentage points and 1.2 percentage points for coal, crude oil and natural gas respectively (*i.e.* by 2020 the tax rate is equal to 50%, 40% and 30% respectively.) It was found that while such a policy would lead to only a very small decrease in chemicals production both in OECD regions and worldwide to 2020, there would be a significant reduction in energy-related emissions from the chemicals industry in OECD countries, particularly in SO_x emissions (Table 19.2).

Table 19.2. **Effects of subsidy and tax policy shock runs on the chemicals industry and its environmental impacts**

(% change from Reference Scenario in 2020)

		Policy simulations	
		Chemicals industry subsidy removal and tax on all chemicals use	Energy subsidy removal and tax on all energy use
Gross production	OECD	20%	–1%
	World	–12%	–1%
CO_2 emissions	OECD	–24%	–7%
	World	–6%	–3%
SO_x emissions	OECD	–24%	–45%
	World	+3%	–4%
Water use	OECD	–23%	–1%
	–	–	–

Note: The energy subsidy and tax policy simulation was applied to all sectors, not just the chemicals sector. See Chapter 12 for more details.
Sources: Reference Scenario and Policy Simulations.

2. This assumes that the increased transfer of chemicals production to non-OECD regions would not result in any additional increases in process efficiency or increased uptake of emissions reduction technologies.

Voluntary agreements

The chemicals industry in OECD countries is continuing with voluntary initiatives to reduce pollution and ensure the safety of its products. Notable examples are the current initiatives of the international chemicals industry to fill data gaps and undertake initial hazard assessments on 1 000 high production volume chemicals by 2 004 (the International Council of Chemical Associations Initiative). Another is the initiative by the US chemicals industry to fill data gaps on 2 800 chemicals in the same time frame (HPV Challenge Programme). Public access to this data and the resultant vigilance of NGOs, as well as government review of the results of these initiatives (*e.g.* through the OECD HPV Chemicals Programme), are critical to their success.

Further efforts need to be made by the industry to increase the usefulness and credibility of programmes on Responsible Care and Product Stewardship. The role of small and medium-sized enterprises in these programmes is important. A key challenge to industry will be to develop performance measures – not only concerning general concepts but also regarding specific actions – which can show stakeholders that progress is being made. To increase credibility in this area, the chemicals industry could move more quickly toward the adoption of environmental management systems such as ISO 14001, which create a framework for self-regulation, and certain provisions for monitoring and enforcement could be added.

With the shift toward greater production of chemicals in non-OECD countries, the industry should be encouraged to implement effective Responsible Care programmes wherever they operate. If Responsible Care really works, then the environmental performance of companies operating in non-OECD countries where effective chemicals regulations do not currently exist should improve. Industry compliance with OECD's new Guidelines for Multinational Enterprises will be a positive step forward. To show progress, companies should be encouraged to develop corporate-wide environmental reporting documents.

Information and other instruments

Policies need to be established to ensure that whatever information is available on chemical hazards and risks (*e.g.* on the Internet) is reliable and presented in a way that is useful to all potential users, including the general public and non-OECD countries. This concerns not only how the information is presented (*e.g.* on the Web, in hard-copy publications), but also the context in which the hazard, exposure and risk information is presented.

Policies for solving the lack of information problem should encourage the development of computer models that can predict the effects chemicals can have on human health or the environment if they are exposed. Such models could reduce the need for animal testing, and generate more data faster than is currently possible.

Each year, more and more chemical companies are producing reports that document emissions from all of their plants worldwide. This information is being posted on Web sites which does allow for some rough comparisons to be made. However, agreed guidelines are necessary so that all companies apply the same standards, making meaningful comparisons and effective management decisions possible.

REFERENCES

Caswell, B. (2000) personal correspondance with Claudia Fenerol, OECD, 7 September.

CEFIC (European Chemical Industry Council) (1998), *Facts and Figures* 1998, CEFIC, *www.cefic.be*.

CIA (Chemical Industries Association) (1999), *Main World Chemical Markets by Geographic Area*, 1996-2010, UK, CIA, London.

CIA (2000), "Responsible Care: UK Indicators of Performance 1990-99", July, *www.cia.org.uk/industry/iop.htm*.

CMA (Chemical Manufacturers Association) (1998), 1998 *Statistical Handbook*, CMA, Washington DC.

CMA (1999), 1999 *Statistical Handbook*, CMA, Washington DC.

IEA (International Energy Agency) (2000), "Energy Balances of OECD Countries and Energy Balances of non-OECD Countries", 1971-1998, on-live services and IEA Secretariat Estimates, IEA/OECD, Paris.

IEA (1999), CO_2 *Emissions from Fuel Combustion*, IEA/OECD, Paris.

Ireland Environmental Protection Agency (1995), *National Waste Database: Report* 1995, Environmental Protection Agency, Dublin.

JRCC (Japan Responsible Care Council) (1999), *Responsible Care*: 1998 *Annual Report*, JRCC, Tokyo.

OECD (1997), *Evaluating Economic Instruments for Environmental Policy*, OECD, Paris.

Sigman, R. (2001), "OECD Environmental Outlook for the Chemicals Industry", background document for the OECD *Environmental Outlook*, OECD, Paris.

Statistics Finland (1992), "Teolissen Toiminnen jätteet 1992", Statistics Finland, Helsinki, *www.stat.fi/tk/yr/ye142.html*

Swift, K. (1999), "Where is the Chemical Industry Going?", *Business Economics*, October; XXXIV, Number 4: 32-41.

UNEP (United Nations Environmental Programme) (1999), *Production and Consumption of Ozone Depleting Substances*, 1986-1998, Ozone Secretariat, UNEP, Nairobi.

US EPA (United States Environmental Protection Agency) (2000), "TRI 1998 Releases", July, *www.epa.gov/tri/tri98/index.htm*.

WEC (World Energy Council) (1995), *Efficient Use of Energy Utilizing High Technology; An Assessment of Energy Use in Industry and Buildings*, WEC, London.

20
Waste

KEY SIGNALS

- The annual generation of municipal waste in OECD countries increased by around 40% between 1980 and 1997, amounting to about 540 million tonnes in 1997. With continued economic growth, municipal waste generation is expected to keep growing, reaching an estimated 770 million tonnes annually by 2020, which would represent a further increase of some 43%.
- In 1997, approximately 64% of municipal waste was destined for landfill, 18% for incineration, and 18% for recycling in OECD countries. Further implementation of waste management policies, however, is expected to help to reduce landfilling and increase recycling in the future. By 2020, about 50% of municipal waste is expected to be landfilled, 17% incinerated, and 33% recycled.
- In the mid-1990s, total waste generated in OECD regions was estimated to be 4 billion tonnes. In the future, both municipal and industrial waste are expected to show high growth rates, while waste generation from the agricultural and mining sectors is expected to grow at a slower rate to 2020.
- Existing waste-related policies – such as regulatory and economic instruments, the dissemination of information, and extended producer responsibility (EPR) – are contributing to more effective management of waste in OECD countries. In order to reduce waste-related environmental problems in the future, not only better implementation of existing policies, but also changes in production and consumption patterns, will be needed.

20.1. Introduction

Along with the increasing affluence associated with economic growth and changes in consumption patterns, per capita waste generation has steadily increased during the last twenty years in OECD countries. While waste recovery (*e.g.* recycling or reuse) has increased considerably, it has not been sufficient to reverse the trend of increasing volumes of waste destined for final disposal. Many countries have experienced adverse environmental and health effects resulting from improper management of hazardous and non-hazardous wastes.

The environmental pressures from waste generation are diverse. First, waste represents a loss of both material and energy resources. Second, the generated waste has to be collected, treated and then disposed of. These processes entail various environmental impacts, including land use, air and water pollution, and greenhouse gas emissions. Reducing such impacts implies not only preventing waste from being generated, but also increasing waste recovery and recycling, and disposing of any remaining waste in an environmentally safe manner.

20.2. Environmental pressures from waste

Waste generation

Municipal waste

Economic growth and changes in consumption patterns in OECD countries have been accompanied by a continuous increase in municipal waste generation.[1] Since 1980, municipal waste generated in OECD countries has

1. The available data on waste generation and management are limited. The only area where data for analysing past trends and developing an outlook for the future are available is municipal waste. Other waste streams, such as industrial waste, construction and demolition waste, and hazardous waste are discussed briefly within the limit of available information.

increased by around 40% in absolute terms, and by around 22% on a per capita basis. In 1997, OECD countries produced 540 million tonnes of such waste annually, corresponding to about 500 kg per person. In recent years, OECD countries have experienced a slight change in municipal waste generation trends. While total amounts have continued to increase, the average annual growth rate slowed down: decreasing from about 3% in 1980 to about 1% in 1990, possibly reflecting the first results of waste minimisation policies put in place in many OECD countries (OECD, 1999*a*).[2] These trends should be interpreted with caution, however, and may also reflect changes in data quality and coverage. Waste data are generally difficult to compare or aggregate as definitions and surveying methods vary considerably across countries and over time.

Figure 20.1. **Municipal waste generation, GDP and population in OECD countries, 1980-2020**

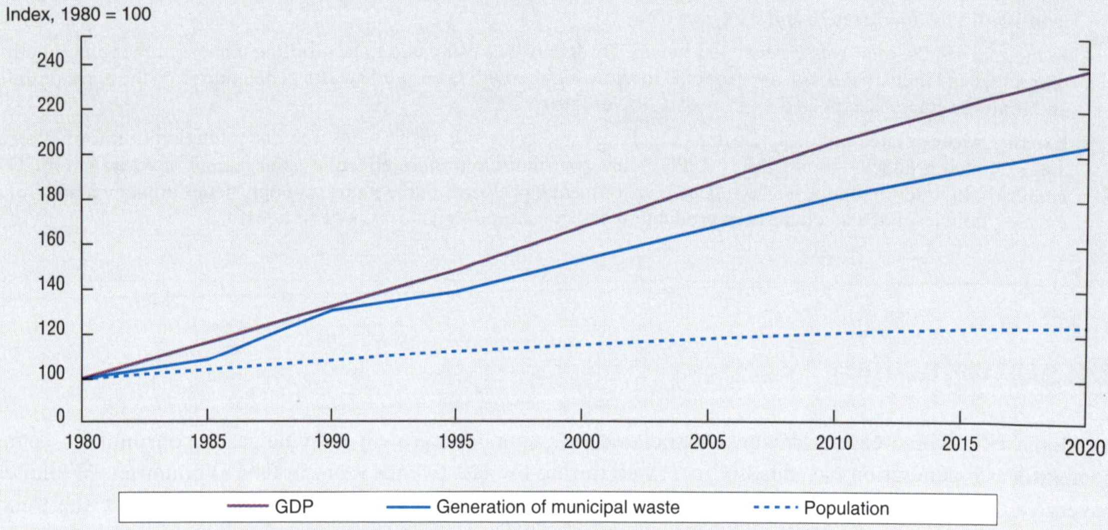

Sources: OECD (1999*a*) and Reference Scenario.

Over the next 20 years, socio-demographic developments such as the trend towards smaller households, and increasing affluence and consumption levels, are expected to lead to further increases in the quantity of waste generated per person in OECD regions (Figure 20.1). For OECD countries, municipal waste generation is projected to grow by about 43%, amounting to 770 million tonnes or 640 kg per capita per year in 2020. This growth rate is slightly lower than the projected GDP growth rate to 2020. In non-OECD countries, municipal waste generation is expected to increase at roughly the same rates as GDP. Thus, by 2020 municipal waste generation in non-OECD regions may more than double compared to 1995 levels (Figure 20.2).

Municipal waste generation in OECD regions is projected to grow by 43% from 1995 to 2020.

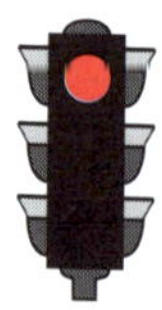

2. Estimate excludes eastern Germany, the Czech Republic, Hungary, Korea, and Poland.

Figure 20.2. **Municipal waste generation, 1995-2020**

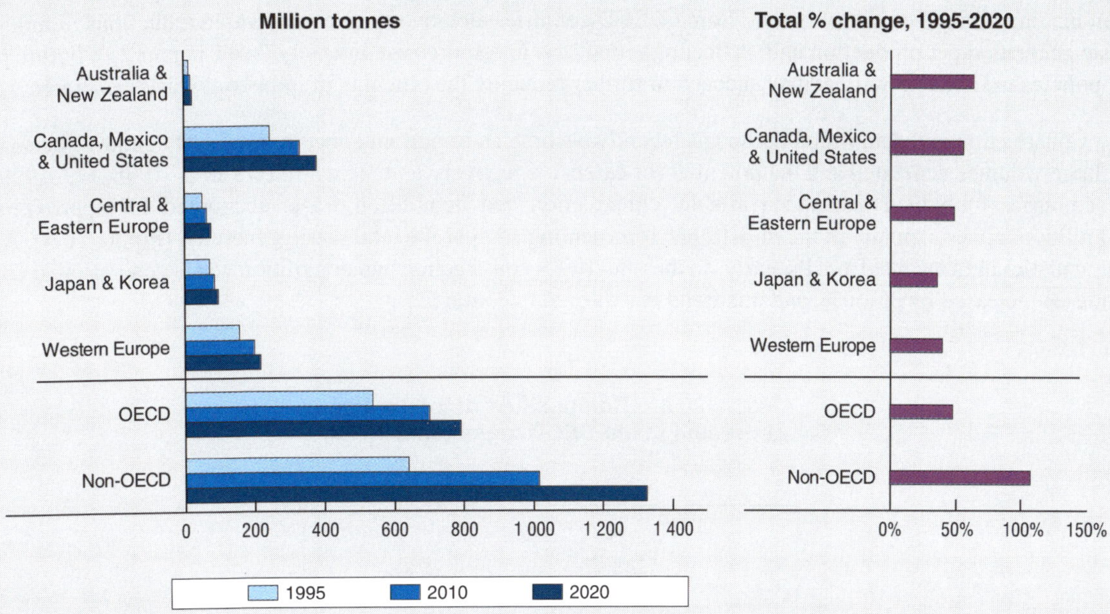

Source: Reference Scenario.
Note: 1995 data for Australia and New Zealand are educated guesses.

Hazardous waste

Total hazardous waste generated in OECD countries in the mid-1990s amounted to approximately 110 million tonnes per year. Of this, North America accounted for about 55 million tonnes and Europe for 43 million tonnes (OECD, 1999*a*). Annual hazardous waste generation is estimated to be about 100 kg per capita in OECD regions.

Because of widely varying and constantly changing definitions of hazardous waste in OECD countries, no clear trends can be provided. However, most OECD countries report increases in hazardous waste generation between 1985 and 1997, with only a very few reporting decreases or stabilisation during this period (OECD, 1998*a* and 1999*a*). The generation of hazardous waste in OECD countries is projected to increase roughly at the same rate as economic growth over the next two decades.

Other wastes

The waste generated by manufacturing industries in OECD regions has been increasing in recent decades. From 1980 to the mid-1990s, it is estimated that this waste increased by about 40%, amounting to around 1 billion tonnes annually in the mid-1990s, approximately 25% of the total waste generated in OECD regions. The generation of industrial waste from manufacturing has increased roughly in proportion to overall industrial production, increasing by about 15% between 1990 and 1997 (OECD, 1991 and 1999*a*), while industrial production levels increased by about 15% and GDP by about 17%. In OECD regions, more than one-third of manufacturing waste is produced in North America and nearly half in OECD Europe. The intensity of industrial waste generation per unit of GDP varies widely between OECD countries, due mainly to differences in waste definitions and in the levels of clean technology in use.

Waste from manufacturing industries is projected to increase in OECD countries over the next two decades as a result of output growth. However, structural changes within industry, as well as the modernisation of plants and equipment and the introduction of cleaner processes, could lead to a gradual reduction in the generation of waste from manufacturing per unit of output. Some OECD countries already report considerable reductions in industrial waste generation per production unit, reflecting reductions in resource use intensity (see Chapter 23). Future product policies in OECD countries may succeed in further reducing the amounts of industrial waste generated.

Construction and demolition waste has recently become an important concern in OECD countries because of the large volumes generated and the potential for extensive recovery and recycling (US EPA, 1998; EC, 1999). In the countries for which data are available, construction and demolition waste accounted for approximately 550 million tonnes annually in the mid-1990s, representing 14% of the total waste generated (Figure 20.3). Available statistics indicate that from the early- to the mid-1990s construction and demolition waste generation in OECD countries increased by 35-40%, and this trend is expected to continue.

Figure 20.3. **Composition of total waste generated in the OECD region, mid-1990s**

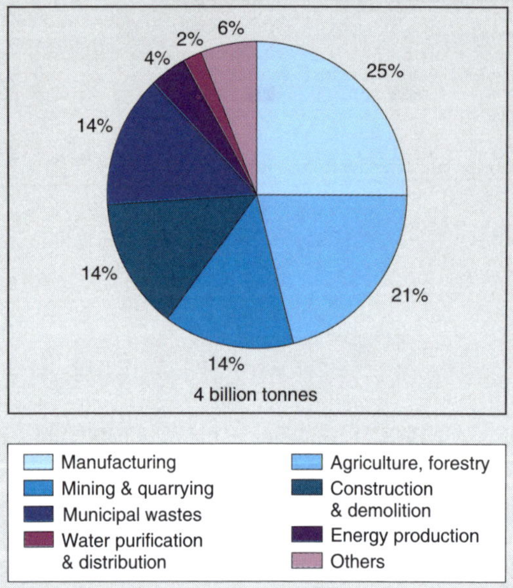

Source: OECD (1999*a*).

In the mid-1990s, about 800 million tonnes of waste (21% of total waste) were generated by the agricultural and forestry sectors, and about 550 million tonnes were generated by the mining and quarrying sector (14% of total waste). Although these two categories of waste represent 35% of total waste generated, such waste is usually managed *in situ* with, for example, agricultural waste often used for compost on farm.

In total, the waste generated in OECD countries in the mid-1990s was estimated to be about 4 billion tonnes (Figure 20.3). With the structure of OECD economies expected to shift over the next couple of decades towards services and high-technology manufacturing, and with a declining share of output from the primary sectors, the composition of total waste is expected to change significantly. The generation of municipal waste and of waste from manufacturing industries and construction and demolition, is expected to increase faster than the average growth rate of total waste, while waste from the primary sectors, such as agriculture and forestry and mining and quarrying, is expected to grow at a slower rate.

Waste prevention and management

Since the late 1980s, OECD countries have been paying increasing attention to waste problems and are shifting strategies from simple collection and disposal to a "waste hierarchy" approach, with the main emphasis on preventing waste generation, followed by reuse and recycling of waste, and then by energy recovery. More attention is also being paid to life-cycle management of products, whereby the materials used are chosen in part for their relative environmental effects and the available disposal options. Extended producer responsibility is gaining ground in a number of OECD countries – under which producers accept some responsibility for the treatment or disposal of their products after consumers have finished with them – and has helped to encourage the development of products and material choices based on lifecycle approaches.

A 1996 survey in OECD countries on the status of waste minimisation[3] showed that although all responding countries prioritised waste prevention over any other waste minimisation method, and many countries had introduced targets for absolute reduction of municipal waste generation, only a very few reported an actual reduction in this waste. In general terms, most performance improvements in waste management have been due to efficiency gains (*e.g.* less waste per unit of GDP) and to increased recycling and recovery. Systematic frameworks for the structured assessment of national of waste prevention programmes are lacking in most countries. Initial work to address this has been undertaken by OECD (OECD, 2001*a*).

Box 20.1. Environmental benefits of recycling

The recycling of many materials has significant environmental benefits. For example, every tonne of iron or aluminium recycled not only replaces a tonne that would have been mined, but also avoids several tonnes of "hidden" material flows associated with the extraction and processing of these metals (Adriaanse *et al.*, 1997). In addition, recycling requires only a fraction of the energy needed to produce these metals from primary ore. The following energy savings have been reported for products using recycled rather than virgin materials: aluminium 95%, copper 85%, lead 65%, zinc 60%, paper 64%, and plastics 80% (BIR, 1998).

Municipal waste

In the mid-1990s, 95% of the population in OECD countries had access to organised municipal waste management services. Approximately 64% of municipal waste was destined for landfill, 18% for incineration and 18% for recycling, including composting (OECD, 1999*a*). Although landfill is still the most widely used method, recycling has increased considerably in most OECD countries. Recycling programmes have included separate collection and voluntary agreements. Recycling rates for paper/cardboard and glass have increased in most countries (Figure 20.4). Some countries have introduced the concept of extended producer responsibility with regard to waste collection and/or recycling of items such as used packaging, batteries and tyres. Composting biodegradable waste has become a major option for reducing the total volume of waste sent to disposal. Incineration of waste is also increasingly used, and energy recovery is gradually becoming an integral part of incineration. In the mid-1990s, over 50% of municipal waste incinerators in OECD regions were equipped with energy recovery systems. Standards for landfilling and incineration have been strengthened in a number of OECD countries.

3. According to the working definition of the OECD, waste minimisation encompasses the following three elements: 1) preventing and/or reducing the generation of waste at the source; 2) improving the quality of the waste generated, such as through reducing the presence of hazardous materials; and 3) encouraging re-use, recycling and recovery (OECD, 1996).

Over the next 20 years, the municipal waste management situation is likely to change considerably. The Reference Scenario projects a significant increase in the share of waste that is diverted to recycling and a decrease in that which is landfilled. In 2020, about 50% of municipal waste is likely to be landfilled, 33% to be recycled and 17% incinerated in OECD regions – compared to 64% landfilled, 18% recycled and 18% incinerated in the mid-1990s. Non-OECD regions are also projected to show significant changes in waste treatment methods, with landfill expected to decrease from about 80% in 1995 to about 70% in 2020, and recycling to increase from about 10% in 1995 to about 20% in 2020 (Figure 20.5).

In OECD countries, a smaller share of municipal waste is expected to be landfilled in 2020, with a greater share diverted to recycling and to incineration.

Figure 20.4. **Recycling rates in selected OECD countries, 1980-1997**

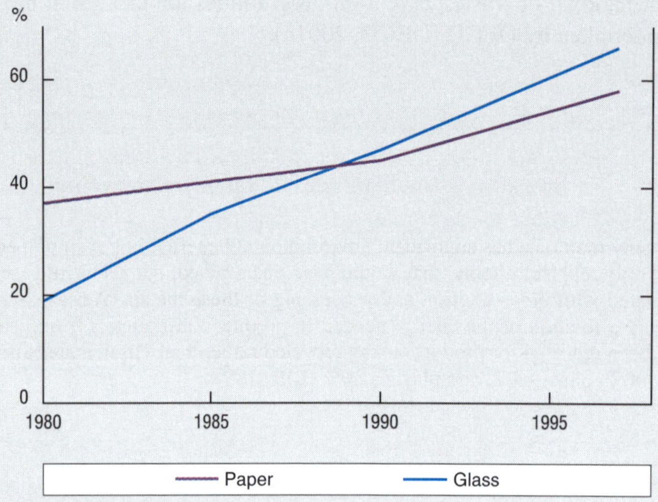

Note: Based on data from Austria, Finland, Germany, Japan, the Netherlands, Spain, Switzerland and the United Kingdom.
Source: OECD (1999*a*).

Hazardous waste

Management of hazardous waste has improved, although the level of progress differs widely among countries. Measures taken include better definition of hazardous waste, the introduction of reporting requirements for waste generation, systems for tracking waste transfers, permits for waste transport and disposal, and the construction of specialised incinerators. Major steps have also been taken to ensure proper control of transboundary movements of hazardous waste. With the stricter regulations that have been introduced over the past two decades on the disposal of hazardous waste, it seems likely that better recovery of the materials and more use of high-temperature incineration will take place in the future.

Hazardous waste generation is projected to increase in OECD countries, but management of this kind of waste is expected to continue to improve.

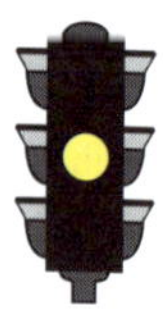

Hazardous waste contained in municipal waste has become a concern in most OECD countries, with some countries introducing separate collection and management systems for this category of waste.

Figure 20.5. **Municipal waste management, 1995-2020**

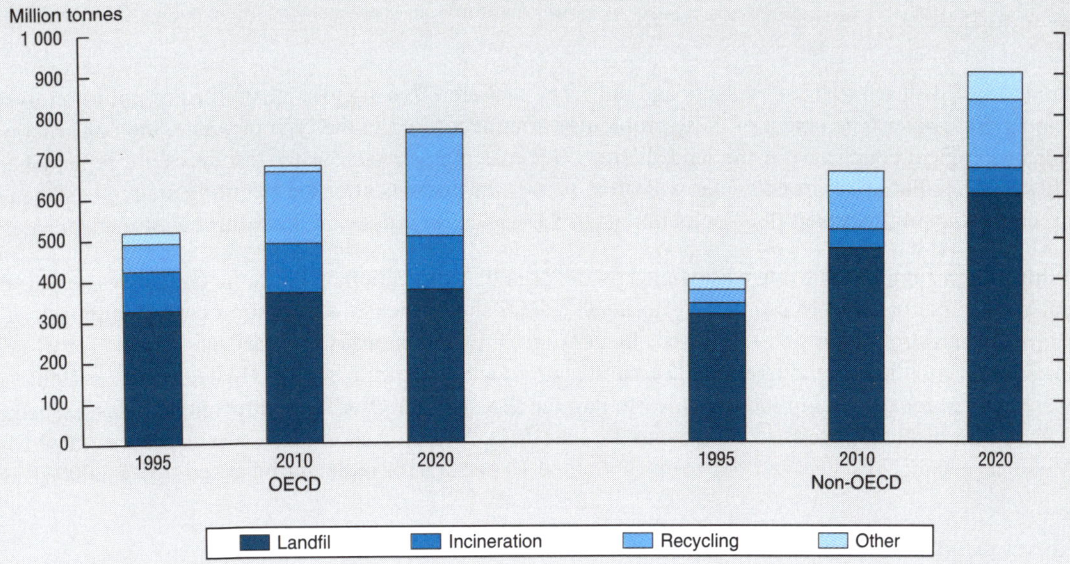

Source: Reference Scenario.

Other waste

Experience from OECD Environmental Performance Reviews shows that practices for managing waste from manufacturing industries vary widely among countries. In some OECD countries, most of the waste generated from manufacturing industries is recycled or recovered, with the remaining waste being incinerated or landfilled. Other OECD countries landfill a major part of their industrial waste. In the EU in the mid-1990s, 30% of construction and demolition waste was landfilled, 25% recycled and 45% treated otherwise (EC, 1999). Other OECD countries were recycling as much as 90% of construction and demolition waste (US EPA, 1998).

20.3. Environmental impacts of waste

Air pollution and climate change

The main air pollutants released in waste processing and incineration are acidic gases, polycyclic aromatic hydrocarbons, dioxins and furans, dust and heavy metals. In the EU, emissions from incinerators were reduced after 1990 due to the closing of many small incinerators, the introduction of cleaning systems, and higher temperature incineration (which reduces the release of toxins, such as dioxins and furans) (EEA, 1999).

Waste-derived greenhouse gas (GHG) emissions comprised 2% of total GHG emissions in OECD regions in 1998. Waste generation and treatment accounted for 34% of methane emissions (mainly from landfills) in 1998. The Reference Scenario projects that waste derived methane emissions in OECD regions will increase by 20% from 1995 to 2020. In non-OECD countries, where waste generation is expected to double with landfilling the main disposal method, waste-derived methane emissions are projected to increase by about 140% over the same period. Waste-derived methane emissions can be mitigated either by avoiding landfilling of organic matter or by collecting and utilising the gas at the landfill sites. Some OECD countries have already introduced, or plan to issue, general bans on landfilling organic waste. Efforts have also been made to establish gas collection facilities at existing and new landfill sites for biodegradable waste.

Water pollution

Some landfills leach toxic substances and nutrients, and are often significant sources of pollution to surface water and groundwater. The extent of these problems varies according to the type of waste, the construction, and the hydro-geological conditions of the landfill sites. Because many groundwater resources are replenished only very slowly, the pollution of groundwater will often persist for decades after the polluting source has been eliminated. Sorting and pre-treatment (*e.g.* incineration) of the waste can reduce the harmfulness of the leachate.

While ocean dumping of waste was banned by the London Convention of 1972, the Global Waste Survey carried out by the International Maritime Organisation (IMO) shows clearly that while ocean dumping by OECD countries is decreasing, it has not yet ceased. In 1991, six OECD countries provided 15 permits for disposal of industrial waste into the sea, and five OECD countries provided 16 permits in 1992 (IMO, 1996). In 1996, no permits were granted for disposal of industrial waste into the sea, but four OECD countries nonetheless issued a total of 35 permits for disposal of sewage sludge into the sea (IMO, 2000*a*). And in 1997, two countries granted 14 permits for sewage sludge disposal and one country granted 16 permits for incineration at sea (IMO, 2000*b*).

Soil contamination and land use

One of the major environmental and health problems related to waste is soil contamination, which often leads to the pollution of groundwater and surface water. In OECD countries, 25-50% of contaminated areas are the result of waste management activities.[4] In 13 OECD countries, 475 000 potentially contaminated land areas have been identified, and the costs of remediating these contaminated land sites have been estimated at about US$330 billion (Harjula *et al.*, 2001).

Soil contamination at existing waste dumping sites is expected to continue, affecting land use and groundwater quality.

Land contamination is often not restricted to the land area concerned, but also affects the future land use of the surrounding areas, the quality of surrounding water bodies, and public health. Furthermore, the remediation of contaminated areas often takes decades and entails significant resources. In the coming years, most OECD countries will need to establish clear liability frameworks and financing mechanisms to address land contamination. Many governments today are implementing integrated policies for contamination management based on precaution (Prokop *et al.*, 2000).

20.4. Policy options

With growing attention being given to waste-related environmental problems, the OECD is encouraging a shift in policy towards the minimisation of waste generation. This includes improving the ability of governments to prevent or minimise pollution and hazards by taking into account the "cradle-to-grave" lifecycle management of activities and substances. Reflecting this, the OECD has adopted several Council Recommendations for economically efficient and environmentally sound waste management, such as the Council Recommendation on the "Comprehensive Waste Management Policy" (1976) and the Council Recommendation on "Integrated Pollution Prevention and Control" (1991).

In order to reduce landfilling, governments in OECD countries will need to intensify efforts to reduce municipal waste generation and increase the recovery of materials. Waste minimisation activities and the implementation

4. Based on OECD Environmental Performance Reviews.

of product take-back programmes will divert increasing levels of waste material from final disposal. In addition, more OECD countries are likely to introduce total or partial prohibitions against landfilling mixed municipal waste and certain specified recoverable waste streams, such as tyres, paper, electronic scrap, etc. The goal is that only inert, non-hazardous waste should be landfilled in the future, justified mainly by requirements to decrease the emission of greenhouse gases from landfills and the pollution of groundwater.

Technological developments and diffusion

Since technological changes influence production and consumption patterns, they also affect the volume and composition of waste generated. The development of clean technologies can contribute to eco-efficient production patterns that not only generate less waste, but also reduce the toxicity of waste. Technological change toward smaller products with improved functions, which is currently typical for electronic equipment, will contribute to generating less waste per unit of output.[5] However, technological changes that shorten product life-cycles, resulting in more frequent product replacement, can result in increased waste generation.

The development of environmentally friendly products and service technologies can contribute to achieving environmentally sustainable waste management. For example, biotechnology can be used to reduce pollutants from waste processing. Micro-organisms can aid in soil remediation, and the use of biodegradable materials can help to reduce waste. In the area of material recovery, sorting technologies make economically profitable reprocessing of discarded products possible by substituting manual sorting which requires higher labour costs. Recently, considerable quantities of waste have been incinerated in cement kilns, steel ovens and industrial boilers with the development of new combustion technologies (EEA, 1999). Although the environmental impacts of such incineration need to be examined more closely, these technologies suggest possibilities for more sustainable waste management.

Regulatory instruments

The installation and operation of waste disposal facilities are generally controlled by legal standards and requirements to minimise the emission of pollutants. The regulations that are applied to landfill sites and incineration plants have been strengthened in a number of OECD countries. Emission standards and operating criteria have been implemented for incinerators of municipal and hazardous waste. Measures to prevent the generation of packaging waste, to limit the heavy metal contents in packaging and batteries, and to safely collect and dispose of waste, have been strengthened in recent decades. Nevertheless, stricter monitoring of waste treatment installations and methods is necessary to ensure that they comply with regulations.

In addition, serious efforts have been made to eliminate, or minimise, the amount of waste that goes to landfill. For example, the European Community has adopted "a directive on the landfill of waste", which introduces a schedule to decrease landfilling of biodegradable material by 65% from the 1995 levels within 15 years (EC, 1999). In some OECD countries, the sorting and/or pre-treatment of waste destined for landfill has gradually become a part of waste treatment processes, reducing the quantities of waste – or eliminating hazardous waste – that is destined for landfill.

Economic instruments

Economic instruments can play an important role in achieving the different objectives of a waste management strategy, ranging from encouraging prevention efforts (*e.g.* waste collection charges) to discouraging the least desirable disposal practices (*e.g.* landfill taxes). Furthermore, economic instruments can be used to change general consumption patterns which affect waste production through encouraging consumer preferences away from products that are neither reusable nor recyclable (see Chapter 5).

5. However, the production of advanced materials generally requires more energy and other inputs, as well as creating more waste in their manufacturing. Furthermore, many advanced materials, such as composites, are generally non-recyclable.

Box 20.2. Effects of using economic instruments for waste management

Charges or taxes on municipal waste have been highly effective in reducing waste generation or landfilling in many countries. In Denmark, a tax on non-hazardous waste has doubled the cost of waste dumping and increased the cost of incineration by 70%. Between 1987 and 1993, the landfilling of household waste fell by 16%, construction waste by 64% and "miscellaneous" waste by 22%. Recycling also increased considerably: 77% for paper and card-board, 50% for glass (OECD, 1997). In the US, about 3 400 local communities in 37 states apply taxes on household waste, which are calculated according to the volumes discharged. The result has been a significant reduction in the volume of waste to be disposed of and a significant increase in recycling.

Almost all OECD countries use economic instruments for waste management. User charges for municipal waste collection and treatment are the most widely applied instruments. Many OECD countries use deposit-refund systems to promote the collection, reuse and recycling of beverage containers, batteries, cars, tyres, and home appliances. Product charges on packaging, batteries, tyres, and home appliances are also used in many OECD countries, mainly to promote the collection and recovery of post-consumer waste.

Voluntary agreements

Voluntary agreements have gained increasing attention as an instrument for waste management. Most of the EU countries include voluntary agreements in their waste management efforts for reducing waste generation and increasing recycling. They can be used as safeguards against potential drawbacks of current regulations, or for testing new policy areas. Existing evidence shows that adding voluntary agreements to a policy mix of traditional command-and-control instruments can improve the flexibility and cost-effectiveness of the policies, as well as reduce administrative costs (OECD, 1999*c*).

Extended producer responsibility (EPR)

Faced with growing waste generation and a shortage of suitable sites for waste disposal, many OECD governments have reviewed available policy options and determined that the responsibility for the post-consumer phase of certain goods should rest with those who produce and distribute these goods. Extended producer responsibility (EPR) is an emerging policy approach under which producers accept significant responsibilities – financial and/or physical – for the treatment or disposal of post-consumer products. EPR programmes change the traditional balance of responsibilities among the manufacturers and distributors of consumer goods, consumers and governments, particularly at the post-consumer stage (OECD, 2001*b*). Through such programmes, producers are encouraged to re-evaluate decisions concerning materials (and chemicals) selection, production processes, product design, packaging, and marketing strategies in order to reduce the costs of takeback requirements.

While the idea of EPR began with a focus on packaging waste, today there is a wide range of EPR programmes in operation for a variety of products. The current trend shows an expansion of EPR policy to more products, product groups and waste streams. EPR has an important role to play in increasing resource efficiency by harnessing materials that would have gone to landfill, and at the same time influencing product designers to reduce raw material inputs and to select materials that are easily reused and recycled.

Information and other instruments

In waste-related policies, the active participation of business, environmental organisations and the general public – in particular the consumer – is very important. Waste prevention can be achieved through encouraging consumers to buy products that pollute less, come from recovered materials, or can be reused or recycled. Promoting information provision and consumer education in this area can contribute to progressive changes in consumption patterns. Eco-audit schemes for economic activities can be used to increase awareness of the amount of waste generated and the costs of its disposal, thereby constituting an incentive to develop waste prevention strategies. Ecolabels and environmental procurement can also be used for this purpose.

Waste-related information based on reliable data is important for the formulation of realistic objectives and for assessing performance. For these purposes, the regular collection of robust data on waste generation and disposal by all sectors and households is needed, and this will require a better harmonisation of coverage and definitions of the different waste streams.

REFERENCES

Adriaanse, A., S. Bringezu, A. Hammond, Y. Moriguchi, E. Rodenburg, D. Rogich, and H. Schütz (1997), *Resource Flows: The Material Basis of Industrial Economics*, WRI, Washington DC.

BIR (Bureau of International Recycling) (1998), personal correspondance with Henrik Harjula, OECD, Paris.

EC (European Commission) (1999), "Construction and Demolition Waste Management Practices and Their Economic Impacts: A Report to DG XI", EC, Brussels.

EEA (European Environment Agency) (1999), *Environment in the European Union at the Turn of the Century*, EEA, Copenhagen.

Harjula, H., J. Stutz, and S. de Tilly (2001), "Waste Outlook", background document for OECD *Environmental Outlook*, OECD, Paris.

IMO (International Maritime Organisation) (1996), "Final Report on Permits Issued in 1991 and 1992" LC.2/Circ.366 and 367, IMO, London.

IMO (2000*a*), "Revised Final Report on Permits Issued in 1996", LC.2/Circ.404, IMO, London.

IMO (2000*b*), "Final Report on Permits Issued in 1997", LC.2/Circ.405, IMO, London.

OECD (1991), *The State of the Environment*, OECD, Paris.

OECD (1996), "Outcome of Session I: Definitions, Terms and Concepts Used in Waste Minimisation" in "Outcome of the International Workshop; Building the Basis for a Common Understanding on Waste Minimisation, 16-18 Oct., Berlin", OECD, Paris.

OECD (1997), *Evaluating Economic Instruments for Environmental Policy*, OECD, Paris.

OECD (1998*a*), *Towards Sustainable Development – Environmental Indicators*, OECD, Paris.

OECD (1998*b*), *Eco-Efficiency*, OECD, Paris.

OECD (1999*a*), OECD *Environmental Data – Compendium* 1999, OECD, Paris.

OECD (1999*b*), "OECD Joint Workshop on Extended Producer Responsibility and Waste Minimisation Policy in Support of Environmental Sustainability, Waste Minimisation through Prevention", ENV/EPOC/PPC(99)11/Final/Part 2, OECD, Paris.

OECD (1999*c*), *Voluntary Approaches for Environmental Policy: An Assessment*, OECD, Paris.

OECD (2001*a*), "*Strategic Waste Prevention: Core Messages from the* OECD *Reference Manual*", OECD, Paris.

OECD (2001*b*), *Extended Producer Responsibility: A Guidance Manual for Governments*, OECD, Paris.

Prokop G., I. Edelgaard and M. Schamann (2000), "Management of Contaminated Sites in Western Europe", topic report No. 13/1ppp, European Environment Agency, Copenhagen.

US EPA (United States Environmental Protection Agency) (1998), *Characterisation of Building-Related Construction and Demolition Debris in the United States*, June, US EPA, Washington DC.

Section VI

SELECTED CROSS-CUTTING ISSUES

This section raises some important cross-cutting issues related to changes in the state of the environment and to the use of natural resources. The chapters on human health and the environment and on the social and environmental interface examine the effects of environmental change on human health, and the distribution of pollution and access to natural resources across different regions and income groups in OECD countries. The chapter on the social and environmental interface also examines issues relating to environmental democracy (the availability of and access to environmental information, public participation in environmental decision-making, and environmental education), as well as the potential distributive effects of some environmental policies. The chapter on resource use efficiency presents a broad perspective of natural resource use across OECD economies, drawing together some of the overall trends and projections discussed in the individual sectoral and environmental issues chapters of the earlier sections of the *Outlook*.

21

Human Health and the Environment

KEY SIGNALS

- Environmental degradation can have a significant impact on human health. Estimates of the share of environment-related human health loss are as high as 5% for high-income OECD countries, 8% for middle-income OECD countries and 13% for non-OECD countries.
- Air pollution and exposure to hazardous chemicals are important causes of the environment-related burden of disease in OECD countries. The transport and energy sectors are major contributors to air pollution, while important sources of chemical pollution are agriculture, industry, and waste disposal and incineration.
- Opportunities for reducing environment-related health risks are considerable. The benefits of many environmental policies in terms of reduced health care costs and increased productivity significantly exceed the costs of implementing these policies.

21.1. Introduction

Concern for health has traditionally underlain much of the political priority given to environmental issues in OECD countries. The impact of environmental risk factors on health are extremely varied and complex in both severity and clinical significance. For example, the effects of environmental degradation on human health can range from death caused by cancer due to air pollution to psychological problems resulting from noise. This chapter attempts to describe the major impacts on human health of environmental degradation and to estimate the associated amount of health loss. A better understanding of the economic costs of environment-related health loss can help to inform environmental policy design.

21.2. Impacts of environmental degradation on human health

Many factors influence the health of a population, including diet, sanitation, socio-economic status, literacy, and lifestyle. These factors have changed significantly during the economic transitions that have shaped present society and resulted in a considerable increase of life expectancy in OECD regions (Ruwaard and Kramers, 1998). Recent studies show that the major determinants affecting life expectancy in OECD regions from 1970 to 1992 were better working conditions, and increased GDP and health expenditure per capita. However, they also indicate that during the same period the negative impacts of air pollution on human health increased in OECD countries (Or, 2000).

Environmental damage is responsible for 2-6% of the total burden of disease in OECD countries.

In order to provide a complete picture of a population's health status, the various aspects which affect it can be combined in a measurement of the "burden of disease", as expressed for example in "disability adjusted life years" (DALYs). They give an indication of how the duration of disease combined with the impact of disease can alter the ability of people to live normal lives as compared to those with no disease (Murray and Lopez, 1996). Figure 21.1 shows estimates for the average total burden of disease, using the DALYs approach, for all OECD countries, for OECD countries grouped by income level, and for non-

Figure 21.1. **Total burden of disease, with estimated environment-related share, mid-1990s**

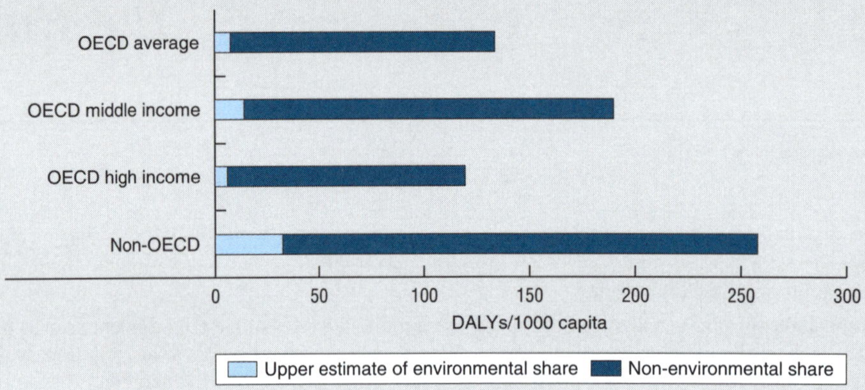

Note: OECD middle income countries: Czech Republic, Hungary, Mexico, Poland and Turkey (World Bank, 1999).
Sources: de Hollander *et al.* (1999), Melse and de Hollander (2001), Smith *et al.* (1999), UNEP/RIVM (1999) and WHO (1999).

OECD countries (Melse and de Hollander, 2001). It clearly shows a significant difference between OECD countries and non-OECD countries (and the influence of income levels on the burden of disease within OECD countries) as regards both total burden of disease and the health conditions related to environmental degradation. The environment-related share of the burden of disease is greatly dependent on income, with higher environmental shares generally occurring in lower-income countries. In OECD countries, this share is estimated to be 2-6% of the total burden of disease.[1]

Figure 21.2 indicates the different patterns of disease in OECD and non-OECD countries, as well as the environment-related share of the various health conditions (Melse and de Hollander, 2001). In non-OECD countries, the majority of the burden of disease can be attributed to communicable disorders (e.g. infectious, maternal, prenatal), while in OECD countries health is lost primarily though the non-communicable (chronic, degenerative) diseases. In OECD countries conditions like heart disease and depression make up a major portion of the burden of disease. In non-OECD countries, diseases in children under four years old account for 50% of the total burden of disease, while in OECD countries the percentage for young children is significantly lower (7% of the total burden of disease).

The large environmental share of health problems in non-OECD countries (diarrhea, TBC, etc.) is primarily due to factors related to poverty, such as limited access to proper food, housing, health care and drinking water. Environmental determinants of human health in OECD countries, on the other hand, are related more to the exposure to air pollutants (particularly in urban areas) and chemicals in the environment than to poor living conditions. Although emissions of many air pollutants have declined in OECD countries in recent years, urban air quality problems related to some pollutants are on the increase, with serious repercussions for human health (see Chapter 15).

Sources of human exposure to chemicals are many and varied. Chemicals can reach the environment, for example, through emissions from industries, anti-fouling paints on marine vessels, pesticides in agriculture, waste incineration and leakage from waste disposal sites. While emissions of chemicals from industries and other point

1. Figure 21.1 only shows the upper estimate of the environment-related share of the burden of disease, with the range (due to uncertainties in risk and the exposure to disease categories used as a basis for the calculations) estimated to be 2-5% for high-income OECD countries, 4-8% for middle-income OECD countries, 2-6% averaged for all OECD countries, and 8-13% for non-OECD countries.

Figure 21.2. **Patterns of disease burden with estimated environment-related shares, mid-1990s**

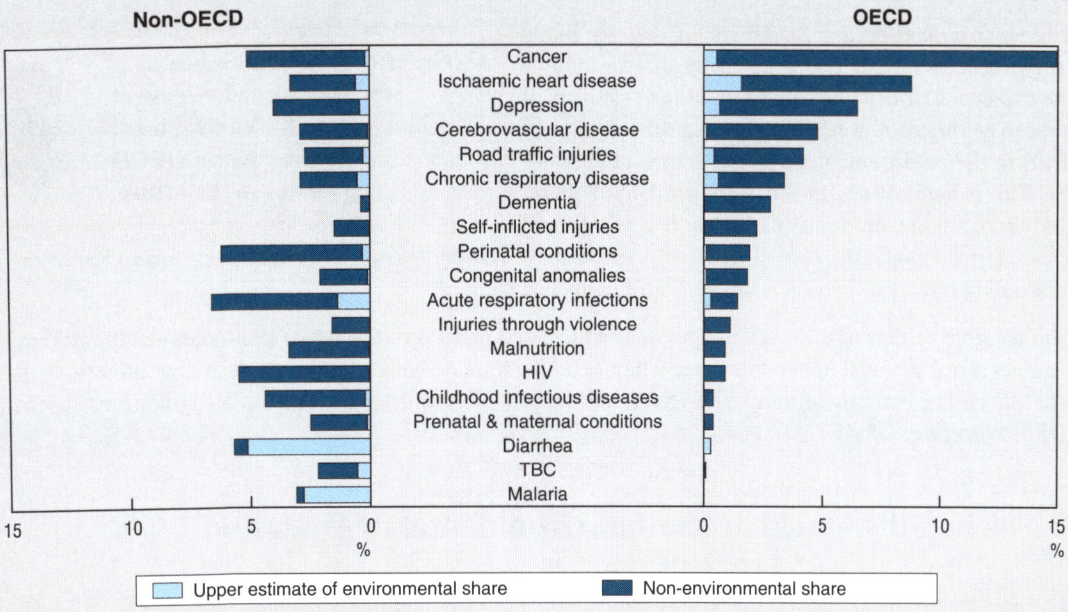

Source: based on Smith *et al.* (1999), UNEP/RIVM (1999), and WHO (1999).

sources of pollution have declined considerably in the past decade in OECD countries, their release to the environment is still significant (see Chapter 19). Uptake through food consumption is another important source of chemical exposure. Intensive agricultural production uses chemicals in pesticides and fertilisers, and in feed additives and medication for livestock (see Chapter 7). Residues remain in fruit, grains, vegetables, meat and dairy products, all of which can reach the consumer. Other sources of chemicals in food include bio-accumulative chemicals in the environment, such as heavy metals and persistent organic pollutants (POPs), which can be found in fish, meat and dairy products.

Other human health risks that have recently received considerable attention include unsafe livestock feeding practices through which toxins reach the food chain unintentionally. Dioxins that have accidentally contaminated poultry feed can move up the food chain to humans, and using feeds that contain diseased animal remains can cause the so-called "mad cow disease" (BSE) in livestock which has been linked to a new form of Creuzfeldt-Jacobs disease in consumers.

The effects on health from exposure to chemicals and air pollutants vary from allergies to cancer. Although the link between exposure and disease is often not clear, a direct causal relationship has been found for some cases. Even at low exposure levels, urban air pollutants can cause asthma, allergies, respiratory diseases and cardiovascular diseases if the exposure is continuous or long term. Heavy metals have been shown to cause neurological disorders and various cancers. POPs can also cause various cancers and are suspected of causing birth defects and reproductive disorders (Colborn *et al.*, 1996).

Environment-related threats to human health that do not result from direct exposure to chemicals or air pollutants are less common in OECD countries, but may still have significant impacts. A well-known example is the effect on the ozone layer of ozone-depleting substances (ODS) used in cooling systems and spray cans. The depletion of the ozone layer has led to increased exposure to UV-radiation and a greater risk of skin cancer. Although the production of ODS in OECD countries has substantially decreased in recent years (see Chapter 19), exposure levels to UV-radiation are still above acceptable levels in many regions of the world.

In addition to physical diseases, environmental contamination can also cause psychological problems. Noise, one of the determinants of the quality of urban life, can have an impact on human health, decreasing the quality of life and potentially contributing to depression. As Figure 21.2 shows, depression is one of the major diseases in OECD countries.

The effects on human health of the widespread release of chemicals to the environment may worsen in OECD countries in the future.

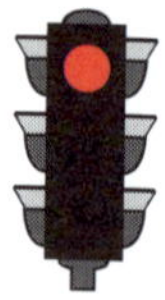

The environment-related health issues that are likely to be prominent in OECD countries in the future include both the expansion of existing threats and the possibility of new ones. The threat of continuing widespread release of chemicals to the environment gives the greatest cause for concern. This is not only a question of the amount of chemicals that end up in the environment, but more a question of their characteristics and effects. Unfortunately, the latter are often unknown, as the recent discovery of the endocrine disrupting effects of certain pesticide ingredients has shown.

The possible effects of climate change are a widely recognised future threat to human health, although their exact impact is not yet well understood (see Chapter 13). Climate change might result in new infectious diseases, as well as changing patterns of known diseases, and loss of life due to extreme weather conditions (McMichael, 1999; Newman *et al.*, 2001).

21.3. The health-related costs of environmental degradation

The impacts on human health from degradation of the environment affect society not only in terms of loss of quality of life, but also in terms of expenditure on health care, loss of productivity and loss of income. Since these impacts are very different, different approaches are required for estimating their magnitude.

Direct expenditure on health care for environment-related diseases can be estimated using the environment-related shares of the burden of disease discussed in Section 21.2 and data on health care expenditures in OECD countries (OECD, 1999). These estimates are fairly rough, but are useful as proxy indicators for current environment-related expenditure on health care and the possible savings that may result from environmental policy interventions. These indicators can therefore be helpful in estimating the economic benefits of environmental policy options.

Table 21.1 shows that direct health care expenditures due to environmental degradation are substantial. These costs may add up to as much as US$130 billion per year for OECD countries, equalling 0.5% of GDP. Both the share of GDP that is spent on health care and the environment-related share of the burden of disease differ from country to country within OECD regions, with the largest differences being found between high-income and middle-income OECD countries. Although the economic benefits resulting from environmental measures seem to be lower in middle-income OECD countries, they can still be significant. For example, the potential economic benefits in terms of health cost saving estimated for measures proposed in Turkey's national environmental action plan to reduce SO_x and particulate emissions are US$125 million annually (OECD, 1999*b*).

Table 21.1. **Total health expenditure in OECD countries and the environment-related share**

Burden of disease	134 DALYs / 1 000 capita
Environment-related fraction (lower and upper estimate)	2.3%-5.8%
GDP	US$22 467 billion (PPP)
Total expenditure on health	9.9% of GDP
Environment-related health costs	US$50-130 billion

Sources: OECD (1999*a*), and Melse and de Hollander (2001).

In order to estimate the indirect costs of environmental degradation in terms of loss of quality of life, a different approach is needed. The monetary values of these indirect costs can be derived through measuring people's willingness-to-pay (WTP) for good health. WTP approaches (Melse and de Hollander, 2001) reflect the value people attach to health and can provide estimates for the fuller economic benefits that could result from implementing certain environmental policies. Willingness-to-pay values are available for a number of high-income OECD countries where environment-related health costs in terms of WTP can be estimated to be approximately 3.2% of GDP and US$750 per capita annually (calculated from Aunan *et al.*, 1998; Melse and de Hollander, 2001).

An example of the magnitude of economic benefits which can be obtained through environmental interventions is presented in a recent study of environment-related health costs resulting from road transport particulate air pollution in three high-income OECD countries (de Hollander *et al.*, 1999). The monetary WTP-valued annual health costs in these three countries for diseases resulting from total particulate matter emissions were estimated to be US$765 per capita or 3.5% of GDP, while costs of diseases from road transport particulate air pollution accounted for US$411 per capita or 1.9% of GDP.[2] The study found that a 5% reduction in road transport particulate emissions could result in a benefit of US$1.5 billion (or US$21 per capita) by averting 1 000 premature deaths and 1 250 cases of bronchitis.

Although data on environment-related health costs and cost-benefit ratios of environment policy interventions will never be as precise as market values for other goods, the available evidence strongly suggests that measures to improve the quality of the environment can prove to be very worthwhile investments. This is true not only because of the intrinsic value of the environment and human health, but also when only the monetary costs and benefits are compared.

21.4. Policy issues

Clearly, the loss of health due to environmental degradation is substantial and calls for interventions. These environmental policy interventions can in turn save money in health care costs. The upper estimate of the current environment-related share of the burden of disease is 6% for OECD countries combined and 13% for non-OECD countries. The cost-benefit ratio for any given policy intervention will depend on the state of the environment and the pattern of disease of the affected population. Since these factors can differ between OECD countries and even within countries, there are only few recommendations that can be generally applied. However, certain priority issues for intervention common to almost all OECD countries can be identified (summarised in Table 21.2).

Table 21.2. **Priority environment-related diseases, issues and sectors in OECD countries**

	High-income OECD countries	Middle-income OECD countries
Diseases	Cardiopulmonary diseases Cancer Depression	Communicable diseases Cardiopulmonary diseases Cancer
Issues	Air pollution Chemicals Noise/liveability	Sanitation/food/housing Air pollution Chemicals
Sectors	Transport Industry/agriculture Housing	Public hygiene Transport/energy Industry/agriculture

Source: Melse and de Hollander (2001).

The most urgent issues to be addressed in OECD countries in relation to limiting health loss from environmental degradation are air pollution and exposure to chemicals. The issue of air quality, and especially urban air quality, emphasises the need for policies resulting in less volume, and cleaner means, of transport (see Chapter 14). The problems related to exposure to chemicals call for policy interventions to limit industry, energy and transport emissions, agricultural chemical use, and to promote food safety (see Chapter 19).

2. Relatively low WTP values and environment-related burdens of disease were used in this study.

A number of the environmental policy instruments discussed in the sectoral and environmental issues chapters of this report – especially those related to decreasing air pollutants and chemicals in the environment – can contribute to reducing loss of health as well as avoiding other effects of environmental degradation (see Chapters 15 and 19). As policies are formulated to address these environmental issues, the health benefits associated with policy interventions should be considered together with the environmental benefits.

REFERENCES

Aunan, K., G. Pátzay, H. Asbjorn Aaheim and H. Martin Seip (1998), "Health and Environmental Benefits from Air Pollution Reductions in Hungary", *Science of the Total Environment* 212: 245-268.

Colburn, T., D. Dumanoski and J.P. Meyers (1996), *Our Stolen Future*, Little, Brown and Company, London.

de Hollander, A.E.M., J. M. Melse, E. Lebret and P.G.N. Kramers (1999), "An Aggregate Public Health Indicator to Represent the Impact of Multiple Environmental Exposures": *Epidemiology*: 606-617.

Melse, J.M. and A.E.M. de Hollander (2001), "Human Health and the Environment", background document for the OECD *Environmental Outlook*, OECD, Paris.

McMichael, A.J. (1999), "From Hazard to Habitat; Rethinking Environment and Health", *Epidemiology*: 460-464.

Murray, C.J.L. and A.D. Lopez (eds) (1996), *The Global Burden of Disease*, Harvard School of Public Health (on behalf of WHO and the World Bank), Harvard University Press, Cambridge.

Newman, J., N. Beg, J. Corfee-Morlot and G. McGlynn (2001), "Energy and Climate Change: Trends, Drivers, Outlook and Policy Options", Background document for the OECD *Environmental Outlook*, OECD, Paris.

OECD (1999*a*), "Health Data 99: A Comparative Analysis of 29 Countries", CD-Rom, OECD, Paris.

OECD (1999*b*), *Environmental Performance Review: Turkey*, OECD, Paris.

Or, Z. (2000), "Determinants of Health Outcomes in Industrialised Countries: A Pooled, Cross-country, Time-series Analysis", OECD *Economic Studies* 30: 53-77.

Ruwaard, D. and P.G.N. Kramers (1998), *Public Health Status and Forecasts Report* 1997, RIVM/Elsevier-de Tijdstroom, Bilthoven/Maarssen.

UNEP/RIVM (United Nations Environment Programme; Dutch National Institute of Public Health and Environment) (1999), edited by Van Vuuren, D.P. and J.A. Bakkes, *Alternative Policy Study: Environment and Energy in Europe and Central Asia* 1990-2010, UNEP/RIVM, Geneva/Bilthoven.

Smith, R.K., C.F. Corvalan and T. Kjellstrom (1999), "How Much Global Ill Health Is Attributable to Environmental Factors?", *Epidemiology*: 573-584.

WHO (World Health Organisation) (1999), *World Health Report: Making a Difference*, WHO, Geneva.

World Bank (1999), *World Development Indicators*, World Bank, Washington DC.

22 The Social and Environmental Interface

KEY SIGNALS

- "Environmental democracy" – including the availability of, and access to, environmental information, opportunities for public participation and partnerships in environmental decision-making, and access to courts for environmental stakeholders – is essential for enhancing public awareness of environmental issues and the development and implementation of widely supported environmental policies.
- A number of international agreements have recently been adopted to strengthen environmental democracy in OECD countries, including the Aarhus Convention (1998) and the OECD Council Recommendations on Pollutant Release and Transfer Registers (1996) and on Environmental Information (1998).
- Participatory decision-making for environmental policy design is on the rise in OECD countries, with more transparent decision-making processes, increasing experience with multi-stakeholder dialogues, and the emerging use of new information and communication technologies for enhanced public participation.
- The impacts of environmental degradation (*e.g.* air pollution) and access to environmental services (*e.g.* city parks) are not evenly distributed across regions and income groups. Poorer households often have the highest levels of exposure to pollutants and the lowest levels of access to publicly provided environmental services.
- Some environmental polices may be socially regressive (*i.e.* lower-income groups may spend a higher proportion of their income on some environmental taxes than higher-income groups). Such effects should be considered in the design of environmental policies, and a number of corrective measures can be used to avoid or alleviate them.

22.1. Introduction

Across OECD countries there are very diverse social conditions, traditions and perceptions, many of which should be considered when assessing environmental pressures and changes, and the policies to address them. As discussed in Section II of this *Outlook*, there are a number of social factors which drive environmental change, including demographic and labour force developments and changing consumption patterns. Conversely, as described in Chapter 21, pollution and changes in environmental quality often have serious impacts on human health and quality of life.

An important aspect of the social and environmental interface which has not been addressed in detail in previous chapters, is the question of "environmental democracy". Access by the public to environmental information and the involvement of stakeholders in the policy-making process is a prerequisite both for the development of widely accepted policies and for the public to take responsibility as consumers and social actors. Another discussion which has emerged from the debate on the social and environmental interface is about "environmental justice", focusing on the spatial distribution of the effects of environmental degradation and access to environmental services. Thus, the population in some local areas, specific social groups or households in OECD countries may be disproportionately affected by environmental degradation, or may not have the same access to environmental services as others. Finally, the effects of environmental policies on income and employment in OECD countries will be considered.

22.2. Environmental democracy

An important precondition for triggering and fostering integrated, pro-active approaches to environmental management and sustainable development is "environmental democracy", understood as encompassing the availability of, and access to, environmental information, opportunities for participation and partnerships of individuals, firms and NGOs in environmental decision-making, and access to courts. The first cycle of the OECD Environmental Performance Reviews has shown that the level of development of environmental democracy varies significantly nationally and locally across OECD countries.

A number of international agreements have been adopted to strengthen environmental democracy in OECD countries. These include: the 1998 Aarhus Convention on Access to Information, Public Participation in Decision-Making, and Access to Justice in Environmental Matters of the United Nations Economic Commission for Europe; the OECD Council Recommendations on Pollutant Release and Transfer Registers (1996) and on Environmental Information (1998); and the 1990 EU Directive on Freedom of Access to Information on the Environment. This section looks briefly at three key themes in OECD country efforts to support environmental democracy: access and assimilation of environmental information; public and stakeholder participation in environmental policy-making; and environmental and sustainable development education.

Availability and access to environmental information

Improved access to quality information can enhance public awareness of environmental issues. Furthermore, it is a prerequisite for the development and implementation of widely agreed and accepted environmental policies. Improved access to information strengthens the potential of information as a vehicle for making the community an active participant in the regulatory process and in reducing environmental pollution and degradation of natural resources (Tietenberg and Wheeler, 1998). This is true not only in the context of public reaction to, or engagement in the development of, environmental policies, including the distributive implications or issues of environmental justice, but also in terms of empowering consumers to make better choices in favour of the environment.

OECD countries have made progress in providing environmental information to the public.

In the context of environmental policy development, significant progress was made in the 1990s across OECD countries in providing environmental information to the public.[1] This progress has been driven by a desire to implement more responsive and cost-effective environmental policies, improve public involvement in policy development, and increase the accountability of all stakeholders. Provision of environmental data, indicators, state of the environment reports and electronic access to environmental information is now often routine. In the area of accident prevention and preparedness, public authorities and enterprises are co-operating to better inform citizens of potential and existing risks and of measures to take in case of an emergency. Enterprises more commonly provide environmental information on a voluntary basis or as part of their obligations (*e.g.* Pollutant Release and Transfer Registers).

In the 1990s, most OECD countries officially recognised the right to environmental information as an enforceable right in national courts and, in some cases, even as a human right that can be enforced internationally. A number of international legal instruments have emerged that establish obligations for different stakeholders to provide environmental information and give valuable reference points. Some practical difficulties remain, however. These include clarifying the legal frameworks governing information provision, raising citizens awareness of their information rights and their ability to exercise them, better implementing these rights, and providing environmental information relevant to citizens in their local context.

Within the marketplace, information is conveyed by labelling and advertising, as well as by retailers, distributors and other "third party" sources (consumer organisations, environmental organisations, and industry associations). The provision of information to consumers serves a number of purposes. It not only protects consumer interests by ensuring accurate information and supporting informed consumer choice by increasing consumer familiarity with different products and their characteristics but it is also a response to the assumption that obstacles to environmentally conscious behaviour (including purchasing) are in part due to a lack of consumer knowledge or inertia through habit.

1. For a fuller discussion of public access to environmental information, see OECD (2000*a*).

Information is one means by which to fill these gaps, to help reconcile personal and social welfare choices, increase the demand for environmentally sound products and stimulate a change in household consumption behaviour. Consumers also seek and receive a vast amount of information outside of the market via both the mass media and public service information. The mass media – television, screen, radio, print, and increasingly the Internet – are major conduits for bringing images and messages into the home. Recent media coverage of the debate over genetically modified organisms (GMO), for instance, has in some countries contributed to heightened public concern and the demand for GMO food labelling. Public service information also influences consumer decision-making and environmental awareness, sometimes directly (*e.g.* information campaigns of the late 1970s and early 1980s to increase energy conservation, or more recent initiatives to change water consumption patterns or to encourage better waste recycling).

Nearly every government, NGO, or private sector strategy for addressing current environmental problems calls for a better-informed public. However, the effectiveness of information in stimulating action is hampered by a number of opposing factors. Among the most important of these are people's reactions to the growing volume and complexity of information on the environment that is available both in the market and in the media, which makes it difficult to interpret environmental information effectively and accurately. Public scepticism *vis-à-vis* the credibility of most information sources is also increasing (see Chapters 5 and 16). Thus, despite reported high levels of public concern about environmental issues in OECD countries today, the number of consumers actively looking for environmental information to guide their day-to-day decision-making seems to be decreasing. Governments, the private sector and NGOs should all find more effective ways of communicating environmental information to citizens.

Information alone, of course, is not sufficient for guaranteeing that the public gives greater priority to environmental protection in their day-to-day decision-making. A number of other factors influence consumer choice, including competing criteria, perceived personal and social costs and benefits (*e.g.* the free-rider dilemma), and the presence or absence of a facilitating environment. A facilitating environment includes relative market prices that internalise environmental externalities, access to alternative products or services in the market, well performing technology and infrastructure (*e.g.* public transport) and a government policy framework that encourages environmentally sustainable behaviour (see Chapters 5 and 16). Turning information into action also requires certain decision-making skills, which underlines the importance of education and learning for achieving environmentally sustainable development. Without this facilitating context, it will be difficult even for well informed and concerned consumers to reduce their environmental impacts.

Public and stakeholder participation

A recurrent theme in the discussion of environmental problems and policy-making is the need to more fully engage the public. At a minimum, engagement means a greater awareness by consumers of the environmental impact of their purchases and behaviour. The primary means for promoting increased awareness is usually through information dissemination. Increasingly, however, engagement is also taken to mean an active participation of the consumer/citizen in public decision-making processes as one of several stakeholders or partners. Here the possibilities for promoting engagement are much more numerous, ranging from traditional public hearings and focus groups to more intensive public "construction" of alternative strategies. Each of these mechanisms is premised on a different set of assumptions and expectations concerning consumer/citizen involvement in the decision-making process and the desired outcomes.

The rise in interest in participatory decision-making, including processes open to the average citizen and not just to major civic representative groups, is part of a changing climate for public policy development (see Chapter 24). First of all, we are seeing an increasing trend across OECD governments for more open and transparent decision-making processes – a trend that is compatible with the movement towards greater decentralisation of government decision-making and the search for partners to work with government to achieve environmental protection. Second, there is growing experience with multi-stakeholder

Many OECD countries are making their environment-related decision-making processes more open and transparent.

dialogues involving the private sector and some representatives of civil society (social and environmental NGOs) who have gained a voice in environmental policy development in OECD countries. Third, international declarations and instruments have been developed in support of participatory approaches (*e.g.* Principle 10 of the Rio Declaration, the 1998 Aarhus Convention). Finally, we are seeing a widening of the stakeholder circle to include the public at large, the "silent" majority whose collective actions or local interests make it a central actor at the negotiating table.

OECD countries have different levels and traditions of civic involvement, and enhancing the participation of civil society is often a challenge (OECD, 2001*a*). Most OECD countries have initiated specific policies to improve consultation, public participation, and citizens' involvement, particularly in the areas of environment, health and social policies. There is a substantial and growing literature and empirical base on multi-stakeholder decision-making for the environmental regulation of industry (*e.g.* zoning laws, waste facility siting, public access to information, workplace environmental health and safety). Moreover, the call for participatory decision-making is increasingly targeting other areas, such as sustainable consumption or sustainable household initiatives (OECD, 2001*b*).

The emergence of new information and communication technologies (ICTs) has opened a number of possibilities for enhanced public participation in environmental policy-making. Some OECD countries have begun to exploit these tools through, for example, e-mail submission of comments or electronic discussion groups on government websites. However, while ICTs have great potential for engaging citizens, they also generate serious policy concerns – for example the creation of a "digital divide" between citizens who have access to such technologies and opportunities and those who do not (OECD, 2001*a*).

Environmental and sustainable development education

In order for information dissemination strategies and participatory development mechanisms to be effective, one of the most important medium- and long-term strategies for environmental protection in OECD countries will be to strengthen environmental and sustainable development education. Education – through both formal schooling and "informal" learning (*e.g.* professional training, adult education) – is a critical complement to social and economic policies. It is crucial for creating public and corporate understanding and for improving dissemination and implementation of new ideas and practices.

Environmental education rests on a foundation of knowledge about social and ecological systems. Education for sustainability includes and goes beyond this, and is concerned with the development of individual and collective competencies for decision-making. Clearly both are essential and embrace respect for individual and social rights and responsibilities, concern for natural and social well-being, critical thinking and skills for civic participation (OECD, 1999).

Despite widespread acknowledgement of the importance of education for attaining environmental sustainability, major efforts are still needed to integrate environmental and sustainability education into standard school curricula. Instead of serving as an integrating discipline through which many different subjects could be better understood, environmental and sustainability education too often remains on the periphery of most formal school programmes and is treated on an *ad hoc* basis in informal settings. One major question is whether environmental education or sustainability issues represent a separate educational agenda from the basic aims of nurturing active learning and responsible citizenship – long-standing educational goals in many countries – or, instead, give these goals new direction and urgency? Another challenging question is how far does education, as currently organised, actually create impediments to sustainable development and environmental protection? Do some education systems, for instance, undermine sustainability either through inhibiting curiosity and critical thinking or through fostering potentially selfish values? The challenge then becomes one of reducing harmful effects as well as optimising beneficial initiatives (OECD, 1999).

Strengthening environmental and sustainability education and learning through formal academic programmes, as well as through professional training and continuing education courses that are easily accessible also to those who are no longer in school, will be key to involving all sectors of the population in environmental decision-making in the future. Workplace and community-based education and learning initiatives are additional areas that offer large potential. It will be increasingly important to tie environmental and sustainability education to practical experience or specific decision-making contexts. Some empirical research has shown that many of today's adults who are active in environmental issues had earlier experiences with both "nature" and some kind of social activism. This argues strongly for simultaneously reinforcing the three pillars of environmental democracy – information, participation and education.

22.3. Social distribution of environmental quality

Neither the impacts of environmental degradation (*e.g.* air pollution) nor access to environmental services (*e.g.* city parks) are uniform across different social groups and households – a situation which has given rise to "environmental justice" discussions. The focus of such discussions is on the socio-economic, cultural and demographic characteristics of those households, income groups or geographical areas which are most exposed to pollution, or which have the least access to environment-related (public) services. The argument is that there is often some "unfairness" in the spatial distribution of environmental pressures and services across income groups and households.[2]

Limited information is available on how environmental quality and access to environmental resources is spread across different communities in OECD countries.

The first cycle of the OECD Environmental Performance Reviews provided examples of environmental problems that were closely related to disadvantaged communities or less developed territories in OECD countries, where implementation of effective and efficient policies was hampered by these disparities.[3] Several recent studies in OECD countries have examined the relationship between the spatial distribution of toxic releases from industrial plants and the socio-economic and demographic characteristics of surrounding neighbourhoods (Johnstone, 2001). Work has also been done on the siting of hazardous waste facilities. In general, studies find that the lowest income groups often have the highest degree of exposure and the highest income groups the lowest degree of exposure. Similarly, poorer households often have the lowest levels of access to environment-related public services, such as public parks and water supply.

These results are hardly surprising since the firms that generate local pollution will have an incentive to locate where the financial costs of environmental damages are lowest. This tends to be in poorer neighbourhoods where property rights (*e.g.* such as effective environmental liability regimes) and income-linked demand for environmental quality are less manifest. Even though rights to the environment are often not well defined, poorer household will still live in more polluted areas and areas that are less well served by environment-related public services because environmental quality affects other markets. For instance, housing close to city parks, and far from polluting plants, tends to be more expensive (Johnstone, 2001). The real estate market in OECD countries therefore reflects an informal "sorting" process where households express their preferences for environmental quality (and other neighbourhood attributes) through what they can afford to buy or rent.

Households can also express their demand for environmental quality through their role as voters and stakeholders in the political process. Implicit in many calls for improved participation is the assumption that increasing the awareness and engagement of the public in decision-making processes for environmental protection will ultimately strengthen that protection and, potentially, also result in behavioural changes by consumers. One essential objective is to ensure good representation of those who do not have easy access to the public debate or are under-represented: minorities, indigenous populations, the unemployed, etc. These vulnerable categories are often those who may suffer most from unsustainable policies while, at the same time, they usually do not have the capacity to lobby for their priorities and concerns.

2. For a discussion of alternative views on what constitutes a "fair" distribution of environmental quality, see Johnstone (2001).
3. Examples discussed in the OECD Seminar on the Social-Environmental Interface (Paris, September 22-24, 1999) were presented by experts from the US, Mexico, Turkey, Australia, Belgium and the UK.

Local environmental policies tend to reflect the preferences of those who get involved because of the competition of political interest groups. The neighbourhoods where demand for environmental quality is greatest also lobby hardest for their interests. There is good evidence to support the claim that waste management firms and industrial facilities pay close attention to the likelihood of facing political opposition when choosing sites (Hamilton, 1993). Finally, there may be differences in the willingness and capacity of different groups to engage in collective action in order to ensure that their demands are met. Thus, even if individuals within groups have the same underlying demand for environmental quality, there may be other factors preventing them from making this demand effective as a group. In terms of public policy-making, this means that:

- First, all households should have access to information which allows them to make an "informed" decision. However, the complexity of environmental problems is such that providing information alone is not always sufficient this means that.
- Second, throughout the decision-making process, all stakeholders should be able to participate in an open and transparent manner. If access is not equal, the political system may generate results that are biased toward those who do have access. Thus open public hearings about waste siting decisions, for example, are important.
- Third, failures in the associated markets (such as labour and housing), which may restrict the mobility of affected households, should be eliminated. For instance, if markets for mortgages are systematically biased against poorer households in a way that does not reflect actual market risks for the lender, then measures may need to be taken to ensure that restrictions on the mobility of these households do not result in ill-health. However, it is risky for governments to assume that mortgage givers are behaving irrationally (*e.g.* overestimating risk for poorer households), and as such interventions in the market should be undertaken with care.[4]
- Fourth, since protecting local environmental quality often necessitates collective action and the capacity to engage in collective action may differ between groups, some households will not be able to ensure that their demand for environmental quality is taken into consideration. In this case, governments can help communities which are less able to engage in collective action – for example, by providing technical assistance with respect to organisational issues. Some OECD governments are trying to overcome this problem by allowing non-governmental organisations and others to launch public interest actions on behalf of diverse groups of victims.

When there are high levels of absolute poverty and an unequal distribution of wealth, it may be deemed socially abhorrent for poorer households to have to "trade off" some minimum level of environmental quality for other goods and services. Such a view implies that governments should mandate minimum levels of access to environment-related public services. In terms of pollution levels, this means that governments should continue to regulate maximum concentration levels consistent with what society views as fundamental to a "reasonable standard of living". Even if environmental quality is considered a "basic need", these levels may be higher than the levels households appear to be willing to tolerate as they trade off environmental quality for other goods and services. Different societies will define such "environmental basic needs" differently. Clearly, factors such as potential damage to children's development are more relevant to defining these conditions than factors such as convenient access to environmental amenities.

22.4. Distributive effects of environmental policies

Another aspect of the social and environmental interface is the distribution of the effects of environmental policies. Who pays the costs or, rather, who pays relatively more? The distribution of environmental protection costs can be observed at two main levels: among firms and among households.

The fact that firms will be unequally hit by pollution abatement costs can be seen as a distributive and/or as a competitiveness issue. Inevitably, polluting firms will have different marginal abatement costs. Firms may also be subject to different requirements according to local conditions. In the case of tradable permits, whether they will

4. Restrictions on the siting of industrial or waste management facilities based explicitly upon the social characteristics of the "receiving" community have also often been advocated. The positive distributional consequences of such measures are likely to decrease in the longer run due to the links between local environmental quality and real estate markets (Johnstone, 2001).

initially be "grandfathered" (free allocation) or auctioned will have marked distributive implications. The distribution of environmental costs between private entities and the general public might also be examined. Efforts to address such effects should primarily be guided by concerns to maximise economic efficiency.

The main distributive issue related to environmental policies, however, concerns the distribution of environmental costs among households. There are two main aspects, one related to income and the other to price effects. First, richer people are generally willing to pay more for environmental quality than lower-income groups. Since environmental quality is by and large a public good, this implies that poorer people should get more environmental protection than they are willing and able to pay for. Second, if the cost of environmental protection were passed on to prices, the impacts would often be bigger for the poorer segments of the population than the richer ones, in particular where prices for necessities like food, water or energy are increased due to environmental protection measures.

The distributive effects of environmental policies need to be considered when choosing the instruments for promoting and implementing these policies, whether they be economic instruments (such as taxes, removal of subsidies and tradable permits), regulatory instruments, or the flanking policies which may need to accompany them. Distributive effects are discussed below, using the case of environmental taxes, and focusing mainly on taxes related to energy. Many of the problems, issues and solutions described here may also be relevant for other policy instruments.

Distributive effects of environmental taxes

The issue of distributive effects of environmental policies is particularly sensitive and subject to debate in the case of environmental taxes. The distributive effects of environmental taxes, especially those related to energy, may be observed in three ways (Smith, 1998). First of all, there is a direct distributive impact related to the structure of household energy expenditure (heating and transport) for different income groups. The larger the proportion of low-income household expenditure devoted to energy, the more regressive[5] the impact of the tax will be. Second, indirect distributive effects will originate from the taxation of production inputs. The more energy intensive the processes, the higher the taxes on the goods produced will be. Finally, an energy tax may affect final consumers, but it may also affect energy producers or production factors (*e.g.* through a fall in wages or lower return on capital). At the same time, part of the tax may be borne by energy consuming countries, and another part by energy exporting countries, according to the elasticities of supply and demand.

In so far as many environmental taxes apply to mass consumption products and services, such as motor vehicles and energy, they may have a potentially higher relative effect on lower-income households. The level of the tax also matters. Relatively low environmental taxes on products such as detergents, fertilisers, batteries and pesticides are likely to have a limited impact, while large-scale and fiscally heavier environmental taxes, such as those on energy, can have more profound implications for lower-income households.

Evidence of the distributive effects of environmental taxes remains scant. There is some evidence that energy taxes tend to be income regressive, but the degree of income regressivity in many cases is weak (Barde, 1997 and OECD, 1997*a*). A study of different energy taxes in eleven EU countries found that the distributional impact of energy/carbon taxes differed by energy use: taxation on transport fuels was weakly progressive, while taxation on domestic energy was weakly regressive (Barker and Köhler, 1998). Overall, an energy/carbon tax was found to be weakly regressive.

Analysis of the distributive effects of environmental policies in OECD countries remains scant.

To avoid regressive distributive effects of environmental policies, in particular environmental taxes, three types of corrective measures can be envisaged: compensation, mitigation, and tax shift. Compensation measures are basically *ex post* and outside the realm of the taxes as such, *i.e.* they do not affect the tax rate or structure. These are corrective measures, such as lump sum compensation calculated on the basis of average tax payments per household. In this case, compensation will have a progressive effect on the assumption that the poorest households

5. With regressive taxes, low-income groups spend a larger proportion of their income on the tax than higher income groups.

on average pay less tax than the richest households. Tax refunds are a typical compensation measure. In several OECD countries, energy taxes are partly repaid to households and/or businesses in the form of subsidies for energy-saving investments/expenditures.

Mitigation is an *ex ante* measure to reduce the rates of specific environmental taxes to alleviate the tax burden on specific segments of the population. This can take the form of reducing tax payments for low-income groups or on specific mass consumption and indispensable commodities such as heating fuels or agricultural inputs. However, the outcome would weaken the desired environmental impact of the tax, and would significantly increase the administrative complexities. Nevertheless, a large number of special tax provisions of this sort are applied in OECD countries. The OECD-EC database on environmentally related taxes indicates hundreds of provisions such as tax exemptions and reduced tax rates.[6]

Tax shifts, *i.e.* the reduction of other taxes, such as labour and income taxes (the case of the "double dividend"), is a widespread form of compensation. It is assumed that the regressive impact of the new environmental tax will be compensated by the reduction of other taxes. However, poorer households are often affected relatively more by environmental taxes while wealthy households will benefit most from lower income taxes. The net distributive implication of this approach is not clear. This form of compensation may even prove to be strongly regressive (Smith, 1998).

22.5. Employment implications of environmental policies

In the early 1970s, with sustained economic growth in OECD countries and full employment, the impact of environmental policies on employment was hardly considered a problem, but the oil crises of 1973-1974 and 1979 brought it centre stage. It was felt at the time that environmental spending was diverting resources from the "productive" sectors, which generated value-added and employment. When the recession started, forcing companies out of business, the blame was sometimes laid on environmental constraints. Between 1990 and 1994, the number of unemployed in OECD countries rose from 25 million to 34 million (8.5% of the workforce). While the unemployment rate has decreased since (it was 6.6% in 1999) it remains high in several countries (8.8% in the EU in 1999). Although the purpose of environmental policies is not to create jobs, the need for a proper integration of environmental and employment policies is widely recognised.

Three main approaches have been used by OECD countries to minimise or compensate for negative employment effects of environmental policies (OECD, 1997*b*). These include using environmental expenditures as a direct tool of macro-economic policy; using environment-based job creation programmes to stimulate employment in specific regions, sectors or occupational groups, and promoting environment-related "active" labour market policies; and combining green tax reform and employment creation.

Some countries have also used environmental expenditures as a demand-inducing, anti-recession, economic policy in the past. However, since the mid-1980s – and for a variety of reasons – OECD governments have become increasingly reluctant to address cyclical unemployment problems through increased spending policies.

Since the early 1980s, many governments have focused instead on job-creation programmes that can be targeted to specific labour markets where high levels of unemployment exist. Many of these projects have included environmental activities, often of a public works nature. Despite their occasional environmental benefits, the primary objective of these programmes has been employment creation. Such programmes have often been criticised because of their temporary nature, their low effectiveness in generating long-term employment, and their risk of competing with other commercial activities. Spending public money on "environmental" job-creation programmes is only one of several possible options.

Rather than emphasising direct short-term employment programmes, several OECD countries have increasingly turned their attention to a more complex set of measures aimed at improving the quality of the labour supply, especially through vocational guidance and education programmes. Because the transition to cleaner industries, products, services, and environmental habitats will ultimately require new skills and/or management techniques, employment and environmental policies may also be integrated at this level.

6. Available on *www.oecd.org/env/policies/taxes/index.htm*.

With regard to combining green tax reform and employment creation, it is often argued that reducing the tax wedge on labour – in particular the employer's social security contributions – in order to lower labour costs would reduce unemployment. Therefore, shifting the tax burden from labour to pollution, *i.e.* compensating additional environmental taxation by a reduction of labour taxation, should provide a "double dividend": reduced pollution and increased employment. The likelihood of such a double dividend is subject to controversy in the economic literature. However, a survey of simulation studies indicates four key conditions for the realisation of a double dividend (OECD, 2000*b*):

- The initial structure of the tax system should be sub-optimal so that a dividend from removing sub-optimal tax provisions can emerge.
- The tax incidence is a crucial issue. If the burden of pollution taxes finally falls upon households through higher prices of the taxed commodities (which is a reasonable assumption), the reduction of the tax wedge on labour will be less effective and the employment effect reduced or cancelled. Since labour is a relatively immobile factor of production, this ultimate tax incidence on labour is likely to occur.
- The degree of substitutability between factors of production is important. If there is a possibility to use more labour instead of energy and capital, increased employment is more likely to occur.
- The mobility of production factors are crucial. In the case of an energy tax, if labour is a better substitute for energy than capital, there will be a shift to more labour-intensive production techniques. If capital is relatively immobile internationally, the tax burden will be shifted to capital. If the capital is internationally mobile, it will move abroad to avoid the tax. In this case a high degree of international co-ordination is required.

The more effective the environmental tax is in changing the behaviour of producers and consumers, the more rapidly the tax base may erode. To maintain the same revenue flow, governments will have to increase other taxes or increase existing environmental taxes with two possible consequences: a further incidence on labour and a possible reduction of pollution beyond the optimal level. In this case, environmental taxes exacerbate tax distortions.

There is still uncertainty about the evidence of a double dividend from environmental taxes. *Ex post* evaluations are needed.

What happens in reality? It has been shown that the results of many models converge to indicate that a carbon-energy tax would yield some double dividend (Majocchi, 1996). The employment effect, however, would be limited (a rise in employment to the order of 2%). Recent simulations show similar results: a real, but small potential for a double dividend (OECD, 2000*b*). The main conclusions from existing studies on the double dividend are summarised in Box 22.1.

Box 22.1. Main conclusions from existing studies on the double dividend

- For most European countries, larger employment effects can be expected if the cuts in social security contributions are targeted on the unskilled labour force.
- Positive effects on GDP can be expected if the revenues are used for a cut in capital taxes (thus favouring investment), and this is gradually implemented.
- Both GDP and employment effects depend on the size of the tax reform. Generally, the results of the simulations show positive effects on GDP and employment when an energy tax is introduced stepwise and the energy price increase does not exceed 4-5% per year. Higher tax rates lead to negative effects on employment and GDP.
- The structure of the labour market also matters. In case of wage rigidities, *i.e.* if the level of wages does not decrease with unemployment, lower social security contributions will reduce real labour costs and increased employment is likely to occur. In the case of a non-competitive market for goods, tax reductions will be captured by increased profits.

Source: OECD (2000*b*).

Most countries which have implemented green tax reforms, or are considering doing so, are in one way or another expecting the double dividend effect. How effective these mechanisms really are, however, still remains to be seen, as no *ex post* evaluations have been carried out so far.

REFERENCES

Barde, J-P. (1997), "Economic Instruments for Environmental Protection: Experience in OECD Countries", in OECD, *Applying Market-Based Instruments to Environmental Policies in China and OECD Countries*, OECD, Paris: 31-58.

Barker, T., and J. Kohler (1998), "Equity and Ecotax Reform in the EU: Achieving a 10 per cent Reduction in CO_2 Emissions Using Excise Duties", in *Fiscal Studies* 19(4), November: 375-402.

Hamilton, J. T. (1993), "Politics and Social Costs: Estimating the Impact of Collective Action on Hazardous Waste Facilities" in *Rand Journal of Economics* 24, 101-125

Johnstone, N. (2001), "The Distribution of Environmental Quality and Resources: Objectives, Evidence and Policy", background paper for the OECD *Environmental Outlook*, OECD, Paris.

Majocchi, A. (1996), "Green Fiscal Reform and Employment: A Survey", *Environmental and Resource Economics*, December: 375-397.

OECD (1997*a*), *Environmental Taxes and Green Tax Reform*, OECD, Paris.

OECD (1997*b*), *Environmental Policies and Employment*, OECD, Paris.

OECD (1999), *Education and Learning for Sustainable Consumption*, OECD, Paris.

OECD (2000*a*), *Proceedings*: OECD *Seminar on Public Access to Environmental Information*, Athens, 5-7 October, OECD, Paris.

OECD (2000*b*), "Greening Tax Mixes in OECD countries: an Assessment", OECD, Paris.

OECD (2001*a*, forthcoming), OECD *Analytical Report on Sustainable Development*, OECD, Paris.

OECD (2001*b*), *Trends and Issues in Participatory Decision Making for Sustainable Consumption*, OECD, Paris.

Smith, S. (1998), "Distributional Incidence of Environmental Taxes on Energy and Carbon: a Review of Policy Issues", *Réforme Fiscale Verte et Instruments Économiques pour une Coopération Internationale: le Contexte Post Kyoto*, Ministère de l'aménagement du territoire et de l'environnement, Paris.

Tietenberg, T. and D. Wheeler (1998), "Empowering the Community: Information Strategies for Pollution Control", Frontiers of Environmental Economics Conference, Airlie House, Virginia, October.

23 Resource Use Efficiency

KEY SIGNALS

- Economic activities are becoming more efficient in their use of natural resources, but the volume effects of total increases in global output on resource use are outweighing these efficiency gains.
- Further improvements in resource efficiency are essential to reduce the environmental pressures of resource use to sustainable levels. A "de-coupling" of environmental degradation and resource use from economic and population growth is needed.
- The means by which resource efficiency can be improved include increased resource saving, resource re-use, resource prolongation, and substitution of resources.
- Policies are needed to provide the appropriate incentives for improving resource efficiency, starting with the removal of adverse incentives to use resources, and policies to ensure that the various environmental and social costs of resource use are reflected in the prices paid by resource exploiters and users.

23.1. Introduction

One of the most widely-discussed environmental problems in the 1960s and 1970s was the perceived threat of increasing resource scarcity, particularly for non-renewable resources such as fossil fuels and metals. More recently, however, the focus has shifted. First, it was found that the available and accessible stocks of a number of key non-renewable resources are actually larger than first believed. In addition, it has been recognised that substitutes or recycling possibilities for many of these now exist, and will be brought into use more and more as the stock of non-renewable resources depletes and the alternatives become more cost-effective. More of a concern for the future will be the better management of renewable resources to ensure that they do not become in effect "non-renewable". The exploitation of capital stocks of renewable resources (*e.g.* wood, water, biodiversity and fish) needs to take place at a rate that does not exceed their regenerative capacity. Indirect pressures on renewable resources, such as pollution and habitat loss, also need to be reduced in order to ensure resource growth and regeneration.

It is clear that the unsustainable management of natural resources will negatively affect both economic activities and the regenerative and assimilative capacity of natural ecosystems. Hence, further improvements in *in situ* resource management and the efficiency with which natural resources are used downstream (*i.e.* resource use efficiency[1]) will be essential to reduce total environmental pressures of resource use. Ensuring that market prices reflect the scarcity of these resources is important for achieving appropriate rates of exploitation. Better management of the associated environmental impacts from the use of natural resources, such as the degradation of ecosystems, pollution emissions, and solid waste generation, will also be needed.

23.2. The effects of resource use on the environment

The environmental consequences of resource use depend on the nature of the resources and the way in which they are used in the economy. In the first instance, they can be distinguished by resource type. Non-renewable

1. Resource use efficiency is usually understood to specifically concern the transformation and the use of resources as "downstream" economic goods and services, rather than management *in situ*.

resources include metallic and non-metallic mineral deposits, as well as fossil fuels. As they are non-living, non-renewable resources have no regenerative capacity. Depending on their nature and use in the economy, however, there can be significant potential for recoverability after initial use. Renewable resources include forests, freshwater, fish stocks, wildlife, etc. Most renewable resources are characterised as living "biological" resources, which reproduce themselves. Thus, as long as the reproductive capacity is maintained properly, consumption today will not affect their availability for potential consumption tomorrow. If consumption today affects the reproductive patterns of the resources, however, it will affect tomorrow's availability. A third class of resource type, non-exhaustible resources, includes resources – such as energy from the sun and tides – where consumption does not increase scarcity.

Box 23.1. **Indicators of resource efficiency**

A variety of research has been conducted on potential resource efficiency indicators, including the "Total Material Requirement" (World Resources Institute), "Factor X" (the Wuppertal Institute), and "Eco Efficiency" (the World Business Council on Sustainable Development).

Aggregated indicators of resource use, however, need to be interpreted with care. Adding the volumes of resources used in a society and comparing them on a country-by-country basis can sometimes be misleading regarding the environmental impacts of resource use, since a tonne of iron ore is not equivalent to a tonne of mercury. Indicators of resource use efficiency should reflect qualitative concerns (both in environmental and economic terms) and indirect (general equilibrium) effects arising from factor and product substitution. Further work on linking resource flows to potential environmental impacts is still needed.

The processing of resources downstream can result in significant pollutant generation. This can arise directly from the dissipation or combustion of the resources themselves. For instance, the combustion of fossil fuels results in a wide variety of local, regional and global air pollutants. However, this pollution can also be caused indirectly as resources are transformed downstream by processing sectors. The generation of biochemical oxygen demand (BOD) from wood pulping is an example (see Chapter 18). Material resource use also generates solid waste. This can either arise at the point of resource exploitation (*e.g.* overburden from mining) and processing, or at the end of the useful life of products which have been manufactured from resources (*e.g.* household solid waste). However, depending on the degree of material deterioration and the cost of collection, sorting and reprocessing, there is potential for significant recovery of such resources (see Chapter 20).

The management of resources is largely determined by their ownership regimes. At one extreme is the open access ownership regime, in which there are no means of excluding others from resource exploitation. A notable example is high-seas fisheries. Because there is no owner protecting his ownership, these types of resources tend to be over-exploited (see Chapter 9). At the other extreme is private ownership, in which rights to the resource are clearly defined. Common property regimes are in an intermediate position, and can take a variety of forms. Access is restricted, but by means of conventions, norms and rules, and not through limits set by the owners *per se*.

The values associated with resources are diverse. Many resources are valued directly as inputs in the production of economic goods and services. However, market prices do not reflect the value of resources properly. Many resources (particularly forests and surface waters) play an important role in supporting ecosystem functions like carbon fixation, water catchments and temperature buffers. Such resources need to be valued not only for their immediate usefulness as inputs, but also for their wider environmental benefits. Thus, unsustainable levels and patterns of resource use can undermine the integrity of broader ecosystems (see Chapters 7 to 11). In addition, resources as components of ecosystems possess an "intrinsic" (or "existence") value which cannot be attributed directly to any of their present or potential future uses as resources for production.

23.3. Resource use efficiency: recent trends and outlook to 2020

> **Despite increases in resource use efficiency, total consumption of resources is continuing to increase in OECD countries.**

Given the environmental impacts of resource exploitation, processing and use, increasing the efficiency with which resources are used in the economy has the potential to generate significant environmental benefits. Figure 23.1 indicates some examples of historical and projected future reductions in resource and energy intensity for OECD economies. These are compared with the net growth in total use of these resources and energy. The Figure reveals that the general decrease in the resource intensity of production that occurred in recent decades is expected to continue to 2020.

In many sectors, the introduction and adoption of advanced technologies and economic restructuring in OECD countries have contributed to significant improvement in resource use efficiency. Despite this, the total consumption of resources is projected to consistently rise to 2020, indicating that in most cases the volume effects of economic growth and increased total production and consumption outweigh the efficiency improvements. For example, while OECD countries reduced the energy intensity of their economies by 31% between 1973 and 1996, they increased total energy consumption by 23% over the same period. This trend is expected to continue, with total energy use in OECD countries projected to grow by a further 35% to 2020 under the Reference Scenario, while the intensity of energy use in the economy declines by almost 20% (see Chapter 12).

Similarly, waste generation, which can be considered to be an indirect indicator of how society uses raw materials, continues to increase, although the average annual growth rate has slowed down: from 3% between 1980 and 1990 to about 1% between 1991 and 1997. The Reference Scenario projects that the generation of municipal waste in OECD regions will increase by a further 43% from 1997 to 2020, amounting to 770 million tonnes in 2020. Despite a continuous emphasis on preventing waste generation and encouraging waste recycling, socio-

Figure 23.1. **Resource and material intensity of OECD economies, total use and intensity of use relative to GDP, 1980-2020**

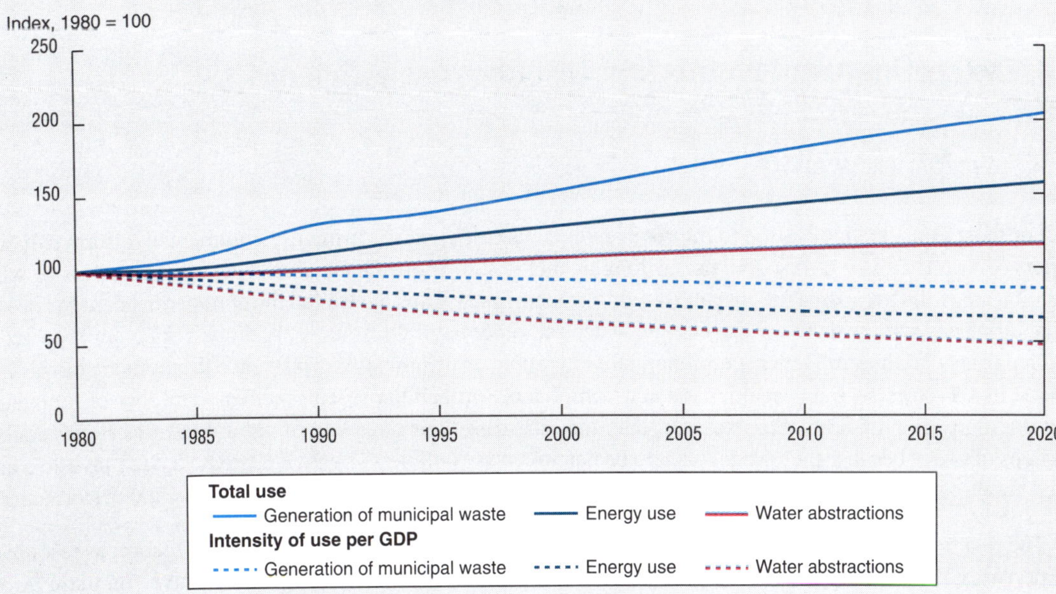

Sources: OECD (1999) and Reference Scenario.

demographic developments – such as the trend towards smaller households associated with increasing affluence and consumption levels – are expected to result in continued increases in municipal waste generation (see Chapter 20).

In some cases, significant improvements in resource use efficiency have resulted in only modest increases in the total amount of resources used and – in some countries – even in reductions in the total use of selected resources. A reduction in per capita freshwater use of 11% since 1980 has meant that total water use by OECD countries rose by only about 5% over that period, and almost one-third of OECD countries realised net reductions in their total water use (see Chapter 8). In other cases, however, there are no signs at all of a de-coupling of environmentally damaging trends from GDP growth. This is true, for example, for transportation, with projections that motor vehicle kilometres travelled in OECD regions will increase by a projected 40% to 2020, and passenger air kilometres are expected to almost triple worldwide (see Chapter 14).

Many of the studies that have been conducted on resource use show similar results: in general, increased total production and consumption have outweighed the efficiency improvements in resource use. For example, the World Resources Institute recently analysed the material output flows for five OECD countries (Austria, Germany, Japan, the Netherlands, and the United States) between 1975 and 1996 (Matthews, *et al.*, 2000). The report shows that the materials outflow intensity of all the studied countries has fallen since 1975. Resource inputs and waste outputs rose relatively little on a per capita basis and fell dramatically when measured against units of economic output. However, overall resource use and waste flows into the environment continued to grow. From 1975 to 1996, the total quantities of wastes, emissions, and discharges in all studied countries increased by between 16% and 29%. For example, flows of fuel-related contaminants to the US environment increased by about 25% between 1975 and 1996. The report projects that these trends will continue into the future. The global throughput of material is likely to triple over the next 50 years, including a nearly 300% rise in global energy consumption and manufacturing activity. Over the same period, the world's population is projected to increase by 50% and global economic activity is expected to increase roughly five-fold.

A scenario analysis of future trends in the use of various resources and materials undertaken by the United Nations show comparable results. With assumed sectoral growth rates of between 1.7% and 2.7% per annum, significant improvements in "material efficiency" are projected even under the most pessimistic scenario (assuming that existing technologies and current trends continue). Nonetheless, absolute consumption continues to rise for all materials under the most optimistic scenario (assuming that technologies which are not currently commercially available will gradually come on stream) (Worrell *et al.*, 1997).

23.4. Determinants of changes in resource use efficiency

Resource prices and scarcity

For most firms, the motivation to improve resource use efficiency is primarily commercial: efforts will be made to improve the efficiency of resource use as long as this will result in financial cost saving, regardless of whatever regulations regarding resource use may exist. For instance, many firms have cited the commercial benefits of improving the efficiency with which they use energy, water and other natural resources. Although price *per se* is one of the most important variables influencing commercial motivation, governments can influence the attitude of firms towards resource use. Factors such as penalties for non-compliance, probability of enforcement, and loss of reputation can affect the motivation of firms. For example, they may increase their recycling of natural resources and reduce their emissions of toxics because they do not want to be perceived as "polluters" by their customers, their investors are concerned about potential future liability claims, or regulators are requiring improved environmental performance.

The relationship between resource prices and resource use is rather straightforward in theory: as resource scarcity increases, prices will rise and resource users and exploiters will have incentives to "save" on these resources. Figure 23.2 shows recent trends in real resource prices for a number of resources and resource-based products. Regardless of whether the prices are for non-renewable or renewable resources, they have shown a general downward trend.

Figure 23.2. **Indices of real primary commodity prices, 1980-1998**

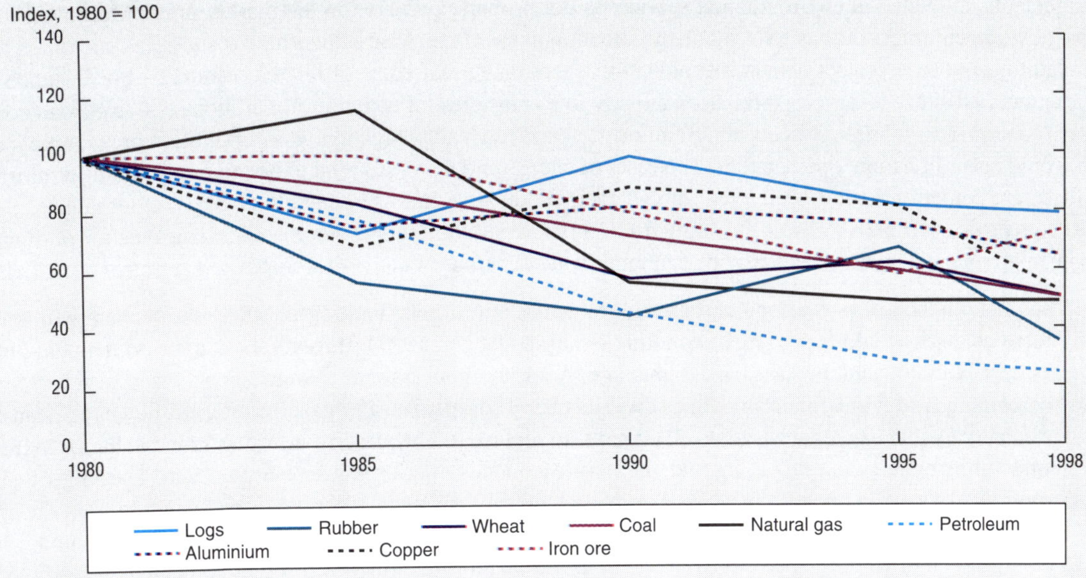

Source: IMF (1999).

Of the primary commodities shown here, only oil is likely to become economically scarce in the foreseeable future. While oil reserve estimates are uncertain, it is expected that many non-OPEC oil fields will have passed their periods of peak production before 2020 (OECD, 1997*a*). As world oil supply is likely to become even more concentrated in the Middle East, the oil market could become more volatile, resulting in unstable oil prices. Declining real resource prices, such as those identified for the other commodities, are likely to generate incentives for a greater substitution of these resources in place of other factor inputs,[2] thus weakening incentives for greater resource efficiency.

For renewables, the future flow of resources will depend on the size of the stock. There is *a priori* no reason to expect that increased exploitation rates today will result in increased prices tomorrow. In forestry and fisheries, "cultivation" technologies (plantation forests and aquaculture fisheries) are being applied that produce bigger harvests. The situation appears to be stabilising in terms of forest resource stocks in OECD countries, with total forest coverage stable or increasing in part because of the growth in intensive plantation forests. However, a continuing trend of net deforestation is occurring outside of OECD regions (see Chapter 10). In the fishery sector, 9 of 16 FAO marine regions are over-fished according to FAO (FAO, 1999). Increases in the amount of fish harvested worldwide are expected to 2020, but all or most of the increase is expected to come from new aquaculture developments.

Barriers to resource use efficiency

Attempts to improve resource use efficiency face several barriers: they can be economic, regulatory, or physical, or they can result from a lack of information. The most recognised of these are market and intervention failures and inefficiencies. Equally important is the "locking-in" of resource use in existing technologies and practices, because of high sunk investment costs and social inertia, that can inhibit change (OECD, 1998). The lack of access to information throughout the economy on resource and energy use efficiency and technological alternatives also represents a "barrier".

2. Assuming the relative prices of the other factor inputs remain stable, decrease less than the natural resource prices, or even increase.

Market failures, such as the non-internalisation of externalities or missing or poorly defined property rights, are a primary barrier to resource use efficiency. If the price users pay for a resource does not reflect both the cost of its exploitation and the associated environmental services, then consumption will be excessive. Under current market arrangements, the value of environmental services is not properly reflected in the market price. Moreover, in many cases, government interventions have made the situation even worse. Many government subsidies and other policies intended to support or protect resource exploitation encourage greater use of related resources. For instance, many governments subsidise resource exploitation directly (*e.g.* subsidies for coal mining) or indirectly (*e.g.* road construction in forested areas). These policies inhibit innovation and can lead to higher levels of pollution or resource intensity than would occur in a more open market. The costs of energy and materials tend to be relatively small for most firms, yet these are sometimes subsidised even though their production and use generate negative externalities. Subsidy reform, green tax reform (*e.g.* shifting taxation from labour to pollution and resource use), and the creation of future markets in environmental goods would provide economy-wide dynamic incentives for improving resource use efficiency.

The phenomenon of technological, behavioural and institutional "locking-in" makes any change look costly, even where changes would bring large economic benefits (Jaeger, 1997). Important changes do happen, but their timing is unpredictable and they are hard to manage. A large organisational change has occurred in recent years in newly deregulated industries, such as airlines and electricity companies. The gasoline engine, however, is an example of a technology that has so far survived all attempts to replace it. The massive infrastructure that has been installed for manufacturing engines and supplying fuel, the low-cost and high performance resulting from "learning-by-doing", and the presence of scale economies makes it extremely hard for any radical technological changes to enter the market. However, electric battery or fuel cell technology could compete with the gasoline engine in the near future, if produced in large enough volumes (see Chapter 14). In this area, no single firm or group of companies is ready to make the massive investments needed for creating a new transportation infrastructure because it is still unclear which technology will prove to be best and what the financial results would be of such an investment. Government initiatives can play a significant role in this field. Governments will need to continue to support investment in physical infrastructure, including transport and communication networks, and long-term research in basic science.

23.5. Policy issues

Table 23.1 shows the potential means by which improvements in resource efficiency can be achieved. These include: resource saving, resource reusing, resource prolonging, and resource substituting (Johnstone, 2001).

Table 23.1. **Channels through which increases in resource efficiency can be achieved**

Channel	Nature of change	Examples
Resource saving	Involves changes in products and production technologies which allow for the use of less inputs per unit of output.	– Increasing use of drip-irrigation systems in agriculture (see Chapter 7). – Downsizing of motor vehicles (see Chapter 14) and electronic goods. – The wider use of electric arc furnaces and continuous casting technologies in steel production (see Chapter 17).
Resource reusing	Involves the increased use of "secondary" resources in production.	– Production of steel from scrap (see Chapter 17). – Increasing use of recovered paper (see Chapter 18). – Increasing energy recovery from waste incineration (see Chapter 20).
Resource prolonging*	Involves the use of processes which allow for relatively more product durability, slowing down material turnover rates.	– The use of new longer-lasting composite materials. – Product (or component) re-design, allowing for longer use.
Resource substituting	Involves the use of alternative resource inputs in manufacturing processes and in products.	– The substitution of renewables or non-exhaustible energy sources in place of fossil fuels (see Chapter 12). – The use of fibre optics in place of copper wiring.

* A possible conflict exists between the environmental benefits of increased product durability and the benefits of more rapid development and diffusion of new, cleaner technologies. Furthermore, some advanced materials like composites are high performing and can be very durable, but they require more energy and other inputs to produce, create more waste in their manufacturing, and are practically non-recyclable.

The diverse nature and utility of natural resources require a comprehensive approach to policies for improving resource use efficiency. The current and future technological changes are expected to have significant environmental spill-overs. The Polluter Pays Principle and the User Pays Principle should play an important role in providing adequate incentives for resource use efficiency to resource users and exploiters. However, prices alone cannot provide sufficient incentives for sustainable resource use in the short run. Stronger pricing signals should be combined with direct regulations, voluntary agreements, and information-based instruments in the relevant policy areas. In many cases, the most important policies bearing on resource use efficiency may not be motivated by environmental concerns but instead by other interests. Policy co-ordination with non-environmental government agencies and the development of synergies with different policy areas are required.

Technological development

In many cases, improvements in resource use efficiency are associated with technological development. Considerable saving and substitution in energy consumption and the use of materials (*e.g.* a reduction in energy consumption or material input per unit of output), as well as a substantial reduction in emissions and waste generation, can be achieved by optimising technological systems. Significant reductions in the input of scarce materials (*e.g.* metals or groundwater) per unit of output can only be expected from improved industrial production systems and technological change.

The net effects on resource use of the rapid development of many new technologies – including information and communication technologies (ICTs) and modern biotechnology – are unclear.

Several technology foresight studies have identified a number of technologies that are emerging. Among these, the development of information and communication technology (ICT) and biotechnology is expected to have significant environmental spillovers (see Chapter 6). Biotechnology, including genetically modified organisms (GMOs), are being proposed as the solution to some of our resource use problems: they have the potential to reduce the amount of damaging inputs (*e.g.* pesticides, fertilisers, antibiotics) used in natural resource sectors (*e.g.* agriculture, forestry, fisheries), and increase production levels to meet human needs. However, they may also have potentially negative effects on human health and on global eco-systems, which should be clarified further. Similarly, progress in ICT can improve the way information is treated, stored, and diffused, and also has the potential to bring about structural changes in other industrial sectors, through better eco-efficiency and de-materialisation. However, they also have the potential for negative environmental impacts, such as greater consumption of electricity and paper for ICT devices, and more waste generation from the short lifecycle of computers (see Chapter 6).

It is important to design and implement adequate environmental policies that provide long-term incentives for developing and diffusing new environmentally friendly technologies, including those that can lead to a de-materialisation of the economy and a reduction in resource intensity. Environmental policies, however, are often insufficient to induce radical and continuous innovations for de-coupling environmental degradation from economic development on their own. Effective co-ordination between environmental policy and technological/innovation policy will be needed to realise more significant changes.

Effects of environmental policy on resource use

The intensity of resource use can be influenced by a variety of environmental policies. The traditional approach of "end-of-pipe" pollution control regulations is often criticised because of its innovation-stifling effects compared with the use of market-based instruments. The actual effect of regulatory instruments on resource efficiency depends on the way they are designed and implemented. Regulatory instruments that specify the use of certain inputs and processes may limit innovation and provide little incentive to change patterns of resources. In some cases, however, regulations are needed to ensure sustainable use of resources, where market-based instruments would not be able to achieve the same effects, or only at very high administrative costs. For example, imposing constraints on the nature of resource exploitation – such as fishing gear restrictions or minimum fish sizes – are often used to ensure the sustainable use of these resources.

Policies to promote resource use efficiency need to address both the supply and the demand side. Environmentally perverse subsidies should be removed to improve market functioning. Economic incentives to address pollution emissions tend also to increase the profitability of reduced energy and materials use through internalising the environmental externalities arising from this use. Charges and taxes on pollutants (*e.g.* SO_x, NO_x, CO_2) have been introduced in a number of OECD countries, and have in many cases proven to be more cost-effective in reducing pollution than regulatory standards. Natural resource management policies which increase the price of primary resources, such as fishing licenses, timber concession fees, and mineral deposit rents, will result in a more efficient use of resources downstream. In some cases, this can also be achieved by creating property rights – *e.g.* tradable fishing quotas – for the resource.

Improved information about environmental impacts and greater consumer demand for green products are likely to make it more profitable for firms to develop environmentally friendly technologies and business operations. In this regard, developing indicators of resource use is important. Assuming that manufacturers will respond to consumer demand, efficiency standards combined with eco-labelling can promote the production of resource efficient consumer durables (IEA, 1997; OECD, 1997*b*). Helping firms to recognise and act upon commercial opportunities – for example improved environmental accounting procedures – can change resource use practices. In the field of resource re-use and recycling, public information sharing plays an important role. The markets for secondary materials and components are growing, but information about them is not sufficient. Government procurement policies and the provision of suitable information to the public through information sharing can also improve the performance of secondary materials markets. Environmental education, focused on training in professions that have large and lasting influences on infrastructure development, institutions and lifestyles, can contribute to improving resource use efficiency.

Policy integration

The most important policies determining resource use efficiency may not be environmental. For instance, the fiscal treatment of capital investment can have a considerable effect on the turnover of the capital stock of a company. Depreciation allowances and tax preferences for new investment will affect the firm's decision about when to scrap capital equipment and undertake new investment. Sustainable resource management can be achieved when policies encourage a longer-term view by decision-makers at all levels. This implies that sectoral ministries at the national level should incorporate the environmentally sustainable management of natural resources in their respective areas.

Although transport, energy and industry ministries in many governments used to view environmental policy as a problem rather than a set of shared objectives, some governments are now beginning to succeed in the integration of environmental, economic and sectoral policies (OECD, 1998). A number of governments have produced joint policy documents on their strategies for sustainable development and climate change. More intensive inter-ministerial efforts at all levels in the various ministries can be a major step towards achieving synergistic effects in improving resource use efficiency.

REFERENCES

FAO (Food and Agriculture Organisation of the United Nations) (1999), *Review of the State of World Fishery Resources*, FAO, Rome.

IMF (International Monetary Fund) (1999), *Indices of Primary Commodity Prices*, IMF, Washington, DC.

IEA (International Energy Agency) (1997), *Indicators of Energy Use and Efficiency*, IEA/OECD, Paris.

Jaeger, C. (1997), "Presentation to OECD Workshop on Fostering Eco-Efficiency: the Role of Government", Paris, 3-4 September.

Matthews, E., C. Amann, S. Bringezu, M. Fischer-Kowalski, W. Hüttler, R. Kleijn, Y. Moriguchi, C. Ottke, E. Rodenburg, H. Schandl, H. Schútz, E.V.D. Voet, and H. Weisz (2000), *The Weight of Nations: Material Outflows from Industrial Economies*, WRI, Washington, DC.

OECD (1997*a*), *The World in 2020, Towards a New Global Age*, OECD, Paris.

OECD (1997*b*), "CO_2 Emissions from Road Vehicles", OECD/GD(97)69, OECD, Paris.

OECD (1998), *Eco-Efficiency*, OECD, Paris.

OECD (1999), OECD *Environmental Data: Compendium* 1999, OECD, Paris.

Johnstone, N. (2001), "Resource Efficiency and the Environment", background document for the OECD *Environmental Outlook*, OECD, Paris.

Worrell, E., M.D. Levine, L.K. Price, N.C. Martin, R. van de Broek and K. Blok (1997), "Potentials and Policy Implications of Energy and Material Efficiency Improvement", UN Department for Policy Coordination and Sustainable Development, New York.

Section VII

INSTITUTIONAL FRAMEWORKS AND POLICY PACKAGES FOR ADDRESSING ENVIRONMENTAL PROBLEMS

Traffic lights have been used in this report to signal key trends and projections to 2020 for environmental pressures, the state of the environment, and some of the societal and political responses to these pressures that have been emerging. The signals table provides a summary overview of some of the main areas of concern for OECD countries. The chapter on institutional frameworks for the environment examines the adequacy of both the international and national institutional frameworks to support the policies needed to address the most serious environmental problems facing OECD countries. The chapter on policy packages to address the main environmental problems suggests packages of policy instruments that could be used to tackle the "red lights" identified in the report, with an analysis of some of their potential effects if they were to be implemented.

Signals of the OECD *Environmental Outlook*

	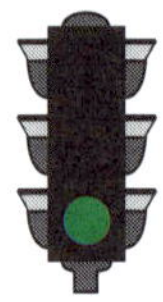	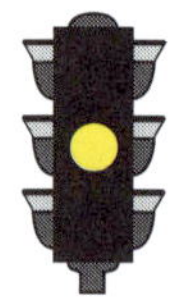	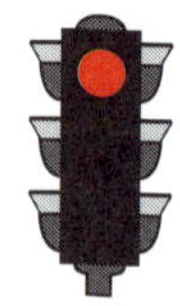
PRESSURES ON THE ENVIRONMENT	• Industrial point source pollution • Some air pollutants (lead, CFCs, CO, SO_x)	• Water use • Toxic emissions from industry • Hazardous waste generation • Energy production and use	• Agricultural pollution • Over-fishing • Greenhouse gas emissions • Motor vehicle and aviation air pollution emissions • Municipal waste generation
STATE OF THE ENVIRONMENT	• Forest coverage in OECD regions	• Surface water quality • Forest quality in OECD regions • Ozone layer integrity	• Biodiversity • Tropical forest coverage • Fish stocks • Groundwater quality • Urban air quality • Climate change • Chemicals in the environment
RESPONSES	• "Green" purchasing • "Green" agriculture • Protected areas • Resource efficiency • Energy efficiency	• Biotechnology • Forest plantations • Aquaculture • Energy and transport technologies • Waste management	

Note: The "green light" signals pressures that are decreasing or conditions in the state of the environment for which the future outlook to 2020 looks positive. It is also used to signal societal responses that are helping alleviate the problems identified. The "yellow light" signals areas of uncertainty or potential problems. The "red light" signals NSpressures on the environment or environmental issues for which recent trends have been negative and are expected to continue to 2020, or for which recent trends have been more stable, but are expected to worsen.

24
Institutional Frameworks for the Environment

KEY SIGNALS

- The environmental problems of the future are likely to be more complex and difficult to resolve than those of the past. They will involve addressing diffuse sources of pollution, and enhancing globally shared resources and sinks. The achievement of environmental objectives will sometimes involve difficult trade-offs with other policy objectives (*e.g.* economic, social); at other times, it will be possible for one set of policy objectives to contribute to another ("win-win" situations).
- Environmental policy-making is likely to increasingly be developed under the guidance of environmental principles or approaches: for example, the Polluter and User Pays Principles, the precautionary approach, the principle of full cost recovery and the complete internalisation of environmental externalities.
- New policy instruments will be required to complement regulations and economic instruments in tackling the new environmental problems. These will include increasing use of information-based tools (*e.g.* eco-labels, civil society consultations, and environmental education), demand-side management policies to influence consumption patterns, and co-operative and voluntary approaches for business and industry.
- Governments are likely to increasingly be seen as facilitators and catalysts of environmental policy development, rather than as providers of environmental protection. Stakeholders will increasingly be considered partners in the process of policy development and implementation.
- At the international level, the need for better integration among existing multilateral environmental agreements (MEAs) and for better co-ordination between existing international environmental organisations and agreements is likely to intensify. The focus will be on the ratification and implementation of existing agreements, and the development of appropriate systems for monitoring and enforcement. Multilateral economic and trade institutions will increasingly be called upon to assess and address the environmental impacts of the activities they administer.
- OECD countries will need to strengthen their dialogues with developing countries and countries with economies in transition, particularly regarding environmental problems of mutual concern.

24.1. Introduction

This chapter provides an outlook for the institutional frameworks which affect environmental policies in OECD countries. A relatively broad understanding of "institutions" is used, extending beyond just "organisations", and including the roles played by specific actors in key decisions that influence environmental performance. It also takes the view that it is not only environmental actors who are relevant – the roles of several non-environmental actors are also discussed, where appropriate. The chapter examines both national and international institutional frameworks, and how stakeholders can contribute to successful policy development. This introduction starts with a discussion of the evolving context in which environmental institutions are developing.

In the early stages of the environmental movement, the main objective was to establish environmental concerns on the public policy agenda. A certain confrontational approach therefore developed, as the pollution control agenda tended to compete with economic interests (transport, energy, industry, agriculture) in an effort to establish adequate environmental controls and standards. Later, as the place of environmental issues at the policy table became more assured, the problem switched to one of finding better ways of communicating environmental needs to economic partners. At this point, the emphasis began to shift to "policy integration".

Awareness also grew that environmental problems were very much part of a system, and should not generally be perceived in isolation of that system. The ecological services provided by the environment therefore gradually came to be more fully recognised, and reinforced the need for more integrative solutions. In recent years, the policy emphasis has been shifting away from essentially an environment/economy focus, to also include social concerns. The links between the environment, the economy, and social development have been emphasised by the concept of sustainable development – an idea given substantial profile by the Brundtland Commission in the mid-1980s (World Commission on Environment and Development, 1987), and reinforced by the 1992 United Nations Conference on Environment and Development (UNCED) meeting in Rio de Janeiro.

Although many of the "old" environmental issues (generally pollution control) remain high on the environmental policy agenda, a range of "new" issues (*e.g.* nature/biodiversity protection, biotechnology, sustainable development, and strategic planning) are also receiving greater attention (Burke, 1998). For example, some OECD countries have recently established a constitutional right to a clean environment, and a few have adopted formal plans for the deeper integration of environmental and economic policies. For the future, it seems likely that the focus on sustainable development will continue to grow, placing increasing emphasis on the interrelationships between the environment, the economy, and social objectives. This implies that environment-related institutions will most likely have to operate in an increasingly complex context over the period covered by this *Outlook*.[1]

The early emphasis on point sources of pollution also led to a focus on the economic actors who could make the quickest (and cheapest) contributions to reducing that pollution – the polluting enterprises themselves. In turn, this caused environmental policy to focus mainly on the supply side of the economy. For example, with the emergence of the Polluter Pays Principle in 1972, firms were told that they would have to bear the costs of meeting environmental regulations imposed on them, and they were increasingly required to conduct environmental impact assessments related to their activities. Certain forms of environmental infrastructure were targeted for upgrading (*e.g.* water and sewage). Contingency plans were developed to help avoid such eventualities as major oil spills. Enterprises were strongly encouraged to use the best available technologies in their production processes.[2]

As it became clear that not all environmental problems could be resolved (at least at reasonable cost) by focusing on the supply side alone, more attention began to be paid to potential contributions from consumers and other parts of the demand system. The User Pays Principle emerged as a logical complement to the Polluter Pays Principle (Smets, 1999); eco-labelling, consumer boycotts, and more elaborate systems for recycling household waste also began to appear. In the future, the demand aspects of environmental questions will become more important. The idea of eco-efficiency at the level of the firm is already broadening out to encompass resource efficiency questions more generally. Life-cycle considerations and the concept of "integrated pollution prevention and control" are gaining acceptance. Emphasis is also shifting toward more strategic environmental impact assessment.

The transboundary or global nature of many of the environmental issues that are of most concern to OECD countries for the future will increase the possibility that these issues may be transformed into potential security threats. Environmental security issues can arise when an environmental problem is considered to be so threatening that extensive and immediate action needs to be taken, outside the domain of traditional environmental policy-making (see Box 24.1). In some cases these can lead to conflict. Examples of potential or actual environmental security threats include water scarcity or pollution (particularly where the available water resources are shared by two or more countries), the depletion of fish stocks in open seas fisheries, changing climate conditions resulting from global warming, and nuclear accidents or major chemical spills.

1. For example, the Amsterdam Treaty of the European Union "...emphasises the integration of environmental protection into Community policies, in order to achieve sustainable development". The 1998 Cardiff European Council then invited "... all relevant formations of the Council to establish their own strategies for giving effect to environmental integration and sustainable development within their respective policy areas...". The Transport, Energy and Agriculture Councils were invited to start this process.
2. Recycling issues were a notable exception to this. Recycling – largely a *demand*-side question – has been part of the environmental agenda for some time. So has demand management in certain economic sectors, notably energy (especially since the energy crises of the 1970s).

Box 24.1. Potential environmental security threats

An environmental problem becomes a security issue when it is "presented as an existential threat, requiring emergency measures and identifying actions outside the normal bounds of political procedures" (Buzan *et al.*, 1998). This *Outlook* shows that some environmental problems are likely to significantly worsen in the period to 2020 (the "red light" issues). Some of these problems have the potential to lead to environmental security threats affecting OECD countries. However, different regions and countries may face different environmental security threats depending on regional factors, such as the scope of the underlying environmental problem as well as the political and administrative capacity to address the problem. Important factors determining the potential for an environmental problem to develop into a security threat in a given region or country include economic vulnerability, technological capacity, cultural and ethno-political factors, political culture and political stability, the assimilative capacity of the environment, and the rate of regeneration of the natural resources.

The linkages between environmental issues and security call for a co-operative and integrated approach towards conflict prevention or resolution. Preventative action is preferred, as solutions can become very expensive and complex once an environmental problem has evolved into a security issue. Agreements such as the Kyoto Protocol, the Montreal Protocol, the Convention on Biological Diversity, as well as numerous bilateral and multilateral water-sharing agreements, are examples of efforts to address potential environmental security threats in a co-operative manner.

It is likely that the future will also see increasing debate about the use of precaution in environmental policy, both at the national and the international level. Given the uncertainty and lack of available data concerning many of the most worrisome environmental issues, some countries have moved away from adopting only those environmental policies that are firmly backed up by a strong scientific basis. Following Principle 15 of the Rio Declaration on Environment and Development, the lack of full scientific certainty should not be used as a reason for postponing cost-effective measures to prevent environmental damage, where there is a threat of serious or irreversible damage. The idea of "precaution" is now embodied in the national laws of several OECD countries, and has been part of the multilateral environmental institutional framework since at least the 1987 Conference on the Protection of the North Sea. It has found its way into the Montreal Protocol on Substances that Deplete the Ozone Layer, the Rio Declaration on Environment and Development, the Framework Convention on Climate Change, the Convention on Biological Diversity and its Protocol on Biosafety. It has also been incorporated into the Treaty of the European Union. The OECD Guidelines for Multinational Enterprises also contain a recommendation to business concerning the need for precaution in environmental matters. Current debates about such questions as the potential impacts of genetically modified organisms and food safety make it likely that this approach will shape future policies in the area of environment and health.

Given the changes that have been occurring in the nature, scope, and focus of environmental policies, it is not surprising that the institutional framework erected to support environmental policies has also been undergoing considerable change. Table 24.1 summarises some of the changes that are taking place. The rows in this Table outline three periods in the evolution of the institutional framework. The first period was from about 1965 to 1985; the second ran from about 1985 to the present; and the third covers the *Outlook* period from now until 2020. These dates are merely indicative, and are intended only to provide a sense of the evolution of institutional changes over time. The columns reflect several important dimensions of the institutional framework as related to the environment.

24.2. National institutional frameworks

Early national environmental institutions were characterised by considerable rigidity and hierarchy. They also tended to have mandates that were limited to single (and narrowly based) issues. Environment Departments, for example, typically had the mandate to protect the environment. Environmental problems were based on "science",

Table 24.1. **Institutional framework – recent evolution and future outlook**

	Evolving Policy Context	National *environmental* institutional frameworks	International *environmental* institutions	International *economic* institutions	Developing countries	General public	Non-governmental organisations	Business community
Previous Stage	• Environment *vs.* economy • Supply focus (PPP, BAT, end-of-pipe)	• Rigid structures • Government as "provider" • Environmental institutions are dominated by scientific considerations and expertise	• Developmental stage (niche-finding) • Environment goals dominate • Policy goals are often expressed in qualitative terms • Economy is viewed in negative "source-of-the-problem") terms	• Economic growth goal dominates • Environment is viewed in negative ("increased cost") terms •	• Institutional power lies mainly with the OECD countries	• Focus on information exchange	• Narrow-based, single issue NGOs • NGOs are often marginalised from key policy debates	• Business objects to environmental rules ("defensive" strategy)
Current Stage	• Environment and economy (*e.g.* sectoral integration) • Demand focus (pricing, UPP, consumption patterns, eco-efficiency)	• More flexible structures, but tradition still dominates • Decentralisation of activities to local level • Government as "negotiating partner" • Environmental institutions increasingly need economics skills	• Partial integration stage (links to economic institutions expand) • MEAs also expand	• Co-existence with environment • Focus is on dismantling rules that are seen to be protectionist • Environment is viewed in more neutral terms	• Institutional power of the developing countries increases • Capacity building, technology transfer, and recognition of "differentiated responsibilities" all become more important	• Focus on consultation and public participation • Increasing emphasis on transparency (access to information)	• More systematic approaches develop (networking, aided by the Internet) • NGOs' influence over routine policy decisions begins to increase (MAI, WTO, WB) • Concerns grow about the NGO "democratic deficiency"	• Business increasingly *co-operates* with governments in defining and implementing environmental policies
Future Stage	• Environment, economy, and social objective (sustainable development) • Integrated supply and demand focus (IPCC, LCA, SEIA, resource efficiency) • Precautionary approaches become more important	• Open/flexible structures • Emphasis on "networking" • Government as "facilitator and catalyst" • Decentralisation continues, but emphasis is increasingly on "functional equivalence" • Role of government does not decline, but it must evolve • Environmental institutions now need integrated analysis skills	• Full integration stage (MEAs are more socially and economically sensitive (*e.g.* Kyoto) • Policy goals are increasingly expressed in quantitative terms (targets, timetables) • Need for international co-ordination intensifies	• Environment is viewed in more positive terms • Recognition that new environmental rules may need to be constructed as part of multilateral economic systems	• Developing countries become full partners in the search for multilateral solutions to environmental problems (and in the work of related institutions) • Institutional governance, mutual commitments and interdependence become more evident	• Focus on dialogue • Accountability, verification, transparency become vital for long-term institutional credibility • Environmental education plays a larger role • Increasing ability of the general public to shape environmental policies/actions (*e.g.* boycotts, company reputations)	• NGOs become "working partners" in key policy debates • Government institutions need to adapt to the emerging NGO presence • Increased networking among NGOs	• Business becomes more pro-active in protecting the environment and ultimately in achieving sustainable development • Role of business in the global environmental effort intensifies (both MNEs and SMEs) • Worldwide operations are increasingly subjected to "best practices" criteria • Privatisation expands, but does not dominate
Integrated action across all the future columns becomes more important								

so most staff in Environment Departments had scientific or engineering backgrounds, and there was little interface with other policy communities – even though these other communities were often directly affected by environmental policies, as well as being important vectors for implementing them. One key result of the broadening nature, focus, and scope of environmental problems was that environment ministries began to expand their internal capacities for economic analysis. Another result was that these ministries began to adopt more flexible (*i.e.* negotiated) approaches to achieving their policy goals. Underlying science was still important (as were the regulations based on that science), but policy trade-offs based on economic or social considerations could now also be taken into account in some circumstances.

A similar shift was occurring within national economic institutions. Recognising the growing political importance of environmental questions, many sectoral economic ministries began to develop environmental units within their organisational structures. Agriculture ministries in particular led this trend, but it was apparent within many transport and energy ministries as well. One important result was that environmental issues came to be more routinely dealt with by these sectoral ministries, rather than having solutions imposed on them by the Environment ministries. Another result was that the sectoral ministries developed many of their internal capacities for dealing with environmental issues, and for explaining their views on these issues directly to the general public, without having to rely on technical expertise from their environmental colleagues.

Several countries also created inter-ministerial working groups or Cabinet-level Committees, Commissions of Enquiry, Task Forces, etc., to more fully examine the interface between the economy, the environment, and society at large. For example, Canada established National and Provincial Round Tables on the Environment and the Economy, drawing together representatives from business, government, and the public to generate strategies for environmentally-sustainable economic development. Canada has also created a Commissioner of the Environment and Sustainable Development in the Office of the Auditor-General. Each government department is required to prepare a sustainable development strategy which, in turn, is evaluated by the Commissioner.

Responsibility for environmental decision-making has also been significantly decentralised in some countries, and in some sectors of the economy. Recent changes in the structure of the Netherlands' Water Boards, for example, will mean that a larger and more diverse portion of the population is now empowered to elect General Councils. This reform is expected to allow a broader range of interests to be reflected in Council decisions (OECD, 1999). Several countries have also made efforts to consolidate their environmental legislation. For example, New Zealand's Resource Management Act (1991) consolidates 75 former statutes into a single law. It also places a duty on policy-makers to assess the costs and benefits of the most efficient and effective means for promoting sustainable land use, in addition to requiring integrated management of pollution and waste.

For the future, the trend toward more local responsibility in institutional structures and operating procedures is likely to continue. This is in line with the principle of "subsidiarity", a basic tenet of institutional behaviour in many countries. However, this trend may be limited by concerns that excessive decentralisation could lead to unacceptable inefficiencies. It will also be limited by regional institutions that have been entrusted with particular (supra-national) mandates by their participants, such as the European Union.

The growing complexity of the sustainable development issue will also make it increasingly difficult for traditional (vertical) institutional structures to deliver credible results. Consequently, there may be a greater reliance on an institutional approach that has been used by the NGO community for some time with considerable success – the idea of (horizontal) "networking" (Reinicke, 1999). This is based on the premise that individuals and groups – not formal organisations – drive innovation. Networking is a bottom-up approach to building institutions, and its use has been greatly facilitated in this global information age. It can help policy-makers deal with three key aspects of modern environmental problems: *i)* managing knowledge flows; *ii)* focusing on particular market and intergovernmental failures; and *iii)* broadening participation in decisions. All of this will be part of a trend toward more horizontal institutional structures – structures that are capable of quick reaction, multidisciplinary analysis, and a significant degree of structural flexibility. It is also likely that business, NGOs, and governments (in both OECD and non-OECD countries) will become more integrated into these networks.

Some observers argue that the role of national governments is set to become progressively weaker as economic globalisation intensifies. This is because the increasing mobility of economic resources in a globalised economy implies less ability to control these flows by (essentially static) governments. But it is also because economic activity is likely to be more complex tomorrow than it is today, and governments are often assumed to not have the internal flexibility to adjust quickly (and adequately) to this additional complexity. Others dispute this view, arguing that "...the state may no longer be able to impose outcomes on all dimensions of policy, but it will remain a key institution in sustaining governance of the international economy... It will make international agreements stick, upwards because it represents a territory, and downwards because it is a constitutionally legitimate power" (Hirst and Thompson, 1995; see also Rosecrance, 1996 and Finger, 1999).

For a number of reasons, the latter view seems more likely to prevail, and certainly over the period covered by the *Outlook*. First, controlling environmental externalities is essential for the provision of public goods. While there is certainly now an expanding role for the private sector (see Box 24.2), the government still has a basic responsibility to protect the public interest in environmental matters. Not only will national governments be necessary to set the operating rules through which economic transactions take place (contribution to sustainable growth), they will also be needed to ensure that market failures do not lead to long-term environmental degradation (contribution to sustainable development).

Second, although national governments will probably agree (for efficiency reasons) to devolve more responsibility for environmental policies to the local level, it will still be important (for equity reasons) to retain some degree of control to ensure that key issues are being handled in a relatively consistent manner. Public opinion will increasingly highlight any gap between environmental policies/laws and their actual implementation, and national governments will become more and more accountable to the public for this "implementation gap".

Third, national governments will probably need to enter into new multilateral arrangements aimed either at resolving specific trans-boundary environmental problems, or at reducing the international competitiveness effects of national environmental policies. In an era where environmental problems (and their economic implications) are becoming increasingly international and there is an increased risk of regional or international environmental security threats, the need for national governments to become more expert with environmental co-operation at the international level is likely to intensify.

Box 24.2. **Increasing privatisation of environmentally related services**

Governments are often criticised for being inefficient economically or financially in their management of public utilities (*e.g.* electricity, urban water and waste management services). These inefficiencies are also often accused of inhibiting the very innovations that could help reduce environmental stresses over the long term. Greater use of private markets can, in some circumstances, make an important contribution to both environmental and economic objectives. However, privatisation is not always feasible. For example, the environmental service involved may be a natural monopoly (*e.g.* water supplies), making it politically difficult to justify the transfer of such a vital resource into private hands, even with the creation of a public regulatory body to supervise it. In this context, the provision of adequate supplies of public goods could be compromised. There are also technical limitations to privatisation. For example, where the capital expenditures involved in developing environmental services are large in comparison with the marginal cost of connecting an additional user to that service, it may be uneconomic to privatise only a portion of that investment.

These limitations notwithstanding, the government role in providing many environmental services has recently been shifting in many countries, from that of being the primary provider of those services, to that of being the creator and regulator of an operating environment which allows communities, the private sector, and non-governmental organisations to become more active in providing these services themselves. For example, substantial privatisation of some aspects of public water supply services has recently been taking place in OECD countries. Even where the water system remains publicly-owned, service management is increasingly being delegated to private operators. This approach seems particularly well-suited to decentralised systems, in which municipalities may see delegation as a useful way of overcoming their own lack of technical expertise and/or financial resources. In the future, the processes of economic globalisation and internal structural reform will probably create pressure for additional privatisation of some activities related to environmental management. However, for the reasons mentioned above, it should not be anticipated that the privatisation of environmental institutions will take place on a large scale during the period covered by the *Outlook*.

24.3. International institutional frameworks

Prior to 1980, some efforts had been made to develop international agreements or institutions aimed at protecting the global commons. However, after the Brundtland report in 1987 public attention began to focus more intently on these global problems. The number of multilateral environmental agreements (MEAs) and the scope and number of regional and global environmental organisations dealing with global commons issues expanded. Many of the MEAs established formal institutions with relatively specific mandates, and focused on specific environmental problems – economic objectives were often not emphasized. Similarly, international economic agreements and the related institutions had previously ignored environmental needs, and the bodies charged with implementing these agreements sometimes seemed to perceive environmental concerns both as a threat to the achievement of their own (economic) objectives, and as the sole responsibility of the environmental community itself.

Both perceptions are now changing. It is increasingly understood that: *i)* poor environmental performance is not a good basis for long-term economic growth and profitability; *ii)* environmental costs are often not a very significant component of economic activity in any event; and *iii)* the key policy concern should be increased welfare, not only increased economic output – with environmental quality being a legitimate component of the former. Economic globalisation is eroding the traditional view that international economic institutions can leave responsibility for environmental problems entirely to their environmental colleagues. Broadly, the public, particularly in OECD countries, has become convinced that economic growth should be accompanied by appropriate policies at the national level to protect the environment. Public pressure is therefore mounting on the international economic and financial institutions (*e.g.* World Bank, Regional Development Banks, WTO, IMF, OECD) to demonstrate that their activities can be reconciled with environmental needs. This broader, more integrative, context for policy-making is however generating a dilemma for international environmental and economic institutions alike. Environmental institutions are finding that scarce resources are having to be applied to the broader goal of sustainable development, and economic institutions are experiencing the same problem with respect to the environment. Changes are likely to be appropriate on both fronts in the future.

Global and regional environmental institutions

A large number of multilateral environmental agreements (MEAs) dealing with global commons issues currently exist, including those that address climate change, biodiversity loss, ozone protection, biosafety, hazardous waste movements, marine pollution, etc. Many of the MEAs established formal institutions with a relatively specific mandate and focused on a single environmental problem. The United Nations Environment Programme (UNEP), established following the 1972 Stockholm Conference on the Human Environment to play a catalytic role in environmental issues in the United Nations system, initiated many of these international agreements. Several other global institutions have also been established, including the United Nations Commission on Sustainable Development (UNCSD) and the Global Environment Facility (GEF).

Five key functions have been identified for global environmental institutions, in particular for MEAs (Waller-Hunter, 1999):

- Where possible, solid grounding in science as the basis for their authority. Without this underpinning, the legitimacy of these institutions will ultimately be eroded.
- Use of quantitative environmental norms and standards in MEAs. MEAs that are confined to dealing with procedural issues will increasingly be questioned on the basis of their actual contribution to improved environmental performance.
- Financial incentives and mechanisms for developing countries to participate actively in MEAs. This will require greater consideration of competitiveness questions alongside environmental issues in the formulation of new MEAs.
- Involvement of the private sector and civil society in contributing to the goals of new MEAs.
- Providing compliance and monitoring systems and dispute settlement mechanisms.

Assessed against these functions, the current system of international environmental institutions can be characterised as somewhat incoherent and incomplete. The scientific function is in some cases under-performed, although in many instances it is clear that full scientific understanding of environmental problems will be a long time in coming. The Framework Convention on Climate Change established by the Intergovernmental Panel on Climate Change is an example of an institution developed specifically to provide a sound scientific basis for the further development of the Convention. MEA-based institutions will also have to learn to accommodate efficiency and cost-effectiveness concerns in their activities. The Kyoto Protocol is a good example of this, inasmuch as it contains "flexibility mechanisms" aimed at increasing the (economic) efficiency of greenhouse gas emission abatement.

In general, MEAs lack effective compliance and dispute settlement mechanisms (see below), and substantive synergies and efficiency gains in the implementation of MEAs are not brought forward as a result of unconnected decision-making processes. This has led to an appeal for a more cohesive system of environmental institutions, involving the major MEAs and organisations like UNEP, the CSD and other parts of the United Nations system dealing with environmental issues (UNESCO, WMO, WHO, etc.). According to some, this could take the form of a World Environment Organisation.

Box 24.3. **Compliance and enforcement mechanisms in MEAs**

At present, there are few examples of MEAs that provide for successful enforcement procedures. Most international environmental agreements already contain provisions for dispute avoidance (conciliation, negotiation, mediation, etc.) and dispute settlement. However, because of difficulties in applying enforcement procedures at the international level, the provisions for dispute settlement currently found in MEAs are generally designed in a way that allows countries to be part of the MEA without submitting themselves to the dispute settlement mechanism. As the obligations for parties under MEAs are becoming increasingly stringent, the need for stronger compliance and enforcement mechanisms at the international level is growing. Already, a number of new or recent initiatives are paving the way for the development of more formalised enforcement mechanisms for international environmental agreements. In the future, compliance and enforcement procedures may be included at an earlier stage in the development of international agreements, building confidence that the MEA will be both effective and fair.

Although there have been several recent high-profile global agreements dealing with environmental issues, awareness is also growing that there are opportunities for more regional approaches to these problems (see Kimball, 1992). The climate change debate (which emphasises that the vast majority of greenhouse gas emissions are produced by a relatively small number of countries), and the ozone depletion debate (where, again, the number of producing countries is relatively small) have both reinforced this regional view.

The possibility of regional co-operation on environmental problems may be particularly important in the future. For one thing, there can be significant transaction costs associated with negotiating and implementing global agreements – costs that may be considerably smaller among more homogenous and like-minded countries. For another, more efficient targeting of the policy response to specific areas is often possible with a regional approach, particularly for issues such as water scarcity, open seas fishing, or regional air pollution for which co-operative regional agreements are already in place or developing in some areas.

Regional approaches can also help pave the way toward wider recognition of environmental goals at the global level. For example, broadening the basis of support for a given environmental standard from the national to the regional level could help reduce concerns within trade/investment institutions about the discriminatory nature of that standard, while still contributing to real progress in environmental performance within the regional grouping. On the other hand, regional approaches may be interpreted in some contexts as attempts to exclude non-participants from the benefits of the agreement. Appropriate care needs to be taken to reduce this concern.

International economic and financial institutions

The ways in which environmental considerations are dealt with by international economic and financial institutions also seems set to become more sophisticated in the future, especially if political momentum to further liberalise global trade and investment is to be maintained within OECD countries. It may not always be obvious how to achieve this result. After all, international economic institutions have until now been largely concerned with dismantling rules which create barriers to trade and investment, whereas international environmental institutions have tended to focus on constructing new rules which sometimes seem to go in precisely the opposite direction. Nor is it obvious that economic institutions are always in the best position to develop efficient international responses to environmental problems. It is still a basic principle of public policy that environmental policies are best enunciated by environmental policy makers, and economic policies are best enunciated by economic ones.

Achieving a common understanding between the economic and environmental communities may ultimately involve the dismantling of some existing rules, but will probably also involve the creation of new ones. Where new rules are appropriate, the challenge will be to create ones that are as efficient as possible (*e.g.* not used as vehicles for disguised protectionism), but that also incorporate criteria to adequately protect key environmental interests. Both communities will therefore increasingly need to view environmental policies as opportunities for economic policy, rather than constraints. "Clean technology" falls into this category, but so do less obvious elements, such as "green consumer markets". There will still be situations where environmental needs imply additional costs for some economic actors, but even here, it is increasingly recognised that there may be ways of minimising these costs, or of spreading them out in a way that allows the underlying environmental improvement to proceed.

Some progress is already occurring toward the kind of integration that is needed. The World Bank has for some time been active in trying to reconcile the simultaneous need for economic growth, environmental protection and, more recently, social development. The WTO established its Trade and Environment Committee in 1994, and the OECD has had a similar Group since 1991. NAFTA contains an environmental sub-agreement. The OECD Guidelines for Multinational Enterprises also contain a chapter on the environment. International financial institutions broadly include: *i)* the international investment and commercial banks; and *ii)* the multilateral development banks. Both groups have already made considerable progress toward the incorporation of environmental goals into their operating procedures, and they represent an increasingly important point of leverage over the environmental behaviour of both investors and investees.

Commercial and investment banks play a key role in conveying new securities to international markets. They also engage in "due diligence" procedures to determine company or project risks. In the US, Superfund liability laws have (since 1990) forced commercial banks to evaluate environmental risks when real estate is used as collateral for loans. This approach (at least for loans backed by real estate) is gradually working its way into lending practices in Europe and other industrialised regions. Investment banks are taking the lead in underwriting government bonds issued by developing country central banks, which accounted for almost half of the US$95 billion in all bond issues from developing countries in 1997 (World Bank, 1998). Outside pressure from multilateral lenders with strong environmental guidelines can sometimes cause these investors to hesitate to invest in projects that may have unacceptable environmental consequences.

Most of the attention being devoted to date by investment and commercial banks to environmental matters has been driven by liability and regulatory compliance issues, and is therefore focused on the "downside" of environmental performance. However, there is some evidence that this may now be changing, especially within the commercial banking sector. For example, as of 1997, NatWest Group in the UK has offered below-market-rate loans to firms with "best of class" environmental management practices. Sumitomo Bank of Japan also provides low-priced loans for investments in Japanese firms that improve the efficiencies of raw material and energy use.

Like the commercial banks, the multilateral development banks (MDBs) have little direct influence in shaping portfolio flows from a financing angle. Still, as the world's most important lenders to developing countries, the standards they set for extending their own credit can have a significant influence via other financing institutions whose own resources are often leveraged by multilateral development bank involvement. For example, the World Bank's environmental guidelines, which became official in late 1997, have been widely adopted by other public institutions. The International Finance Corporation (IFC) also recently updated its Environmental Appraisal

Checklist. These standards, as well as similar ones from the European Bank for Reconstruction and Development and the Asia Development Bank, are commonly referenced by both public and private financiers, especially in developing "due diligence" protocols.

Private sector finance is also increasingly affected by multilateral development bank policies, via the expanding role of the latter in providing insurance and guarantees. The World Bank's Multilateral Investment Guarantee Agency, for example, lends relatively little money directly, but leverages considerable private investment by insuring against political and currency risks. The IFC operates in a similar manner. In addition to its own direct investment activities, the IFC has launched several environment-specific funds aimed at increasing environmental investments in developing countries. In the first instance, this helps to leverage money from the World Bank/ UNDP/UNEP Global Environment Facility for biodiversity preservation, climate change mitigation, protection of the ozone layer, and water pollution control. In turn, this contributes to leveraging private capital into new projects. In cases where these projects would not move forward without IFC's involvement, its influence on environmental content is crucial, since the IFC will not invest its own funds if a project does not meet its environmental guidelines. On the other hand, multilateral development banks may have less environmental leverage than they used to, given that the relative importance of private capital flows has increased substantially in recent years. Among the least developed countries, however, multilateral development bank involvement and overseas development aid lending remain the most significant sources of foreign capital, so any environmental rules imposed on these flows are still likely to be important.

Investment institutions probably hold important keys to environmental protection in the future. Consistent oversight of investment activities is therefore critical, as are transparent regulatory frameworks to govern the finance-environment linkage. Governments in both source and host countries need to work together to engage investors and investees in ensuring that investment contributes to real economic growth as well as appropriate levels of environmental protection.

Co-operation with developing countries

The globalisation of both environmental problems and economic activity has increased the need to take the views of developing countries into account within the international institutional framework. This implies that OECD countries need to strengthen their dialogue with developing countries about environmental problems that are becoming more and more a mutual concern. It also implies some effort to reconcile often fundamentally different perceptions about environmental problems, as well as fundamentally different capacities to do something about them.

One important institutional innovation in this area has been the development of the notion of "common but differentiated responsibility". Under this concept, developing and developed countries agree to work toward the resolution of common environmental problems, but the developed countries acknowledge their special responsibilities in contributing to their resolution. The Framework Convention on Climate Change and the Kyoto Protocol provide practical applications of this concept. The former applies the principle to the climate change debate; the latter establishes quantitative abatement targets for developed countries, but not for developing ones.

The idea of "capacity-building" has also become an important reference point for relations between developed and developing countries. This idea is based on the premise that nations should co-operate to strengthen endogenous capacities for sustainable development by exchanging scientific information, including the development and transfer of innovative technologies. UNCED recognised in 1992 that, although Agenda 21 provided guidance on the changes needed for sustainable development, most developing countries lacked the capacity to actually put these changes into effect. Capacity 21 was therefore created as one way of overcoming this deficiency. The Capacity Development for the Environment Programme of the OECD Development Assistance Committee is another initiative with similar objectives. The development of capacity for addressing environmental problems is an increasing focus of both bilateral and multilateral aid.

In the future, the role of developing countries in contributing to the resolution of trans-border environmental problems (or their trans-border economic implications) is likely to grow. This will probably increase the bargaining power of developing countries in negotiations toward possible solutions – a situation that may ultimately benefit both developing and developed countries. The developed/developing country relationship seems set to intensify over the period to 2020 along at least three key axes.

First, trade and investment liberalisation can be a positive force for environmental protection, provided that environmental governance systems are in place and adequate when liberalisation occurs. The risks of poor governance would seem, at first glance, to be higher in the developing countries than in the developed ones (although "poor governance" should not be confused with low environmental standards that reflect high assimilative capacities, or low values for environmental externalities). This suggests that it may well be in the interests of OECD countries to contribute to improved environmental governance capacities in developing countries, as a way of maintaining political momentum toward economic liberalisation. Such an approach would increase the likelihood that developing countries would benefit from trade and investment liberalisation, thereby increasing their support for that liberalisation in the first place.

Second, developing countries will clearly have a major role to play in reducing environmental pressures at the global level. More innovative ways will have to be found to actively involve developing countries in this effort. The Clean Development Mechanism, Permit Trading, and Joint Implementation under the Kyoto Protocol are examples of steps in this direction. Biodiversity protection is another area in which the co-operation of developing countries is likely to become more important in the future.

The third area in which closer co-operation between developed and developing countries seems set to increase over the next two decades is that of technology transfer. It is clearly inappropriate from an environmental perspective for developing countries to have to go through the same stages of technology experimentation that developed countries have undergone over the past half century. If a technology exists that would reduce international environmental pressures, it makes sense for that technology to be applied wherever it may be able to benefit the environment, including the environment of developing countries. Issues related to the definition of intellectual property rights (and the conditions under which these rights can be transferred) therefore seem likely to grow in importance over the period to 2020.

24.4. Stakeholder involvement

General public

Much discussion of modern environmental problems concerns the need to engage the public more directly in solving those problems. Partly, this is due to a recognition that better environmental policies are more likely when public preferences are taken into account; partly, it is due to the fact that engaging the public in an open and transparent manner is a useful insurance policy against poor implementation of environmental policies. Promises made in public are open to later questions about why those promises have not been kept. Transparency and public involvement in decisions are therefore basic elements of environmental democracy (see also Chapter 22).

The growing interest in participatory environmental decision-making reflects the following developments, among others:

- Decentralisation of government decision-making processes, and the resulting need to seek out partners with whom to work on environmental issues. Globalisation is probably enhancing these trends.
- Growing experience with multi-stakeholder dialogues involving the private sector, NGOs and other components of civil society.[3]
- More use of negotiated arrangements as a tool for environmental policy. These agreements sometimes directly involve civil society as a co-operating partner. They can play a useful role in building consensus.

3. UNCED defined "civil society" in terms of the "major groups" highlighted in Chapters 24-32 of Agenda 21. These included: Women; Children and Youth; Indigenous Peoples; Non-governmental Organisations; Local Authorities; Workers and Trade Unions; Business and Industry; the Scientific and Technological Community; and Farmers (UN, 1992).

- Many OECD countries have made progress in integrating environmental issues into their public education processes, and many have active information campaigns on a variety of topics, including: energy efficiency/ conservation, environmentally friendly driving habits, and environmental education programmes for farmers.
- Development of several international declarations and instruments which in themselves support participatory approaches. Three examples are: Principle 10 of the Rio Declaration; the Aarhus Convention on Access to Information, Public Participation, in Decision-making, and Access to Justice in Environmental Matters; and the OECD Pollutant Release and Transfer Registers (PRTRs) (established in the context of "Community Right-to-Know" programmes).

Non-governmental organisations (NGOs)

NGO inputs to policy discussions on the environment were originally provided by a fairly narrow base of special interest groups. These groups were often very successful in achieving their individual goals, but the process was relatively *ad hoc* and sporadic. It was also fundamentally confrontational. The result was that many NGOs tended to be single-issue-oriented. Later, in many countries NGOs became more and more involved in policy development and implementation, and in environmental education and monitoring.

Greater use of the Internet, coupled with the growing globalisation of economic activity, are now facilitating a more systematic, and more powerful, voice on the part of civil society. Public consultation processes for environmental issues have the potential to expand significantly as a result, with some OECD governments already opening electronic discussion groups on their Web sites. NGOs are also making use of new information and communication technologies to spread their messages and influence further. As indicated in Chapter 10 (Box 24.3), the World Resources Institute now uses an Internet-based forest management monitoring system, Global Forest Watch, to place environmental compliance and monitoring systems more directly in the public sphere.

NGOs are also focusing more on a wider range of environmental problems. The result is improved integration of civil society views into the mainstream of environmental policy-making. NGOs play an important role in alerting public authorities about irregularities or infringements of environmental obligations by individuals, firms, or even public authorities. In fact, provisions encouraging NGO participation in monitoring compliance with environmental objectives can be found in some national legislations or regional agreements. For example, the North American Agreement on Environmental Co-operation (the environmental side agreement to NAFTA) provides for a procedure under which any non-governmental organisation or person can alert that a Party to the Agreement is failing to effectively enforce its environmental law.

Some NGOs adhere to a strategy of confrontation with governments and industry, while others have opted to channel their activism in more collaborative directions with governments and with business. Both approaches can generate benefits. When co-operation is the objective, the NGOs usually bring a fresh perspective on environmental problems to the table. Where confrontation remains the standard, NGOs can often exert pressure on public opinion, resulting in more effective incorporation of environmental concerns into government and business decisions. The result of both approaches is that contacts among government, business and NGOs on environmental topics are more frequent and more fruitful today than ever before. In years ahead, the deeper participation of civil society in environmental discussions will probably place increased pressure on the resource capacities of individual NGOs. It will also place additional pressure on governments to find new ways of assuring a productive dialogue with these NGOs. In particular, questions concerning the "democratic deficiencies" of some individual NGOs will probably become more important.

Despite the expanding role of NGOs, the institutional links between governments and NGOs are still fairly rudimentary. In the future, it may be appropriate to seek new ways of strengthening this linkage, for example through such measures as expanded use of the Green G8 process (a consortium of green NGOs from the G8 countries), formal access for NGOs to key meetings and/or policy discussions, and/or a standing committee of "eminent" environmental institutes. Furthermore, arrangements for financing their contribution to meeting governments' policy needs remain to be worked out.

The business community

In the early days of environmental policy, environmental externalities were relatively easily attributable to individual business activities (*e.g.* "point source" pollution), and business generally was seen as profiting excessively from policies which failed to internalise these costs into their profit and loss statements. One result was that business came to react quite defensively to suggestions for new environmental policy initiatives by governments. Often, this also involved active opposition by firms to these initiatives.

In later years, recognition grew that active opposition to environmental policies was not a very good business strategy. Instead of overtly opposing environmental policies, firms therefore began to emphasise the need for more flexibility in how environmental objectives were actually going to be met (Carrod, 1997). Regulations came to be seen as cost-ineffective ways of meeting environmental goals, suggesting a preference for voluntary approaches. For the same reason, business tended to argue against eco-taxes, and in favour of tradable permit approaches. Firms also began to emphasise the possibility that they might relocate their activities to countries where the environmental costs of doing business were not so high.

For the future, it is anticipated that the business community will recognise more and more that a minimalist approach to environmental behaviour is not likely to be very productive over the longer-term, especially in the face of an increasingly sophisticated marketplace. Many firms are now moving toward more proactive approaches – approaches which involve the private sector operating more in partnership with government, customers, NGOs, and other interested parties.

Market pressure is increasingly being used as a vehicle to change corporate environmental practices. The public is now more capable of launching effective information campaigns aimed at firms that do not conduct their environmental affairs in a manner deemed to be environmentally (and socially) acceptable. Several recent high-profile cases involving large multinationals have resulted in a damaged public image, in boycotts of their products, and ultimately in a reduction in their profits and share values. Consequently, business is taking these problems more seriously, and is actively working to reduce them. One important manifestation of that effort has been the recent development of voluntary codes of conduct for enterprises, and the growing adherence to environmental management systems related to such international standards as ISO and EMAS.

One of the criticisms often levied against these codes of conduct is that they are little more than public relations exercises on the part of the companies involved. While this criticism may be valid in some circumstances, there is growing evidence that firms are using these instruments as a way of diffusing an ethic of continuous improvement in environmental performance throughout their operations. For example, recent OECD work finds that firms are focusing on the implementation phase of their codes – including their integration into internal management and control systems, the build-up of institutional capacity to pool the costs of developing such systems, and certification/verification systems.

Multinational firms are also more likely to consider using standardised environmental practices in all plants of their operations (*i.e.* regardless of location) as one way of improving overall corporate environmental performance. Another result has been the tendency for a growing number of firms of all sizes and origins to engage actively and transparently in consultations with civil society. These trends are likely to continue.

REFERENCES

Burke, T. (1998), "Globalisation, the State and the Environment", in OECD, *Globalisation and the Environment: Perspectives from OECD and Dynamic Non-member Countries*, OECD, Paris.

Buzan, B., J. de Wilde, and O. Waever (1998), *Security: A New Framework for Analysis*, Lynne Rienner, Boulder and London.

Carrod, B. (1997), "Business Strategies, Globalisation and the Environment", in OECD, *Globalisation and Environment: Preliminary Perspectives*, OECD, Paris.

Hirst, P. and G. Thompson (1995), "Globalisation and the Future of the Nation State", in *Economy and Society*, 24(3): 408-442.

Kimball, L. (1992), *Forging International Agreement: Strengthening Inter-governmental Institutions for Environment and Development*, World Resources Institute, Washington, DC.

OECD (1999), *The Price of Water: Trends in OECD Countries*, OECD, Paris.

Reinicke, W. H. (1999), "The Other World Wide Web: Global Public Policy Networks", in *Foreign Policy*, Winter 1999/2000.

Rosecrance, R. (1996), "The Rise of the Virtual State", in *Foreign Affairs* 75(4): 45-61, July/August.

Finger, M. (1999), "Globalisation and Governance", in *Policy Matters* 6, December.

Smets, H. (1999), personal communication.

UN (United Nations) (1992), AGENDA 21: *Programme of Action for Sustainable Development*, United Nations, New York.

Waller-Hunter, J. (1999), "Appropriate Institutions for the 21st Century", in IUCN, *Imagine Tomorrow's World: 50th Anniversary Symposium Proceedings*, Gland, Switzerland.

World Bank (1998), *Global Development Finance* 1998, The World Bank, Washington DC.

World Commission on Environment and Development (1987), *Our Common Future*, Oxford University Press, Oxford.

25
Policy Packages to Address the Main Environmental Problems

25.1. Introduction

Comprehensive packages of policy instruments will need to be adopted in OECD countries to successfully address the "red lights" identified throughout this report – *i.e.* the environmental pressures and conditions that most urgently need attention by OECD countries. Because of the complexity of many of the most urgent pressures on the environment, their often inter-connected nature, and the limited understanding of some of their causes and effects, single policy instruments will seldom be sufficient to effectively resolve these problems. Instead, combinations of policy instruments will be required which target the range of actors affecting the environment, draw on synergies for realising the different environmental policy objectives and avoid policy conflicts, and which address any social or competitiveness concerns about the policy instruments.

The appropriate policy packages will need to be carefully designed in order to achieve the desired environmental objectives in the most cost-effective manner. Policies will be needed which deal with intervention failures, such as inefficient subsidies or poorly designed regulations, and with market failures that stem from the non-internalisation of the environmental costs in the market or the lack of markets for particular resources or environmental services, such as biological resources. The policy packages required to address the identified "red light" problems will in general include a combination of:

- *economic instruments* – taxes, charges and fees, the reform of environmentally damaging subsidies, transferable permits;
- *regulatory instruments* – standards, licenses, permits, regulations, restrictions, etc.;
- *voluntary approaches* – negotiated agreements, unilateral commitments, public voluntary commitments;
- *incentives for technological development and diffusion* – appropriate pricing signals, regulations, support for research and development, etc.;
- *information-based instruments* – data collection and dissemination, indicators, information provision, valuation, education and training, eco-labelling, etc; and
- *other policies* – zoning and land-use planning, infrastructure provision, etc.

Where it is possible to apply them, economic instruments will often be the preferred policy instruments as they can be both efficient (*i.e.* least-cost) and effective (*i.e.* achieve the desired environmental objective) in the right circumstances. However, in some cases the administrative costs associated with their use may be excessive or political barriers to their use may exist, such as concerns that these instruments may be socially regressive or may affect the competitiveness of particular sectors. In addition, their apparent advantages depend upon the precise nature of the environmental problem. One of the main advantages of economic instruments – ensuring the efficient distribution of abatement efforts – generally does not apply for site-specific environmental impacts. Furthermore, it is often not possible to directly target a particular environmental problem due to high monitoring costs, as is the case with monitoring the diffuse run-off of agrochemicals from farms. In some cases an indirect economic instrument – such as a tax on chemical fertiliser or pesticide purchases – may be a reasonable proxy for addressing the environmental problem. However, there will certainly be a continued role for some regulations and restrictions, and perhaps an increased role for "newer" instruments (*e.g.* information-based instruments and voluntary approaches), to complement economic instruments.

Direct regulations will continue to be needed, in particular to deal with issues for which environmental impacts are potentially non-reversible or which exceed the carrying or assimilative capacity of the environment. In such cases, direct regulations or restrictions can be used to set absolute limits on the production or use of certain substances, or the type and level of use of environmental resources, thus assuring the desired environmental outcome as directly as possible. Regulations are also often useful where impacts are very site-specific or for which monitoring costs are high (*e.g.* regulating emissions of air pollutants from motor vehicles). However, in many cases, direct regulations will not be the most cost-effective instruments available because, among other things, it is generally difficult to ensure that the marginal costs of environmental improvements are equalised across the regulated actors.

Voluntary agreements may have an important role to play as well, particularly when political considerations prevent the use of other instruments. They may also provide the conditions for a relationship in which valuable information about abatement costs and technological opportunities are shared between firms and the regulatory authorities. While voluntary agreements are generally not considered to be effective as a stand-alone instrument, in combination with and as a supplement to other policy measures they may contribute to improvements beyond the minimum requirements of regulations. They can work well if they are carefully designed and implemented with clear objectives, and with targets and timeframes for achieving them.

Information-based instruments are usually most appropriate for overcoming information barriers related to the environmental characteristics of products (*e.g.* eco-labels for CFC-free refrigerators) or processes (*e.g.* certification schemes for sustainable forestry management) about which consumers feel strongly. In such cases, they can be used to inform the public and enable consumers to choose (and pay for) products or processes with the environmental qualities they desire. Information-based instruments are most successful when there are close links between the "private" (*e.g.* financial, health) and the "public" (*e.g.* environmental impacts) attributes of the products or processes – allowing consumers to choose products or processes which benefit both themselves personally and the public good. Fuel-efficient cars and organic agriculture are two such examples. Strengthened environmental data collection and appropriate indicator development and use will also be essential for the shaping of public opinion concerning environmental issues, the development of environmental policies, and monitoring the progress of these policies in reaching the environmental objectives and targets that have been set.

While all the policy instruments discussed above can be used to drive technological developments and diffusion in environmentally friendly directions, they can also hamper beneficial technological innovation if they are poorly designed. Thus, regulations specifying the use of particular technologies or processes, or subsidies that are directly tied to certain inputs or processes, can lock-in the use of the technologies, processes or inputs that are supported or required, hindering the potential development of new and improved technologies. Direct support to technological development can be appropriate when there are "positive" spillovers associated with the development of environmentally preferable technologies. In such cases, the market prices may not provide strong enough incentives for innovation to address the environmental problem successfully or quickly enough. The support can be designed as an internalisation of the positive public good aspect of the technological innovation.

The key is to find the right mix of environmental policy instruments, and not to superimpose instruments which send conflicting (or redundant) signals. Some examples of complementary policy packages for addressing environmental problems include the use of:

- information-based measures to increase the speed and magnitude with which firms and households respond to taxes and charges;
- a voluntary agreement as the basis upon which to "grandfather" permits under a tradable permit system; or
- direct regulations applied as a "safety-net" in cases when particular environmental conditions are exceeded under other instruments (*e.g.* to limit car use in urban areas).

This chapter examines some environmental policy packages that could be adopted to effectively address the "red light" issues identified in the report. These packages are not limited only to those policy instruments that are available to environment ministries. Many of the instruments discussed here would need to be implemented either by, or in co-operation with, authorities outside environment ministries. The co-operation of finance authorities is needed to implement environmental taxes, and the co-operation of sectoral authorities is needed to implement policies which affect particular sectors, such as agriculture, fisheries, forestry, energy, transport, and selected industries. This chapter presents the results of model-based simulations of some of the economic policy instruments included in the packages, indicating the potential economic, sectoral and environmental impacts, accompanied by a more qualitative discussion of other instruments with which they could be combined. Further analysis of the full benefits and trade-offs of these proposed policy packages is of course needed. Because implementation is often the weak link in the environmental policy cycle, some policy implementation issues and potential solutions are also discussed here.

25.2. A policy package for the primary sectors and selected natural resources

The pressures on the environment arising from primary sectors, and the possible changes in the state of the environment to 2020 for selected renewable natural resources, were described in Section III of this report. For a number of these pressures and issues, OECD countries have already implemented, or are starting to implement, policies that can successfully address the problems. Such policies have so far resulted in increased forest area and volume in OECD regions and reductions in per capita water use, although the latter are not yet sufficient to reduce total water use. Some of the policies and societal responses that are helping to address environmental pressures from primary sectors include the expansion of environmentally sound management practices in agriculture, increased development of plantation forests and aquaculture fisheries, and the growth in the number and total area of protected lands.

Despite these advances, significant threats to ecosystems remain. Over-fishing of marine fisheries is an increasing problem worldwide, with the majority of the world's capture fisheries already exploited to their full extent, over-fished, or recovering from over-fishing. Projections indicate that fish catch from world capture fisheries is unlikely to increase and may even decline in the period to 2020, as stocks are depleted. While the growth of environmentally sound agricultural practices (*e.g.* organic farming, integrated pest management) is a positive sign in OECD countries, their use is still very limited. Groundwater pollution is expected to continue to worsen in OECD countries to 2020, with the main source remaining the high levels of agricultural pollution through run-off of farm chemicals and nutrients. Furthermore, while efforts of OECD countries to improve conditions at home – for example through aforestation programmes and the expansion of protected areas – are steps in the right direction, they are generally insufficient to change the overall global trends. Global deforestation and biodiversity loss are thus continuing at an alarming rate.

This section presents a package of policies OECD countries can adopt to help address these pressing issues relating to primary sectors and natural resource use – in particular, agricultural pollution, the contamination of groundwater resources, over-fishing of the world's capture fish stocks, tropical deforestation, and global biodiversity loss.

Economic instruments

Subsidy removal or reform

Subsidies to natural resource use – particularly to fishing, agricultural production, forest logging and irrigation water usage – remain high in both OECD and non-OECD countries. Many of these subsidies contribute to over-exploitation of the resource base, ecosystem degradation, and biodiversity loss. As such, an important first step in tackling the "red light" issues identified for the primary sectors and natural resources is the removal of subsidies that encourage environmentally damaging activities.

A simulation has been undertaken with the *Outlook* modelling framework to indicate the potential effects of removing subsidies to primary sectors and the use of natural resources in OECD countries. All the subsidies to the agriculture, forestry, fisheries, and water supply sectors, or to the use of products from these sectors, were removed in the simulation.[1] A total of US$81.5 billion in subsidies to these sectors is specified in the GTAP database for

1. These correspond to subsidies listed in national accounts, as summarised in the GTAP database (Version 4) used in the JOBS model (see Annex 2). Because of a lack of disaggregation of the subsidies in the database, it was not possible to simulate the results of removing environmentally damaging subsidies alone. Instead, all the subsidies listed were removed for the exercise.

1995, amounting to almost 7% of the gross value of output from the sectors. The vast majority of the identified subsidies (US$81 billion) were given to the agriculture sector (Table 25.1). Despite the large levels of support that are listed as going to agriculture, the database still does not reflect all the support to the sector, in particular it does not reflect market price support.[2] An additional exercise was therefore undertaken to approximate the effects of removing market price support to agricultural production, in addition to removing the agricultural subsidies listed in the database. The levels of support to the forestry, fisheries, and water sectors identified in the database are likely to be underestimates of the support to these sectors as well. The indirect nature of most forestry subsidies makes their quantification difficult, but estimates are available of direct government support (*i.e.* not including market price support) to fisheries in OECD countries. These indicate that OECD governments transferred US$6.3 billion to their fisheries industries in 1997, while under US$20 million in subsidies are listed in the database for 1995 (Table 25.1). Given the underestimation of total subsidy levels in the model, the results of the policy simulation should be considered only as an indication of the direction, and rough minimum level, of the effects that might be obtained from subsidy removal in the sectors concerned.

Table 25.1. **Subsidies to primary sectors and natural resource use in OECD countries according to actual estimates and as reflected in the Reference Scenario, 1995 or most recent year, US$ billion**

	OECD estimates	de Moor and Calamai estimates	Reflected in the Reference Scenario
Agriculture	362	335	81*
Forestry	..	..	0.18
Fisheries	6.3	..	0.02
Water use	..	42-47	0.32

* In addition to simulating the removal of the US$81 billion in subsidies to agriculture listed in the GTAP database for 1995, a further exercise was undertaken to simulate the removal of market price support to the sector in OECD countries. In essence, this simulation applied a tax on agricultural production with a tax rate equal to estimates of the market price support in the OECD regions in 1997. For practical reasons, this tax was phased-in over a 5-year period. When fully phased-in, it generated around US$312 billion in "extra" tax revenue. This "tax" was combined with a "subsidy" to the household consumption of agricultural products (meat and other food) of an equivalent amount.

Sources: OECD (1999*a* and 2000), de Moor and Calamai (1998), and Reference Scenario.

While variations in the environmental effects that might result from removing subsidies to primary sectors would be expected to occur between and within regions, the effects of the subsidy removal are estimated to be positive for the environment in OECD regions overall (Figure 25.1). Thus, total irrigation water use in OECD countries is expected to be about 11% lower in volume terms in 2020 compared with the Reference Scenario projections, while nitrogen loading of waterways from agriculture (through fertiliser use and livestock waste) would be about 6% lower. Methane emissions (both from rice and livestock production) are also estimated to be lower in 2020 than projected under the Reference Scenario, but only marginally. Finally, the subsidy removal simulation indicates that some "leakage" could be expected in terms of environmental effects, with global demand for agricultural products increasing as a consequence of removing market price support measures in OECD regions and some shifting in production to non-OECD countries. As a result, global methane emissions and nitrogen loading to waterways from agriculture are estimated to increase marginally in the subsidy removal simulation in 2020 compared with the Reference Scenario. These results highlight the need for both co-operation with non-OECD countries in the reduction of support measures and ensuring that the correct framework conditions (and any necessary accompanying policies) are in place to guarantee that the potential positive environmental effects of these policies are realised.

Where removal of the subsidies is difficult for political or social reasons, it may be possible instead to reform them so that they are no longer tied to environmentally damaging activities – a process that is already occurring in some OECD countries for agricultural and fishery subsidies. Many countries have started to replace agricultural subsidies tied to production levels (*e.g.* market price support) or input use (*e.g.* irrigation water, agrochemicals) with agri-environmental measures that provide payments for activities such as land set-aside schemes, tree planting

2. Market price support measures guarantee a minimum price level for the producer above the market price and provide accompanying measures to ensure guaranteed sales of a certain level of production in excess of demand.

for wind breaks, soil protection and conservation schemes. Similarly, efforts are underway in many OECD countries to reform environmentally distorting subsidies to the fishery sector, particularly subsidies that encourage excess fleet capacity.

Figure 25.1. **Effects in 2020 of removing subsidies to primary sectors and applying a tax to agrochemicals in OECD regions**

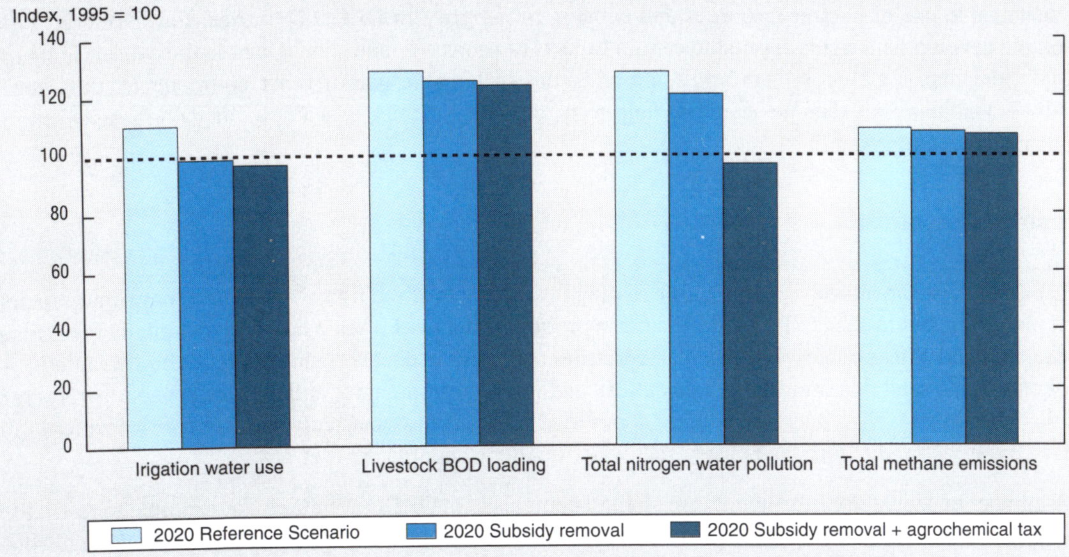

Source: Reference Scenario and Policy Simulations.

Introducing or modifying environmental taxes and charges

In addition to removing subsidies which encourage the over-use of natural resources, OECD countries also have significant scope for increasing the use of economic instruments (*e.g.* taxes, charges, fees, tradable permits) for natural resource management. Many OECD countries already levy charges for water use or abstraction, stumpage charges for timber harvesting, hunting licenses or fees, individual transferable quotas for fishing, or park entrance fees (OECD, 1999*b*). However, these are often set at very low levels which do not fully reflect the external costs of using these resources, and thus do not comply with the User Pays Principle. A more limited number of countries have implemented taxes on pesticide and fertiliser use, and forest clearing. Effective use of such instruments can help ensure that the full environmental costs of using natural resources are internalised in the decision-making process of natural resource users – thus creating incentives for the optimal level of use. In addition to providing incentives for reduced or more sustainable use of resources, these instruments can help generate government revenues, allowing the possibility for reducing other taxes in a revenue neutral tax reform.

Another model simulation was made reflecting the introduction of a tax on chemical inputs used by agriculture (*e.g.* fertilisers, pesticides), in addition to the subsidy removal described above. The tax was set as an *ad valorem* tax on chemical inputs to agriculture in OECD regions that increases by 2 percentage points each year, such that it would be equal to a 50% tax on these inputs by 2020. This simulation led to particularly strong effects in the emissions of nitrogen from crop production in OECD regions, indicating that such a tax might lead to decreases in nitrogen-loading of waterways from agricultural run-off and nitrous oxide (N_2O) emissions (a greenhouse gas) from agricultural soils compared to 1995 (Figure 25.1). Although not quantified in the modelling framework, one can assume that such a tax would also lead to reductions in pesticide use. The estimated effect on real GDP in OECD regions of removing the primary sector subsidies as described above and adding this tax was negligible. A very small negative impact on the value added of the agricultural sectors in OECD economies was found, amounting to a reduction of less than 0.1% of value added of these sectors in 2020 compared with the Reference Scenario.

Economic instruments also have an important role to play in ensuring the conservation or sustainable use of biological diversity and its components. Greater use of park fees or charges, resource extraction charges, hunting and fishing fees, subsidies for conservation efforts, and taxes on unsustainable activities could help to ensure the more sustainable use of natural resources and ecosystem integrity in OECD countries. Furthermore, the use of transferable development rights applied to certain aspects of protected lands could enable the conservation of some essential elements of the ecosystem while allowing the sustainable use of other components, thus generating incomes. Tradable permit schemes can also help to meet water scarcity or water quality concerns within a river basin management approach.

Regulatory instruments

Economic instruments have a strong role to play in tackling the "red lights" identified for primary sectors and natural resources, but to successfully address these problems they will often need to be complemented by regulatory measures and other policies. In particular, countries may wish to enact regulations regarding the allowable toxicity of chemicals used in agricultural, aquaculture and forestry production, and their effects on the environment and human health, as well as address concerns about food safety, control particular site-specific activities (*e.g.* limit intensive livestock production), and ensure that water quality standards are met.

A number of regulatory instruments are already being used in OECD countries, for example to restrict fishing activities to sustainable levels and practices. Regulations are set to limit fish catch (through setting Total Allowable Catch levels), govern the methods used to catch fish and the allowed catch characteristics (*e.g.* age, size), and limit by-catch discard and the degradation of marine ecosystems. In many fisheries, however, these regulations are not sufficient to address the problems of over-fishing, and instead lead to a race-to-fish or to shifting the problem to unregulated fisheries. Furthermore, a lack of adequate systems for monitoring compliance with regulations and other policy instruments has resulted in poor enforcement of fisheries policies in many cases.

The use of regulations and access restrictions to protect biodiversity or biological resources is also fairly widespread in OECD countries. They usually take the form of the creation of natural parks and conservation zones or the enactment of prohibitions on the use or hunting of wildlife species. Because the implementation and enforcement of access or use restrictions tend to be expensive both to government administrations and local communities, they are often accompanied by positive incentives for sustainable use of the resources, or economic instruments that help raise government revenues or apply charges to cover the costs of conserving the resources.

While maintaining forest coverage and meeting demand for forest products is not likely to be a problem for OECD countries in the coming years, policies will be needed to ensure that the right balance is struck between the protection of the remaining old growth forests, the intensification of managed forests, and the development of plantation forests. Appropriate regulations will need to be developed to limit any negative environmental pressures associated with the intensive production of plantation forests.

Technological development and diffusion

Considerable scope exists in the agriculture sector for the adoption of new technologies and management practices beneficial to the environment. The diffusion of drip-irrigation systems and precision application systems, which allow the accurate matching of agrochemical inputs to local conditions and needs, have significant potential to increase input efficiency, while alleviating the pressures on water resources exerted by the agriculture sector. More significant increases in input efficiency and productivity may also be realised through the development of biotechnologies (*e.g.* GMOs) that require less chemical inputs or water, increase crop growth, and are resistant to adverse environmental conditions. However, some of the new biotechnologies under development may lead to negative impacts on the environment as well. The extent of the future use of GMOs is likely to depend on the analysis of their potential health and environmental effects, and on consumer acceptability.

Similarly, the development of aquaculture fisheries and plantation forests allow for rapid increases in output – removing some of the pressures on the natural resource bases in the wild – but can also have harmful effects on the local environment because of the intensive nature of their operations. Policies will need to be designed to address the negative environmental effects of these practices, while still encouraging the development of technologies and practices which can ensure more efficient production.

Fishing technology developments are generally geared towards improving the efficiency and profitability of fishing activities or fish processing, and consequently often contribute to over-fishing of fish stocks. New technologies can, however, also be adapted to support sustainable fishing practices by the use, for example, of satellite-based vessel monitoring systems to increase compliance by vessel operators with fisheries management rules and to collect data on catch and fishing, or by the use of turtle excluder devices on fishing nets.

Many environmentally benign technologies are already on the market, and generally the most effective encouragement for their adoption will be economic incentives. For example, drip irrigation techniques can significantly reduce the amount of water used by farmers, but the costs of installing such a system are relatively high. Given the relatively low savings that could be earned in water costs with current water prices, the use of such systems is limited. However, reducing the substantial subsidies to irrigation water use in OECD countries would make the use of drip irrigation techniques more profitable for farmers. In cases where subsidy removal or tax increases is either politically not possible or would not provide sufficient incentives for the uptake of existing or new environmental technologies, governments that wish to further their use can do so through other measures. In particular, they may choose to directly support research and development of the technologies, engage in farmer education and training programmes, or introduce fiscal incentives to make up the cost difference between the use of traditional techniques and the less environmentally damaging alternatives.

Voluntary agreements

Voluntary agreements in the field of agriculture can complement other policy instruments by involving farmers and farmer associations in the implementation of more environmentally friendly practices. By doing so, voluntary agreements may enhance co-operation between authorities and community groups, and raise farmer awareness of the environmental issues at hand. To ensure that the best use is made of such initiatives, however, governments will need a strong regulatory and economic framework to support them. Voluntary approaches have already been developed in a number of OECD countries in response to various natural resource management concerns. These include voluntary agreements to encourage water users (especially industry and agriculture) to take proactive steps towards reducing water abstractions and emissions to waterways, and to encourage sustainable forest management or fishing practices.

In general, such approaches will be most effective when they are combined with the removal of conflicting policies (*e.g.* subsidies to unsustainable resource use), or with the availability of some project funding to support group initiatives, and the provision of training regarding the available solutions to the environmental problem (OECD, 1998*a*). Close co-ordination with other stakeholders (*e.g.* local residents and recreational land users for voluntary agreements in the agriculture sector) will be needed to strengthen support for, and the accountability of, voluntary approaches.

Information and other instruments

In response to growing concerns by consumers, agricultural producer associations are developing eco-labels based on environmental and health criteria (*e.g.* organic produce) and animal welfare criteria (*e.g.* eggs from free-range chickens). Eco-labelling schemes are also developing in the fisheries sector, certifying that fish and fishery products have been harvested and processed in an environmentally responsible manner. Certification of forests as meeting sustainable management standards has been gaining importance in recent years as well, and some of the largest furniture producing and wood purchasing companies worldwide have decided to only source wood that is certified as coming from sustainably managed forests.

With proper monitoring and independent certification, eco-labelling schemes can allow consumers to directly influence the extent of sustainable activities in place. As a result, such schemes for forests, fisheries and agricultural products or processes can help foster the conservation or sustainable use of biological resources. For many of these activities, a number of national or international eco-certification schemes are being developed or applied, often covering different species or regions, but sometimes overlapping. Where possible, overlapping schemes should be harmonised or combined, with transparent and rigorous monitoring and certification procedures agreed, in order to build trust in the schemes and ensure that they are comprehensible to consumers. Care must also be taken that such schemes do not develop into trade barriers.

For many of the renewable natural resources examined, there is inadequate information on the state of the resources. In some cases this is the result of poor scientific understanding of the resources and their interactions, while in others this may be due to the cost of gathering and processing the relevant data. The collection, processing and dissemination of physical and economic information regarding biodiversity values is of paramount importance for designing appropriate policies, in particular because the complex nature of biodiversity and ecosystem functions means that a wide range of economic activities have an impact on them.

25.3. A policy package for climate change, air pollution, and the energy and transport sectors

Under current policies, OECD countries are likely to increase greenhouse gas emissions by approximately 33% to 2020, far from the overall Kyoto Protocol target of achieving by 2008-2012 a 5% reduction in emissions across the larger group of Annex I countries from 1990 levels. Similarly, urban air quality is expected to continue to worsen in OECD countries, although reductions in a number of pollutants (such as sulphur dioxide and carbon monoxide) are expected in most countries. The main contributors to greenhouse gas emissions and declining air quality are transport and other fossil fuel-based energy use. Pressures from transport and energy use trends will continue to increase to 2020. Motor vehicle kilometres travelled in OECD countries are expected to increase by over 40% from 1997 to 2020, and passenger air kilometres are expected to triple. Similarly, OECD energy use is expected to increase by roughly 35% to 2020. As a result, both motor vehicle and aviation air emissions fall under the "red light" categories, as do urban air quality and the emission of greenhouse gases.

Moving these sectors and issues out of the "red lights" will require a comprehensive policy package. Such a package should include a combination of economic instruments (subsidy and tax reform, introduction of new taxes or charges, and wider use of tradable permits systems), a strong regulatory framework (particularly for setting air quality targets or standards), the promotion of voluntary or negotiated agreements, and the use of information-based policy tools to encourage more sustainable energy consumption and production patterns. Targeted measures to stimulate the development and uptake of cleaner energy and transport technologies – such as research and development funding and support for demonstration projects – can help to accelerate the use of new technologies. In many cases, international co-ordination may be necessary to ensure support for and successful implementation of these policies, for example to overcome resistance to subsidy reform or to the introduction of carbon taxes. Careful monitoring of policies and their effects is also needed to strengthen and refine policies over time. Opportunities exist to combine policies in these sectors, and optimise across what are often seen as separate policy objectives. Air quality, climate change, and protection of the ozone layer, for example, can often be considered simultaneously in the design of policy instruments. This should lead to outcomes that deliver greater social and environmental benefits for a given cost. While the policies discussed here focus on national actions, policies to address climate change will of course continue to be formulated in the context of evolving international commitments, such as the UN Framework Convention on Climate Change to reduce greenhouse gas emissions and the Montreal Protocol on the protection of the ozone layer.

Economic instruments

Subsidy removal or reform

Subsidies are widespread in the energy and transport sectors, although they are often less explicit than in the primary sectors. Governments influence energy and fuel prices through measures that include preferential tax rates, government ownership of energy production or distribution systems, subsidised loans for capital, obligatory purchase commitments for certain fuels, direct and indirect budgetary transfers, trade barriers, and set prices (*i.e.* market price support). Governments also often subsidise public transport infrastructure and operating costs, road building and maintenance, and provide preferential low tax rates for aviation and ship fuels. Furthermore, the full external costs of transportation and fuel use (*e.g.* the environmental and social costs, such as human deaths and illness associated with air pollution) are rarely reflected in the prices users pay for these services.

Partly because of the broad range of measures in place that distort energy and fuel markets, it is particularly difficult to estimate the total level of subsidies that benefit these sectors. The GTAP database used in the modelling framework includes subsidies to fuel production and use, and to electricity and oil refining in OECD countries, that amount to US$4.7 billion. Within the OECD, the only energy-related subsidies that are systematically estimated are the subsidies to coal producers in selected countries. As indicated in Table 25.2, 1997 estimates for coal producer subsidies in five OECD countries amounted to almost US$8 billion, substantially more than the estimated total energy-related subsidies in the GTAP database, yet still an order of magnitude smaller than external estimates of total energy subsidies in OECD countries.

Table 25.2. **Subsidies to energy and fuel production and use in OECD countries according to actual estimates and as reflected in the modelling exercise, 1998 or nearest year, US$ billion**

	OECD estimates	de Moor and Calamai estimates	Reflected in the Reference Scenario
Coal production	7.8*	–	3.3
Total energy subsidies in OECD countries	–	70-80	4.7

* Estimate for five OECD countries (Germany, Japan, Spain, Turkey, UK) only.
Sources: IEA (1999), de Moor and Calamai (1998), and Reference Scenario.

Because of the limited estimates of energy and transport subsidies reflected in the model, the effects of a policy simulation to remove these subsidies are understandably also small. However, previous OECD analysis has shown that a targeted reform of energy and transport subsidies in OECD countries could significantly reduce greenhouse gas emissions without increasing overall economic and social costs (OECD, 1998*b*). OECD case studies indicate that energy subsidy reform could reduce CO_2 emissions in the range of 1-8% of energy-related emissions at the national level, while also improving economic performance (OECD, 1997). The studies indicated that transport subsidy reform could reduce emissions from that sector by 10-15%.

At a minimum, subsidies that support the use of fossil fuels over other, less environmentally damaging, fuels should be reformed to "level the playing field". As the share of nuclear fuel in the fuel mix of OECD countries is expected to decrease over the coming decades, actions will need to be taken to ensure that this is replaced with environmentally sound alternatives. The removal of obligatory purchase commitments for fossil fuels by electricity generators will help achieve a less environmentally harmful fuel mix, as will continued research and development efforts to bring costs down for renewable and other environmentally benign alternatives. A gradual removal of the restrictions on prices that electricity users pay would also contribute to limiting total energy demand.

Regulatory reform is accompanying, and even preceding, subsidy reform in a number of OECD countries. Regulatory reform of OECD energy and transportation sectors can help to establish competitive markets in these sectors, laying the foundation for increased economic efficiency. However, the effect of regulatory reform on the environment in general, and on greenhouse gas emissions in particular, is unclear. As a result, other policies to protect the environment will need to accompany market or regulatory reforms in these sectors.

Introducing or modifying environmental taxes and charges

As was demonstrated during the oil crises of the 1970s, fuel and electricity price increases have a significant influence on producer and consumer behaviour. Higher prices will help to shift patterns towards less wasteful ways of consuming energy and considerably reduce the energy intensity of the economy. Currently the vast majority of environmentally related tax revenues in OECD countries are linked to vehicles and vehicle fuels, while taxation of other fossil fuels (*e.g.* coal) and other energy consumption is very low.[3] Taxes could be restructured so that prices better reflect relative carbon content of fuels or products. While a few OECD countries have introduced carbon or energy taxes in recent years, the tax rates are in many cases quite low, and often exemptions are granted to energy-intensive industries for competitiveness reasons. As a result, the environmental benefits of the taxes are severely limited. In many instances, the exemptions and low tax rates reflect political difficulties of raising energy and fuel prices, as was demonstrated by protests in a number of OECD countries against the oil price increases during 2000.

A model simulation was undertaken reflecting the removal of the subsidies to energy and transport listed in the GTAP database, combined with the application of a yearly increasing *ad valorem* tax on fuel use in OECD countries whose increase is linked to the carbon content of the fuels. The tax was thus set to increase by 2, 1.6 and 1.2 percentage points per annum for coal, crude oil and natural gas respectively. The tax rate thus reached the level of 50%, 40% and 30% of pre-tax prices respectively in 2020.[4] According to this simulation, implementing the combined subsidy removal and fuel tax policy package in OECD countries would reduce CO_2 and SO_x emissions in OECD regions in 2020 by as much as 25% each compared with the Reference Scenario. The simulation indicated only a small leakage effect of emissions to non-OECD countries, such that worldwide there would be a net decrease of 11% in CO_2 emissions and of 9% in SO_x emissions in 2020 compared with the Reference Scenario. From 1995 to 2020 (see Figure 25.2), this

Figure 25.2. **Effects in 2020 of removing energy subsidies and applying a fuel tax in OECD regions**

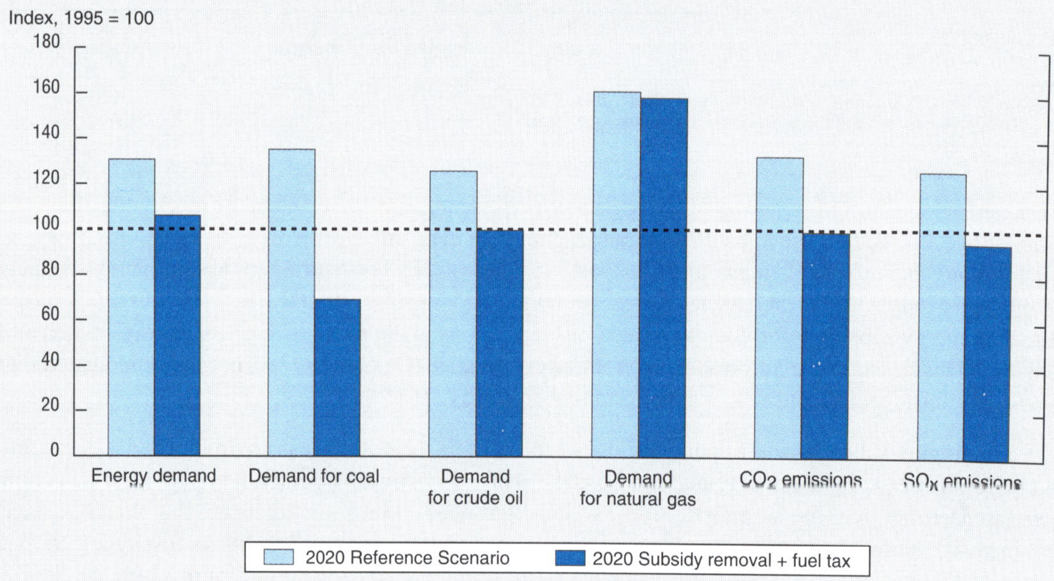

Source: Reference Scenario and Policy Simulations.

3. See the OECD Environmentally Related Tax Database at *www.oecd.org/env/policies/taxes/index.htm.*
4. For technical reasons, the policy simulation is related to the pre-tax prices of the fuels, and not directly to the carbon content of each fuel, and so does not, strictly speaking, reflect a "carbon tax".

combination of policies is estimated to result in a net decrease in CO_2 and SO_x emissions in OECD countries, driven primarily by the large decreases in coal use.

The subsidy removal and energy tax policy package was found to result in slightly reduced economic growth in OECD regions. Thus, real GDP for OECD regions might be roughly 0.1% lower in 2020 under this policy simulation than the level projected in the Reference Scenario. While the CO_2 reductions achieved under this simulation would not be sufficient to achieve Kyoto targets agreed by OECD countries, another OECD study recently indicated that the achievement of such targets by 2008-2010 might cost 0.2% of GDP, with expected costs dropping by a third if mitigation occurs not only for sources of CO_2, but also of CH_4 and N_2O (Burniaux, 2000). Costs were found to be reduced even further to 0.1% of GDP if efficient markets for emission trading among Annex I countries were to be combined with multi-gas mitigation.[5]

A removal of energy subsidies and introduction of new energy taxes would affect certain economic sectors more than others. Under the policy simulation discussed above, value added from the coal sector in OECD countries decreased by almost 33% in 2020 compared with the Reference Scenario. Where subsidy removal or environmental tax use negatively impacts on the competitiveness of particular economic sectors (*e.g.* the coal industry, commercial transportation sector), or raises energy of fuel prices to households, opposition to the reforms will be likely. In such cases, there is a risk that – without complementary measures to address any equity issues which might arise – the reforms might either be blocked or that large exemptions would be granted to the affected sectors, thereby reducing the effectiveness of the policy. In such cases, measures are available which can accompany the environmental policy and address many of these concerns (see Section 25.6). For example, international co-operation for the harmonisation of energy taxes can help to reduce the need for tax exemptions for internationally competitive industries.

Restructuring existing fiscal measures in a more environmentally friendly way could complement this policy package. For example, in the transport sector, the structure of taxation of motor vehicle use based on vehicle ownership could be altered to reflect the pollution potential of different vehicles, or changed to taxation linked directly to fuel use. Other demand-side management tools, such as road pricing charges, can also contribute to providing the right incentives for achieving multiple objectives simultaneously, for example reducing local air pollution, traffic congestion and greenhouse gas emissions. Furthermore, international discussions on the taxation of fuels used in air transport should continue with an aim to adopting tax levels that reflect the external costs of aviation fuel use.

The use of tradable permit schemes

With a few exceptions, use of tradable permits for addressing air pollution or greenhouse gas emissions by OECD countries is limited. However, there is considerable scope for their introduction. Tradable permit schemes can lead to substantial efficiency gains in the achievement of pollution reduction targets compared to regulatory standards, provided that the transactions costs of implementing the schemes are kept low. Success also depends upon clear rules for ownership rights and obligations, together with effective monitoring, reporting and legal enforcement of private entity limits. Limited experience in OECD countries already exists, with some countries using transferable permit schemes to address CO_2, SO_2, NO_x, VOC emissions, and ozone depleting substances (OECD, 2001*b*). For instance, the acid rain programme in the US has contributed to significant reductions in SO_2 emissions at much lower costs than originally estimated. Averaging schemes[6] for fuel efficiency objectives in motor vehicle production are another promising example. Such schemes are particularly cost-effective where they cover a large number of emitters with widely varying marginal costs of pollution reduction, and when the environmental benefits do not vary with the location of the emission reduction.

Tradable permit schemes are also being discussed at the international level, most notably in the context of reducing global greenhouse gas emissions under the Kyoto Protocol. The introduction of the three Kyoto mecha-

5. However, these simulations disregard any short- to medium-term adjustment costs associated with the policy changes.
6. Averaging schemes involve the competent authority setting average limit values for emissions from motor vehicle engines, which vehicle manufacturers have the option of exceeding for some of the products they sell on the market provided that these overshoots are offset by lower average levels with regard to other products.

nisms (international emissions trading, the Clean Development Mechanism and Joint Implementation) would enable emissions to be cut at least cost. As mentioned above, an OECD study found that making use of these mechanisms would have the potential to reduce the costs of achieving Kyoto targets considerably (Burniaux, 2000).

Regulatory instruments

While economic instruments such as charges, taxes and tradable permits tend to be the most economically efficient policies for achieving reductions in air pollution and greenhouse gas emissions, they often need to be complemented by other types of policies. In some cases, their use may simply not be viable. For example, for some air pollutants, the level of tax or charge that would be needed to achieve the agreed emission reduction target may be too high to be acceptable in a political context, or there may not be a suitable base for levying the tax or charge. Regulations may also be the best instrument when the emissions to be controlled are toxic. Thus, regulatory measures are used in OECD countries to restrict the use of highly polluting fuels (*e.g.* heavy fuel oil, residual fuels) or to limit the content of certain elements (*e.g.* sulphur, heavy metals, toxics) in fuels. Clearly defining air quality targets based on acceptable concentrations of pollutants according to their effects on human health and ecosystems is often an essential step towards developing appropriate policy instruments to achieve these targets. Emission standards for motor vehicles are in place in most OECD countries, although further emission reductions – especially for diesel vehicles and motorcycles with respect to CO, NO_X, VOC and particulate emissions which are harmful to human health – will generally be needed to achieve agreed air quality targets.

Regulations may be increasingly used in the future to meet the challenge of reducing total energy demand and greenhouses gas emissions. They can be employed to direct energy production towards a more environmentally friendly fuel mix and conversion efficiency through requiring a greater share of renewables in the fuel mix and the use of high efficiency applications, including electricity co-generation and low emission motor vehicles. Some countries are also adopting policies that can reduce the environmental impacts of energy and transport use by requiring that a certain percentage of new power generation or new motor vehicles be fuelled by renewable or non-fossil fuel energy sources. While regulations can accelerate development and the uptake of new, clean technologies, OECD analysis indicates that they may be relatively costly policies for reducing greenhouse gas emissions when compared to economic instruments (OECD, 2001*a*).

Technological development and diffusion

Programmes supporting the development and diffusion of "clean" technologies through research and development (R&D) policies are a traditional focus of energy and transport policy in OECD countries, and they are becoming an increasing component of domestic climate change policies. Historically, OECD government R&D expenditures relating to energy have been heavily weighted to coal and nuclear technologies (IEA, 1999). Although the share of R&D expenditures devoted to renewables and energy efficient technologies has been increasing in recent years, it is still a relatively small proportion of overall energy R&D expenditures. Governments should take a careful look at the balance of funding for these competing technologies, given the environmental objectives of the future.

Government partnerships with industry can facilitate development, demonstration and more rapid diffusion of clean transport and energy technology. While existing technologies can help to reduce many air pollutant and greenhouse gas emissions, their adoption is often not profitable. In the field of transport, there are combinations of advanced emission control and fuel efficiency technologies which could reduce the environmental impacts of motor vehicles significantly, but diffusion is very slow. Governments can facilitate their adoption through increased fuel prices, the gradual tightening of performance standards, or through sector target-setting. Meanwhile, natural gas and bio-diesel could be further promoted, in particular for urban buses and corporate vehicle fleets. The use of hybrid and electric vehicles are beginning to spread slowly in OECD countries, and research is ongoing into fuel cell vehicles which can reduce the use of fossil fuels and many of the environmental impacts of transportation. Government support for demonstration projects may be critical in the near term to give a boost to new technologies and to familiarise the public with them.

In addition to direct government expenditures on environmental R&D and demonstration projects for the energy and transport sectors, governments can also foster industry research and development investment through a variety of other policies. As already mentioned, these can include fiscal measures, such as fuel charges or taxes which encourage greater fuel-efficiency, and – where they are linked to fuel carbon content – substitution towards lower carbon fuels. They can also assist in the development of niche markets for environmentally friendly alternatives through the removal of barriers to market entry (*e.g.* off-grid power), or the development of eco-labels or other information schemes which enable consumers to make informed choices regarding their energy or fuel use.

Voluntary agreements

Voluntary approaches or agreements as a stand-alone instrument to address greenhouse gas emissions and air pollution can be subject to a number of drawbacks. If they are not properly designed, implemented and monitored, their effectiveness in reducing energy use by the industrial and power sectors is questionable. However, in combination with other policy measures, they can be useful to encourage industry to make improvements beyond the minimum requirements specified in regulations. In the transport sector, long-term agreements on fuel efficiency improvement have been made between governments and vehicle and aircraft manufacturers, airlines and rail operators in OECD countries. Similar forms of agreements exist in many OECD countries for energy-intensive industries. Voluntary agreements may further be useful in exploring mitigation action in areas where more knowledge about technical mitigation options and costs is essential, such as with industrial process emissions of the other greenhouse gases (SF_6, HFC and PFC).

Information and other instruments

Education and public awareness policies are also an essential component of a policy package that aims to reduce urban air pollution and respond to climate change. Information transmitted to the consumer through labels – such as energy efficiency labels for appliances and equipment, or fuel efficiency labels for motor vehicles – can help consumers to identify and choose more environmentally friendly products. Although experience with labelling of fuel consumption has shown that simple consumer information has limited effects on purchasing decisions, higher fuel and energy prices can increase the effectiveness of these instruments. Some power supply systems are also now offering customers the option to choose (and pay for) the power source they wish to supply their electricity (*e.g.* renewables, nuclear).

In terms of reducing the negative environmental effects of transportation activities, land use and transport planning are essential instruments for modifying demand for travel in the long term. Experience has generally shown that increasing the supply of public transport alone is not sufficient to attract more users and reduce private transport. Instead, the promotion of public transport needs to be complemented by more direct incentives to not use private vehicles, for instance appropriate taxation of private vehicle use, reducing the capacity of main urban thoroughfares, and by prohibiting traffic at certain times. Furthermore, policies designed to restrain private car use may be ineffective if at the same time incentives are offered for dispersed housing development.

Finally, understanding of the full environmental or health effects of various air pollutants, as well as the long-term impacts of climate change, should be improved. Further research in these fields, as well as the development of appropriate indicators for measuring progress and the collection of comparable data, will require greater efforts in OECD countries in the future.

25.4. A policy package for households, selected industries, and waste

Many of the pressures and issues examined under Section V in this report have been placed under "green" or "yellow lights". The three industries examined in the section – the pulp and paper, steel and chemicals industries – have all increased resource and energy efficiency in recent years, and reduced the waste and pollution emitted per unit of product. For example, in the steel sector, the greater use of alternative sources of energy (*e.g.* slag and waste used as fuels) and the development of more efficient processes, such as electric arc furnaces, have contributed to a decrease in energy use by the sector of almost 20% since 1970. Similarly, pulp and paper mills have reduced water use per tonne of production significantly, from 80 m^3 in 1970 to 10-15 m^3 today. Increased use of recovered paper has also led to significant energy and resource savings, while the development of new technologies (*e.g.* improved washing of pulp and oxygen prebleaching) has reduced the emission of toxic organochlorines from the pulp and paper sector. Technological changes in the chemicals industry continue to contribute to a decline in energy use and in emissions of substances which promote the formation of tropospheric ozone, acid rain and chemicals that deplete the ozone layer. The significant progress made in phasing out production of ozone-depleting chemicals has resulted from the successful adoption and implementation of the Montreal Protocol.

However, two overall "red light" pressures remain from these issues and sectors in particular – municipal waste generation and the widespread release of persistent and toxic chemicals to the environment. Municipal waste generation has continued to grow in recent decades, closely following economic growth in most OECD regions, increasing by about 40% since 1980, and it is expected to increase by a further 43% to 2020. Although waste treatment systems have improved significantly, and more municipal waste is being diverted to recycling, increasing municipal waste generation per capita remains a problem in most OECD countries. De-coupling waste generation from economic growth will be a major challenge. To a large extent, addressing this problem will require policies that address both production processes (particularly packaging practices), and consumer behaviour.

Similarly, while the release of chemicals from industry has been reduced significantly in recent years because of more effective regulations and enforcement, the release of chemicals through downstream use (*e.g.* agriculture, households) continues to increase the level of persistent and toxic chemicals present in the environment, degrading ecosystems and harming human health. Again, tackling this problem will require a comprehensive package of policies that provides the right incentives for the various downstream actors to reduce their release of chemicals.

Economic instruments

Economic instruments, such as user charges on municipal solid waste collection and treatment, are already widely used in OECD countries but their rates are, in general, too low to influence consumer behaviour. Furthermore, for municipal waste, the rates that are charged are rarely based on the amount of waste produced, and thus provide no economic incentive for households to reduce waste generation. Unit pricing ("pay as you throw") policies using price structures that penalise higher levels of waste generation are starting to be introduced in a few communities in OECD countries. These policies often need to be accompanied by stronger incentives against waste dumping in the environment. Product charges on packaging, batteries, tyres and home appliances are also being introduced in some countries, helping to promote the collection and recovery of post-consumer waste. Such measures can be complemented by positive economic incentives for waste reduction or recycling, such as deposit-refund schemes, payments for goods delivered for recovery, or recycling credits.

Economic instruments for addressing problems of toxic and persistent chemicals in the environment are less widely used in OECD countries, and their potential applications are more limited. Some OECD countries levy charges on fertiliser purchase as a proxy for the run-off of agro-chemicals through fertiliser and pesticide use. As indicated in Section 25.2 above, a model simulation of applying a tax on chemical inputs to the agriculture sector shows that significant reductions in farm nitrogen run-off, and probably also in pesticides use, could be achieved through such an instrument. Economic instruments can also complement the implementation of regulations aimed at phasing-out the use of toxic substances, such as lead in petrol (through tradable permit averaging schemes or differentiated taxes on leaded and unleaded petrol), the production of CFCs (through a tax) or for the recovery of toxic products through deposit-refund schemes (*e.g.* lead batteries or cadmium cells).

The widespread existence of toxic and persistent chemicals in the environment is a serious problem in OECD countries, mostly arising from downstream use and the disposal of chemicals. A wider application of taxes or charges on the use of particularly toxic or persistent chemicals could help to encourage manufacturers to substitute these pollutants with cleaner and safer substances. While it was not possible in the modelling framework to apply a tax to the use of toxic or persistent chemicals only, a simulation was undertaken to examine the effects of a tax applied across all chemicals use, combined with subsidy removal for all the manufacturing sectors. While the direct effects of the diffusion of toxic and persistent chemicals in the environment could not be estimated in the model, estimates of the effects on some other key environmental variables could be made (Table 25.3). The environmental benefits of the combined policies are most noticeable in the chemicals industry, but they would affect the other manufacturing industries as well. Water use by the steel and the pulp and paper industries are estimated to decline by 3% (compared with a decline in gross production in the sector of only 2%), while CO_2 and SO_x emissions are estimated to be 4-5% lower in 2020 as compared with the Reference Scenario for both the steel and pulp and paper industries in OECD countries. A more significant 22-24% reduction is expected for the chemicals industry for the pressures under the policy simulation.

Table 25.3. **Effects of applying an *ad valorem* tax that increases by 2 percentage points each year on all chemicals use in OECD countries and removing all subsidies to manufacturing industries, difference in 2020 from Reference Scenario**

	Energy use	CO_2 emissions	SO_x emissions	Water use
Steel industry	–5%	–5%	–5%	–3%
Pulp and paper industry	–3%	–5%	–4%	–3%
Chemicals industry	–23%	–24%	–23%	–22%

Sources: Reference Scenario and Policy Simulations.

Regulatory instruments

Regulations, such as bans or restrictions on the use of certain substances, appear to be the most appropriate instruments for addressing problems of hazardous substances or toxic chemicals in the environment, as they can directly prohibit or limit the use of these chemicals in particular. However, gathering – and making effective use of – the indispensable information on the characteristics and exposure of all chemicals and chemical products, and on their effects on the environment and human health, remains a formidable challenge. Yet for designing such policies, this information is essential.

Regulations are also needed to specify the required treatment and handling of hazardous waste materials. In addition, regulations can be used to specify targets for the recovery or recycling of a given proportion of municipal or industrial waste, with the targets based on cost-benefit analyses or other decision-making principles.

Voluntary agreements

Voluntary agreements are already commonly used in waste management in OECD countries, primarily in the form of agreements with producers on the recycling of packaging wastes. Partly in response to threats of stronger regulations, voluntary initiatives have been taken by industry to reduce pollution and ensure product safety. However, in the absence of adequate provisions for monitoring and enforcement of these agreements, further efforts need to be made to increase the usefulness and credibility of programmes, such as Responsible Care and Product Stewardship.

Information and other instruments

Achieving waste prevention targets can be supported through information and labelling programmes that encourage consumers to buy products which pollute less, or come from recovered materials, or which can be reused or recycled. Promoting consumer information through education and campaigns can stimulate household awareness and has been shown to increase participation in recycling initiatives in OECD countries. Another waste management policy instrument that is being used more and more in OECD countries is extended producer responsibility (EPR), under which producers are made responsible (financially and/or physically) for the treatment or disposal of post-consumer products. Through extending their responsibilities, producers can be encouraged to re evaluate decisions concerning materials selection, production processes, design, packaging and marketing strategies in order to minimise their costs. EPR can play a role in increasing resource efficiency by encouraging the recovery of materials that would otherwise have gone to landfill or been incinerated, and by influencing product designers to select materials that can easily be reused or recycled.

As indicated above, gathering and disseminating information on persistent and toxic chemicals in a meaningful way is a major challenge for governments and industry. Information generated by industry and government on the characteristics of the chemicals, their potential risks, and any reasons why their use should be restricted, needs to be conveyed to various stakeholders and the general public in a way that can be understood by all. As such, there is a need to make greater use of pollutant release and transfer registers (PRTRs), which can help to monitor and assure reductions in the release of hazardous substances to air, water, soil and waste. As inventories of toxic releases can only capture a small proportion of the total flow of chemicals dispersed into the environment, life-cycle management of chemicals and products should also be promoted, together with integrated management schemes which include increased producer responsibility to ensure that products are used and disposed safely.

25.5. Assessment of a combined policy package

It is clear that a comprehensive package of instruments will be needed to address the "red light" environmental problems that are likely to be of greatest concern to OECD countries over the next two decades. Because of the complex nature of many of these issues – the diffuse sources of pollution, the over-exploitation of common property or shared natural resources – single policy instruments are unlikely to be sufficient. Instead, a combination of economic incentives, a strong regulatory framework, increased technological development and diffusion, and voluntary and information-based policies will be required.

A range of policies that could be included in such a policy mix were discussed above, and potential environmental and economic effects were assessed where information was available. But what would be the net effects of adopting the full package of policies? An important question for policy-makers is whether some of these policies would be complementary – allowing for the adoption of less stringent policies while still meeting the desired objectives or targets – or whether they might provide conflicting incentives instead. A policy simulation was undertaken to examine the net impact of adopting the main economic incentives described above, applied in two stages through a combination of:

1. removing all subsidies (as listed in the model used) to production and consumption across OECD regions in the first stage; and then
2. applying an *ad valorem* tax on fuel use in OECD regions that increases annually by 2, 1.6 and 1.2 percentage points respectively for coal, oil and natural gas use; and applying an *ad valorem* tax on chemical use in all sectors of OECD economies that increases by 2 percentage points per year.

The simulation for subsidy removal alone was found to have limited environmental benefits (Figure 25.3). In part this reflects the low level of subsidies listed in the GTAP database when compared with other support estimates. It is also in part a result of the removal of market price support to agriculture, which – through the lowering of agricultural prices – could lead to greater consumer demand in OECD countries for agricultural products and higher global agricultural production. These results reinforce the need for strong international co-operation in the removal of environmentally damaging subsidies and the implementation of accompanying measures to ensure that the full potential benefits of such a policy can be realised.

The combined policy package – with both subsidy removal and the application of taxes on energy and chemicals use – would have more considerable environmental benefits. CO_2 emissions in OECD regions were found to be 15% lower in 2020 under this simulation as compared with the Reference Scenario, while SO_x emissions were 9% lower, nitrogen loading to waterways 27% lower (largely due to the effects of the chemicals tax on fertiliser use), and methane emissions 3% lower. While not explicitly included in the modelling exercise, reductions in OECD municipal waste generation and agricultural pesticide use could also be expected. Despite the significant decline in greenhouse gas and air pollution emissions that could be achieved under this combined policy package, the economic costs were found to be small, with a reduction of projected GDP in 2020 for OECD regions overall of less than 1% compared with the Reference Scenario.

Figure 25.3. Effects in 2020 of removing subsidies, applying a fuel tax and a chemical use tax in OECD regions

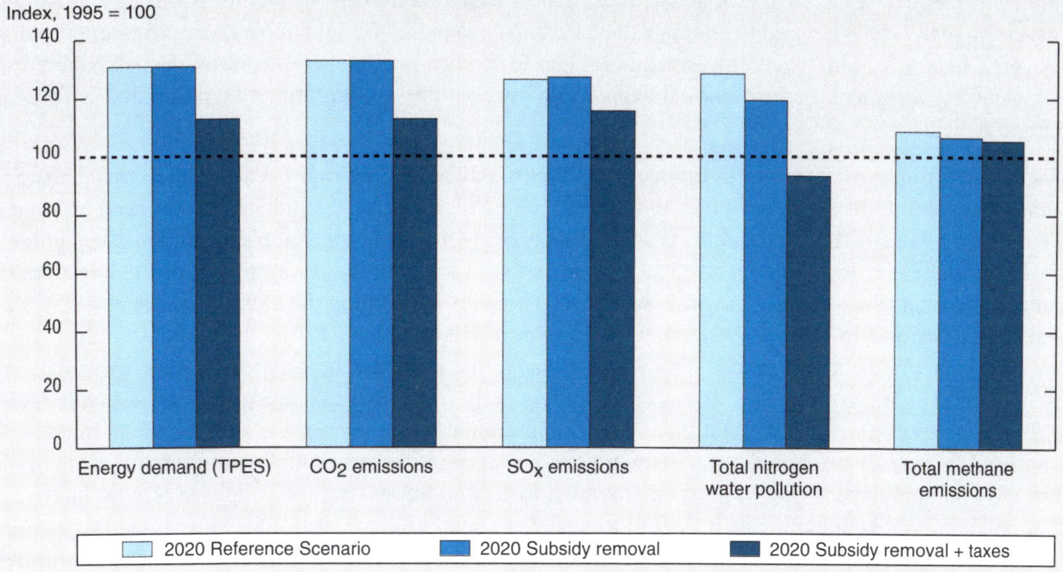

Source: Reference Scenario and Policy Simulations.

25.6. Implementing the combined policy package

Few of the policies discussed in the sections above are new, and many of them have been recommended for a number of decades. As shown throughout the *Outlook*, OECD countries have already implemented, or are starting to implement, policies to address some of the "red lights" identified. However, their use is still limited in most countries. Particular efforts need now to be devoted to the implementation of policies that have already been agreed, as well as to the adoption of new policies to ensure comprehensive policy packages for dealing with these urgent issues. As has been shown over and over again in this *Outlook*, implementation of environmental policies is the weak link in the environmental policy cycle. In some cases, economically efficient environmental policies have not been adopted out of fear that their implementation could have potentially negative social impacts. In other cases, the implementation and enforcement of adopted environmental plans, strategies or agreements have been slow due to loosely defined objectives, lack of involvement of the main stakeholders, weak support of the general public and/or poor enforcement mechanisms.

To ensure successful policy implementation at the national level, measures that address potentially adverse social effects of some environmental policies (*e.g.* on incomes or employment) will need to be developed. In many cases, the implementation gap is the result of concerns that applying a given environmental policy will result in hardships for a particular region, sector, or for particular income groups. Thus, large exemptions are often in place for fuel taxes, with the aim of protecting energy-intensive industries from adverse effects on competitiveness at the international level. Similarly, environmentally damaging support to coal production, agriculture, and fisheries is often difficult to remove because of the potential effects on employment in the regions concerned, and again because of the fear of a loss of competitiveness in the world market for sectors where support is common practice. In both of these cases, stronger international co-operation can be the key to reforming the environmentally damaging policies. Measures can be taken to phase in the reforms over time, in co-operation with the affected stakeholders, and by ensuring that the adjustments in the sectors concerned go smoothly. Where reluctance to implement an environmental policy reflects concerns that the policy may have socially regressive distributional effects – as with the maintenance of low water tariffs for households in most OECD countries – direct measures (*e.g.* raising income support) can in many cases be used to alleviate the hardships in a more efficient and less environmentally distorting manner (see Chapter 8). Finally, an implementation gap for environmental policies often persists where there is a lack of scientific, technical, or institutional capacity to support the design, implementation and enforcement of appropriate policies.

Successful implementation of environmental policies and policy packages will require consideration of the following issues. First, the setting of clear and realistic objectives and targets that can be checked against a set of reliable indicators is essential to record the progress that is being made in policy implementation. Strengthened data collection and indicator development and use will be essential for both the development of the appropriate environmental policies, and monitoring the progress of these policies in reaching the environmental objectives and targets that are set. Second, to achieve these objectives, support will often have to be secured not only from the public at large, but also from the main stakeholders who are part of the implementation process. One way to secure such support is to inform and consult the various stakeholders early in the decision-making process, while facilitating public access to environmental information and reporting regularly on the progress being made in implementation (see Chapter 22). Third, care should be taken to limit potential negative social implications arising from implementation of environmental measures, such as employment loss or additional burdens imposed on low-income groups of the population. Where the policies may lead to negative social effects, careful use of temporary compensation measures, such as direct income payments to affected workers or retraining schemes, can help to smooth the transition. Finally, effective compliance and enforcement mechanisms at the national and international levels will need to be adopted.

Because a large number of the "red light" environmental pressures or conditions identified are of global concern (*e.g.* climate change, biodiversity), increased and more effective international co-operation will be needed to successfully address them in the future. In addition to concluding and ratifying existing and any new international or regional agreements to deal with these issues, mechanisms are needed to ensure their enforcement and provide incentives to encourage compliance. Increased international co-ordination will also help to address concerns of industry with respect to potential loss of competitiveness due to unilateral action taken by one country.

In the future, it is likely that the institutions that support environmental policies will need to increasingly address these issues within the broader framework of sustainable development (encompassing economic, social and environmental objectives) (see Chapter 24). More decentralised and co-operative policy decision-making and implementation (*i.e.* working with the business community and non-governmental organisations, greater public accountability) will be needed, and greater commitments to internationally agreed targets for addressing global and regional environmental concerns. While some of the institutions and the capacity to face these growing challenges are in place, adaptations will need to be made in most national governments, as well as in the international institutions that directly deal with environmental concerns or have an impact on the environment (*e.g.* international trade or investment agreements).

REFERENCES

Burniaux, J-M. (2000), "A Multi-Gas Assessment of the Kyoto Protocol", Economics Department Working Papers No. 270, OECD, Paris.

de Moor, A. and P. Calamai (1998), *Subsidizing Unsustainable Development: Undermining the Earth with Public Funds*, the Earth Council, San José, Costa Rica.

IEA (International Energy Agency) (1999), *The Role of IEA Governments in Energy: 1999 Review*, IEA/OECD, Paris.

OECD (1997), *Reforming Energy and Transport Subsidies: Environmental and Economic Implications*, OECD, Paris.

OECD (1998*a*), *Co-operative Approaches to Sustainable Agriculture*, OECD, Paris.

OECD (1998*b*), *Improving the Environment through Reducing Subsidies: Part II, Analysis and Overview of Studies*, OECD, Paris.

OECD (1999*a*), *Agricultural Policies in OECD Countries: Monitoring and Evaluation*, OECD, Paris.

OECD (1999*b*), "Economic Instruments for Pollution Control and National Resources Management in OECD Countries: A Survey", ENV/EPOC/GEEI(98)35/REVI/FINAL, OECD, Paris.

OECD (2000), *Transition to Responsible Fisheries: Economic and Policy Implications*, OECD, Paris.

OECD (2001*a*), "Instruments and Technologies for Climate Change Policy", ENV/EPOC/GEEI(99)15/FINAL, OECD, Paris.

OECD (2001*b*), *Design and Use of Domestic Transferable Permit Systems for Environmental Policy Making*, OECD, Paris.

Annex 1
Background Documents

The information and analysis of this OECD *Environmental Outlook* is supported by the following background documents, which can be made available upon request to the OECD Secretariat (E-mail: *env.contact@oecd.org*).

Chapter/Topic	OECD *Environmental Outlook* background document	Author(s)	Person responsible in the OECD Secretariat
Modelling and Assessments	Selected papers presented to the OECD Working Party on Economic and Environmental Integration on the development of the JOBS model, 1999-2000	OECD	Nils Axel Braathen
	Global Environmental Scenarios	Eric Kemp-Benedict and Paul Raskin, Stockholm Environment Institute (SEI) Boston, consultants	Helen Mountford
	Abstracts of Selected Environmental Outlook Reports and Assessments	J.A. Bakkes, D.P. van Vuuren and E. Smeets, RIVM, consultants	Lars Mortensen
2. Demographic and Labour Force Developments	Demography and Employment, Urbanisation and Land Use: Patterns, Trends and Perspectives	Voula Mega	Voula Mega
3. Globalisation, Trade and Investment	The Environmental Implications of Trade and Investment Liberalisation	Duncan Brack, consultant	Tom Jones
5. Consumption Patterns	Consumption Patterns	Elaine Geyer-Allely and Hoe-Seog Cheong	Hoe-Seog Cheong
6. Technological Change	Technological Change, Public Policy and the Environment	Nick Johnstone	Nick Johnstone
7. Agriculture	Agriculture and the Environment	Nikos Alexandratos, consultant	Elaine Geyer-Allely
8. Freshwater	Freshwater Outlook	Helen Mountford	Helen Mountford
9. Fisheries	Fisheries Trends: A Background Report	Gordon Munro, consultant, Carla Bertuzzi and Chris Chung	Carla Bertuzzi
10. Forestry	Environmental Outlook in the Forestry Sector	Gerard Bonnis	Gerard Bonnis
	Draft Summary Record, OECD Workshop on the Environmental Outlook and Strategy for the Forestry Sector (Jedlnia, Poland, 15-16 February 2000)	Gerard Bonnis	Gerard Bonnis
	Environmental Outlook: The Forestry Sector	Roger A. Sedjo, consultant	Gerard Bonnis
11. Biological Diversity	Biodiversity	Dan Biller	Dan Biller
	Biodiversity Indicators for the OECD Environmental Outlook and Strategy	Ben ten Brink, RIMV, consultant, with contribution from WCMC	Dan Biller
12. Energy	Energy and Climate Change: Trends, Drivers, Outlook and Policy Options	John Newman, consultant, Noreen Beg, Jan Corfee-Morlot and Gene McGlynn	Jan Corfee-Morlot
13. Climate Change	Energy and Climate Change: Trends, Drivers, Outlook and Policy Options (see Chapter 12)	John Newman, consultant, Noreen Beg, Jan Corfee-Morlot and Gene McGlynn	Noreen Beg

Chapter/Topic	OECD *Environmental Outlook* background document	Author(s)	Person responsible in the OECD Secretariat
14. Transport	Environmental Outlook to 2020 for the Transport Sector	Peter Wiederkehr	Peter Wiederkehr
15. Air Quality	Environmental Outlook to 2020 for Air Quality	Peter Wiederkehr	Peter Wiederkehr
16. Households	Consumption Patterns (see Chapter 5)	Elaine Geyer-Allely and Hoe-Seog Cheong	Hoe-Seog Cheong
17. The Steel Industry	Iron and Steel Industry	Alain Rajotte, consultant	Ponciano Villafuerte-Zavala
18. The Pulp and Paper Industry	Pulp and Paper Industry	Alain Rajotte, consultant	Ponciano Villafuerte-Zavala
19. The Chemicals Industry	Environmental Outlook for the Chemicals Industry	Richard Sigman	Richard Sigman
20. Waste	Waste Trends and Outlook	John Stutz, consultant, Henrik Harjula, Soizick de Tilly	Henrik Harjula
21. Human Health and the Environment	Human Health and the Environment	Johan M. Melse and Augustinus E.M. de Hollander, RIVM, consultants	Niels Schenk/Dian Turnheim
22. The Social and Environmental Interface	Background Paper on the Social-Environmental Interface	Christian Avérous	Christian Avérous
	The Distribution of Environmental Quality and Resources: Objectives, Evidence and Policy	Nick Johnstone	Nick Johnstone
23. Resource Efficiency	Resource Efficiency and the Environment	Nick Johnstone	Nick Johnstone
24. Institutional Frameworks for the Environment	Institutional Frameworks for the Environment – Outlook to 2020	Tom Jones	Tom Jones

Annex 2

Modelling Framework Used for the Reference Scenario and Policy Simulations

1. Introduction

The Reference Scenario and the policy simulations described in this *Outlook* were made using a global general equilibrium model (JOBS) which was developed by OECD's Development Centre.[1] Results from these simulations – *e.g.* concerning developments in value added and relative price – were fed into the PoleStar framework, developed by the Stockholm Environment Institute Boston. PoleStar then enabled us to calculate a number of environmental impacts, as described in the chapters of this *Outlook*. This Annex provides a brief description of the assumptions used in the Reference Scenario to 2020 and the policy simulations, and explains the structure of the JOBS model and the PoleStar system. The final section provides a brief description of the MOVE II model, which was used to project emissions from motor vehicles, as described in Chapters 14 and 15.

2. Underlying assumptions

2.1. Population

The growth of total population and labour supply is exogenous in all the simulations presented in this *Outlook*, and the assumptions used are presented in Chapter 2. The assumptions are based on the medium fertility version of the 1998 UN population projections. The labour force in each region is assumed to constitute a fixed portion of the population in the age group 15-64 years. This means that – on balance – no major net changes are assumed to take place in factors such as the rate of unemployment, male and female labour force participation rates, etc.

2.2. GDP in the Reference Scenario

A specific development of total GDP[2] is assumed for each region up until 2020 in the Reference Scenario, *cf.* Figure 4.1 in Chapter 4 and Table A1. The assumptions used in the Reference Scenario are, as far as possible, the same as those used in the climate change policy analyses done by the OECD on the "GREEN" model, *cf.* Burniaux (2000). However, due to *inter alia* some differences in the regional groupings used in the GREEN model and the JOBS model, the assumptions are not identical. In any case, it is underlined that the growth assumptions do not represent prognoses, merely a starting point for exploring possible impacts of changes in policy assumptions, etc.

While the GDP growth rates are exogenous in the Reference Scenario, capital productivity is endogenous. Furthermore, while the supply of labour is exogenous, the labour productivity parameter (which is uniform across sectors) is calibrated so as to ensure that the ratio of capital (in efficiency units) to labour (in efficiency units) is constant over the simulation period.

In the policy shocks, the growth rates of capital and labour productivity which were calculated endogenously in the Reference Scenario, are used as exogenous assumptions. In these simulations, changes in real GDP and in capital-labour ratios are endogenous.

1. The JOBS model was further developed from 1998 to 2000 for the purposes of this exercise. Papers describing its development were discussed in meetings of the OECD Working Party on Economic and Environmental Policy Integration during this period, including "Macroeconomic Model Simulations: Assumptions in a "Baseline Scenario" [ENV/EPOC/GEEI(99)8], "Macroeconomic Model Simulations: Some Policy Scenarios" [ENV/EPOC/GEEI(99)14], and "Consumption and the Environment: Exploring the Linkages with Economic Globalisation" [ENV/EPOC/GEEI(99)2/REV2]. Based on the discussion in these meetings and other comments, a number of the assumptions in the model were refined for the Reference Scenario and policy shocks presented in this *Outlook*. This Annex describes the final model assumptions and structure used.
2. While the total GDP in each region is given exogenously, the distribution of this total production among the 26 sectors in each region of the model is determined endogenously, reflecting *inter alia* the relative producer prices and the relevant substitution elasticities.

Table A1. **Historical and assumed levels of real GDP, 1995 US$ billion**

	1980	1995	2000	2005	2010	2015	2020
NFT	4 960	7 347	8 476	9 590	10 588	11 463	12 410
WEU	5 861	8 071	9 087	10 331	11 129	11 813	12 539
CEU	370	449	536	668	817	975	1 163
JPK	2 980	5 084	5 277	5 942	6 528	6 998	7 539
ANZ	232	372	432	503	561	611	667
FSU	626	447	394	468	583	710	863
CHN	150	653	942	1 237	1 578	1 995	2 522
EAS	280	784	882	1 116	1 410	1 757	2 159
SOA	153	359	466	578	711	873	1 072
MEA	281	463	512	570	636	709	790
LAT	893	1 252	1 365	1 591	1 835	2 107	2 419
ARW	452	646	740	866	1 004	1 164	1 350
OECD	14 403	21 323	23 808	27 033	29 623	31 860	34 318
Non-OECD	2 834	4 604	5 302	6 426	7 757	9 315	11 175
World	17 237	25 927	29 109	33 459	37 380	41 175	45 493

3. The JOBS model

3.1. Basic characteristics

JOBS is a neo-classical general equilibrium model that was initially constructed to assess the economic impacts of globalisation on individual regions of the world. JOBS is a version of the LINKAGE model, used in the OECD Linkages II project, which *inter alia* resulted in the publication "The World in 2020: Towards a New Global Age" (OECD, 1997*a*). The LINKAGE model was in turn derived from the GREEN model that has been used in a series of analyses of policies to combat climate change.[3]

JOBS is designed for the analysis of dynamic scenarios, which are solved as a sequence of static equilibria. The periods are linked by exogenous population and labour supply growth, capital accumulation and productivity developments. For this *Outlook*, a Reference Scenario was developed, and impacts of a number of policy shocks were compared to this. The simulations were based on data from Version 4.0 of the Global Trade Analysis Project (GTAP) database, developed by Purdue University,[4] with 1995 as the base year. This database contains consistent data for 50 sectors and 45 regions. The JOBS model is implemented with GAMS software, and includes a flexible aggregation facility which may be set by the user up to the maximum dimensions of the GTAP data set. For purposes of the *Environmental Outlook*, a Reference Scenario was developed and policy simulations were undertaken for 12 geographical regions and 26 economic sectors as described in Tables A2 and A3 respectively.

Figure A1 describes the production structure used in JOBS. The inputs used to produce a given output have been divided into several distinct components, namely non-energy intermediate inputs, energy intermediate inputs, one category of labour, one type of capital, land (in agriculture sectors only) and a natural resource factor used in the Forestry, Fisheries, Minerals, Coal, CrudeOil and NaturGas sectors. A nested set of CES functions (Constant Elasticity of Substitution) was used to emulate the different degrees of substitution and complementarity between the various inputs, and a brief description of the substitution elasticities[5] (the σ's in the shaded boxes) used is included in the figure.

At the top of Figure A1, an aggregate bundle of non-energy intermediate inputs is combined with a value-added and energy bundle, with a very low elasticity of substitution (0.05). Hence, these two bundles will always be used in almost fixed proportions. The aggregate non-energy intermediate bundle is decomposed into demand for individual intermediate goods, with no substitution possibility between the different goods. In many cases, this is a reasonable approximation, as the substitution possibilities in reality are often limited. However, in some cases, this technical assumption imposes unrealistic limitations on the substitutions that can take place when relative prices change, *e.g.* as a result of increases in taxes on certain products.[6] This should be borne in mind when interpreting the results of some of the policy shocks simulated.

3. See, for instance, Burniaux (2000).
4. For further details, see *www.agecon.purdue.edu/gtap*. Version 5.0 of GTAP now includes an extension and update of the database compared to Version 4.0. This version was, however, not available at the time the simulations for this publication were made.
5. A substitution elasticity describes the change in the relative use of two factors if the relative price between these factors changes by one per cent. A substitution elasticity equal to zero means that two factors are always used in fixed proportions. A high substitution elasticity means that a small change in relative prices will cause a significant change in the composition of inputs used.
6. One could, for example, expect an increase in the use of *Wood products* as input in the *Construction* sector, if the relative price of *Iron and steel*, or *Non-ferrous metals*, increased. In the current version of JOBS, such a substitution is not possible.

Table A2. **Regions used in the model simulations**

	Region name	Countries
NFT	Canada, Mexico & United States	Canada, Mexico and United States
WEU	Western Europe	Austria, Belgium, Denmark, Finland, France, Germany, Greece, Ireland, Italy, Liechtenstein, Luxembourg, Netherlands, Norway, Portugal, Spain, Sweden, Switzerland, United Kingdom and Iceland
CEU	Central & Eastern Europe	Czech Republic, Hungary, Poland, Slovakia and Turkey (OECD Members), Romania and Bulgaria
JPK	Japan & Korea	Japan and Korea
ANZ	Australia & New Zealand	Australia and New Zealand
FSU	Former Soviet Union	Armenia, Azerbaijan, Belarus, Estonia, Georgia, Kazakhstan, Kyrgyzstan, Latvia, Lithuania, Republic of Moldova, Russian Federation, Tajikistan, Turkmenistan, Ukraine and Uzbekistan
CHN	China	China (including Hong Kong)
EAS	East Asia	Indonesia, Malaysia, Philippines, Singapore, Taiwan, Thailand and Vietnam
SOA	South Asia	Bangladesh, Bhutan, India, Maldives, Nepal, Pakistan and Sri Lanka
MEA	Middle East	Bahrain, Iran, Iraq, Israel, Jordan, Kuwait, Lebanon, Oman, Qatar, Saudi Arabia, Syrian Arab Republic, United Arab Emirates and Yemen
LAT	Latin America	Anguilla, Antigua & Barbuda, Argentina, Aruba, Bahamas, Barbados, Belize, Bermuda, Bolivia, Brazil, British Virgin Islands, Cayman Islands, Chile, Colombia, Costa Rica, Cuba, Dominica, Dominican Republic, Ecuador, El Salvador, Grenada, Guatemala, Guyana, Haiti, Honduras, Jamaica, Nicaragua, Panama, Paraguay, Peru, Puerto Rico, St.Kitts & Nevis, St.Vincent & Grenadine, Suriname, Trinidad & Tobago, Uruguay and Venezuela
ARW	Africa & Rest of the World	Afghanistan, Albania, Algeria, Angola, Benin, Bosnia and Herzegovina, Botswana, Brunei, Burkina Faso, Burundi, Cambodia, Cameroon, Cape Verde, Central African Republic, Chad, Congo, Croatia, Cyprus, Djibouti, Egypt, Equatorial Guinea, Eritrea, Ethiopia, Fiji, Gabon, Gambia, Ghana, Guinea, Guinea Bissau, Ivory Coast, Kenya, North Korea, Laos, Lesotho, Liberia, Libya, Madagascar, Malawi, Mali, Malta, Mauritania, Mauritius, Mongolia, Morocco, Mozambique, Myanmar, Namibia, Niger, Nigeria, Papua New Guinea, Rwanda, Senegal, Seychelles, Sierra Leone, Somalia, South Africa, Sudan, Swaziland, Tanzania, The former Yugoslav Republic of Macedonia, Togo, Tunisia, Uganda, Yugoslavia, Zaire, Zambia and Zimbabwe

Table A3. **Sectors used in the JOBS model**

Name	Content
Rice	Paddy rice
OthCrops	Wheat, cereal grains n.e.s., vegetables, fruit, nuts, oil seeds, sugar cane, sugar beet, plant-based fibers, crops n.e.s.
Livestock	Live animals, Raw milk, wool, silk-worm cocoons, etc.
Fisheries	Fisheries
Forestry	Forestry
Minerals	Minerals
Coal	Coal
CrudeOil	Crude oil
NaturGas	Natural gas extraction
RefOil	Petroleum, coal products
GasDistr	Gas manufacture and distribution
Elect	Electricity generation and distribution
Meat	Meat from all types of animals
OthFood	Vegetable oils and fats, dairy products, processed rice, sugar, food products n.e.s., beverages and tobacco.
Chemicals	Chemical, rubber, plastic products
I_S	Iron and steel
NonFer	Non-ferrous metals
WoodProd	Wood products
PPP	Pulp Paper Publishing
MotorVehi	Motor vehicle manufacturing, including parts
OthManu	Textiles, wearing apparel, leather products, metal products, transport equipment n.e.s., electronic equipment, machinery and equipment n.e.s., manufactures n.e.s..
Construc	Construction
Water	Water supply
TradeTran	Trade and transport services
Service	Finance, business, recreational services, public administration, defense, education, health
Dwellings	Dwellings

n.e.s. = "not elsewhere specified".

Figure A1. **Nested structure of production in the JOBS model**

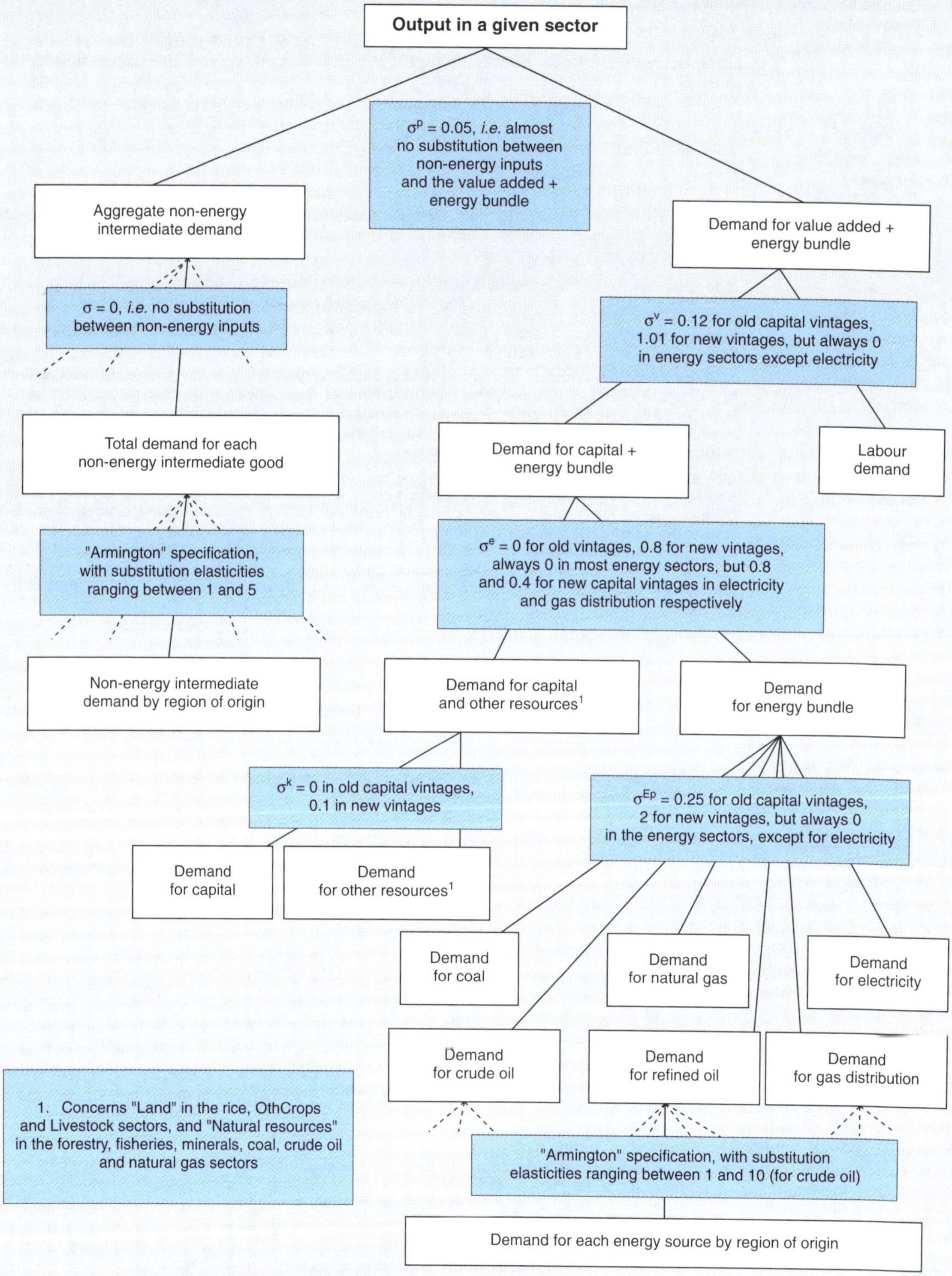

Source: Author.

In the agriculture sectors JOBS does, however, allow additional substitution possibilities: in the Rice and OthCrops sectors, it is possible to substitute between the use of Chemicals inputs and the capital, energy and land bundle, and in the Livestock sector, one can substitute between the use of Land and the use of purchased feedstock. Hence, for example a tax increase on chemicals used in Rice production can lead to a substitution away from the use of such chemicals towards more use of capital, land or energy.

It is important to note that JOBS distinguishes between two vintages of capital – old and new (the latter being largely equal to each year's investment) – with the substitution possibilities (*e.g.* across energy sources) being higher concerning new capital than for old. New capital is perfectly mobile between sectors, and its allocation insures a uniform rate of return in each sector. The rate of return to old capital in expanding sectors is also equal to the economy-wide rate of return on new capital. Declining sectors are assumed to release capital, which is added to the stock of new capital in a given year. The rate of return on old capital in declining sectors will be lower than the return achieved in other sectors, and it is determined by sector-specific supply and demand conditions.

The model does not include an investment function which relates the overall level of investment to the expected rate of return. Indeed, there is no forward-looking investment behaviour incorporated in the model. Instead, the value of investment in each year and region is equal to the value of aggregate saving in the region. Aggregate saving, in turn, is derived from household behaviour.

Household consumption demand is modelled through the use of a so-called "extended linear expenditure system"(ELES). This consumption demand includes a demand for future goods, represented through the demand for savings, which in turn is determined as a residual, as the difference between household disposable income and current expenditures. The demand for each category of goods and services consists of two components: a population-adjusted subsistence minimum and a component reflecting *inter alia* the relative prices of the different categories.[7]

The volume of government expenditure is assumed to be a constant share of real GDP at market prices. Real government saving is exogenous, thus the government is assumed to have a target for the net fiscal position of the public budget. The direct tax rate on household income is endogenous, and the household tax schedule shifts over time to accommodate the given level of real government saving.

One important aspect of the JOBS model is that domestically produced products and imported products of the same type are assumed to be imperfect substitutes, which *inter alia* implies that their prices may differ in a given market. The elasticities describing the degree of substitutability between domestic and foreign products are called "Armington elasticities". The higher the Armington elasticity, the easier it is for users to substitute between domestic and foreign products of a particular type.[8] In the simulations made for this *Outlook*, the Armington elasticities are assumed to be equal between all regions, and between all end-users within each region, but they vary between different sectors. Total demand for a given type of product within a region is called "Armington demand". This total demand is split into a demand for imported and domestically produced products via the Armington elasticities.

One implication of the "Armington specification" of international trade is that each region faces a downward-sloping demand curve for its exports. The more their production is to increase, the lower their relative export prices need to be. Hence, regions that expand their production more rapidly than other regions will tend to experience a term-of-trade loss, *i.e.* their export prices will decrease compared to the prices of their imports.

3.2. Productivity and factor supply assumptions

Exogenous assumptions have been made in JOBS concerning overall agricultural productivity growth, the supply and productivity growth of land (used in the Rice, Other crops and Livestock sectors) and the productivity of Chemicals inputs used in Rice and Other crops and feedstock (used in Livestock). Table A4 below details the assumptions used.

JOBS also includes a sector-specific natural resource factor used in the Forestry, Fisheries, Minerals, Coal, Oil and Natural Gas sectors. For this *Outlook*, it has been assumed that the supply curve for these factors remain unchanged over the simulation period in the Forestry, Fisheries and Minerals sectors, while they shift outward 0.5% per year in the Coal, Oil and Natural Gas sectors.

7. Income elasticity estimates needed to calibrate the ELES system are taken from the GTAP database. A problem with an ELES demand system is that all income elasticities over time converge towards 1, meaning that the budget shares of each product category become constant. That would certainly be wrong for most goods. To alleviate this problem, the demand parameters are re-calibrated between each period, so that the income elasticities remain more or less constant over time.
8. The Armington elasticities assumed in the Reference Scenario are highest for commodity-based sectors such as Oil (10.0), Coal, Refined oil, Iron and Steel (5.0), and lowest (1.0) for service sectors such as Electricity, Gas distribution and Dwellings. Their empirical foundation is not strong. However, sensitivity analyses indicate that, for instance, a doubling of all of these elasticities does not alter the results of the simulations fundamentally. It is technically possible, and it would in principle be of interest, to vary the Armington elasticities between regions and/or between end-users within each region. This would, however, complicate the solving of the model considerably, and would require assumptions to be made regarding a very large number of parameters, with little empirical foundation.

Table A4. **Assumptions regarding the agricultural sectors, annual percentage changes 1995-2020**

	Reference agricultural productivity	Land supply growth	Land productivity growth	Fertiliser productivity growth	Feedstock productivity growth
Canada, Mexico &US	2.5	–0.1	1.5	0.5	0.5
Western Europe	2.0	–0.2	2.3	0.1	0.1
Central & Eastern Europe	1.0	–0.1	1.6	0.0	0.0
Japan & Korea	3.0	–0.2	0.6	1.0	1.0
Australia & New Zealand	2.5	–0.1	2.7	–0.1	–0.1
Former Soviet Union	1.7	–0.1	1.1	0.0	0.0
China	1.8	–0.5	2.0	0.0	0.0
East Asia	2.5	0.8	1.8	0.0	0.0
South Asia	1.8	0.2	1.5	0.0	0.0
Middle East	1.8	0.2	1.5	0.0	0.0
Latin America	2.5	0.6	2.3	0.0	0.0
Africa & Rest of the World	1.8	0.2	1.5	0.0	0.0

Sources: Based on Martin and Mitra (1999), Alexandratos (1995) and OECD (1995).

It is further assumed that the supply-elasticity of the sector-specific factor is 0.25 in Forestry and Fishery, 0.5 in Minerals, 2 in Crude oil and Natural gas and 10 in the Coal sector, where supply in the time horizon of this *Outlook* is believed to be rather flexible.[9]

Finally, JOBS includes assumptions concerning autonomous energy efficiency improvements (AEEI), that is, improvements in the efficiency in use of the 6 energy goods in all sectors of the economy which are assumed to take place independently of any response to changes in relative prices. In the Reference Scenario, these efficiency parameters are assumed to improve 0.75% per year in all sectors and regions.

4. The PoleStar framework

The PoleStar framework was developed by Stockholm Environment Institute Boston to describe developments for a number of environmental parameters. The framework has previously been used in a number of scenario analyses, for instance in the publication *Bending the Curve: Toward Global Sustainability* (Raskin *et al.*, 1998). A number of modifications have been made to the framework for this *Outlook*, so that the results from the JOBS simulations could be used as drivers for the environmental impacts simulated in PoleStar. This section provides an overview of the assumptions used in the present analysis.

4.1. Fuel demand

JOBS estimates economic transfers between different economic sectors. These transfers include those between energy-producing and energy-consuming sectors, which – for the purposes of this analysis – are taken as proxies for the transfer of fuel. The fuel-producing sectors tracked in JOBS are Coal, Crude oil Natural gas, Refined oil and Electricity. Within PoleStar, trends in household biomass consumption and district heat are also estimated. Biomass consumption is estimated based on a cross-sectional analysis, using International Energy Agency (IEA) energy data, of per capita biomass consumption against income. District heat consumption per capita is held at base year levels in all regions.

The JOBS output used in the PoleStar analysis is expressed in nominal US dollar terms. Because values are in nominal terms, trends in raw output do not correspond directly to physical flows. However, JOBS also estimates prices for goods paid for by different consuming sectors, allowing an estimate of transfers in real terms (after correcting for changes in efficiency of resource use in consuming sectors). As a result, using JOBS it is possible to estimate trends in physical fuel consumption, as indices calculated relative to consumption levels in the base year.

Energy use by the energy sectors is also estimated. Three energy transformation sectors are included: oil refining, electricity generation and district heat. Of these, two are treated in JOBS: oil refining and electricity generation. Furthermore, within electricity generation a subset of feedstock fuels is considered: petroleum, coal and natural gas. Other sources of electricity are omitted: nuclear power, hydroelectric and renewables. For the Reference Scenario, trends in refining activity and use of petroleum,

9. The supply elasticity of a sector-specific factor indicates by what percentage the supply of that factor would change if the price changed 1%, for a given choice of supply-curve.

coal and natural gas are derived from JOBS. Real output from the oil refining sector drives refinery production in PoleStar, while real transfers from the refining, coal production and natural gas distribution sectors to the electricity generation sectors are used to generate trends in use of these fuels for electricity production. Trends in electricity generation from nuclear, hydroelectric and renewable plants are based on separate analyses from IEA (IEA, 2000). Fuel shares for district heat production are held at base year levels. Trends in real supplies from energy sectors to other sectors were applied to base year consumption estimates based on IEA energy statistics (IEA, 1999).

4.2. CO_2 emissions

Carbon emissions are estimated from energy consumption data and emissions factors based on IPCC (1995). Carbon dioxide emissions are estimated for fossil fuel combustion and feedstock use. Carbon emission intensities are applied to fuel consumption in all sectors. Industrial process emissions are not included.

4.3. SO_x emissions

Data on sulphur emissions for the base year are based on Posch *et al.* (1996) and Kuylenstierna (1998). Sulphur emission intensities are applied to fuel consumption in all sectors. Additionally, industrial process emissions for the Nonferrous Metals sector are included.

In the industrial sector, a gradual reduction of sulphur emission factors for fuel combustion and process emissions is assumed in most regions in the Reference Scenario, reflecting the fact that these can be affected by end-of-pipe cleaning technologies and fuel switching. In all regions, emission coefficients converge linearly to a value of 0.65 for coal, 0 for crude oil (used as a feedstock in the Chemical sector) and 0.0005 for petroleum, all other fuels staying at the base year value. The values would converge completely in 2050; otherwise they change linearly with time. Some regions start out with emission coefficients below these target values. In that case, the emission coefficient remains at the base year value.

The air pollution projections provided in Chapters 14 and 15 of the *Outlook* were primarily based on the more detailed analysis undertaken using the MOVE II model (see description below), rather than the PoleStar Framework.

4.4. BOD and nitrogen loading

Households and services

Sources of domestic pollution include pollution from both the household and service sectors. For this analysis, pollutant generation is represented by biochemical oxygen demand (BOD). Estimates of BOD are constructed separately for urban and rural areas. Loadings are estimated by multiplying population by an emission factor, and then taking treatment into account. Separate emission factors are used for sewered and unsewered waste. A fraction of sewered urban wastes is assumed to be treated to some degree, while in rural areas it is assumed that the population is not connected to central sewer systems, and no waste is treated. BOD loadings are determined by the size of urban and rural populations and by regional GDP per capita, as described below.

Data on BOD loading factors (kg BOD/person) are scarce, but they are known to vary regionally and by sewered and unsewered population. Values for the US and for representative developing countries are shown in Table A5. In this *Outlook*, the US value is used for OECD regions and the developing countries value is used for non-OECD regions.

Table A5. **Typical loading rates for untreated domestic sewage**

Country or region	BOD [BOD5] (kg/capita/year) Sewered population	BOD [BOD5] (kg/capita/year) Unsewered population
US*	54.9 [45.5]	11.0 [9.1]
Developing countries**	23.7 [19.7]	8.3 [6.9]

* Chapra (1997) and ** WHO (1982).

The average BOD removed with each degree of treatment (primary, secondary, tertiary) was applied to average wastewater treatment levels in four of the regions in this analysis, to obtain the average per cent of BOD removed by region. The relationship between treatment level and income (GDP per capita, measured in Purchasing Power Parities) assumed is as follows: below an income level of around US$2 400, it is assumed that only primary treatment is used, in which 25% BOD is removed. Above an income of US$7 700, treatment levels are assumed to remain at 80%.

Estimates of the percentage of the urban population connected to sewerage and the percentage of sewered wastewater that is treated are based on data from the Global Urban Observatory (UNCHS, 1998). The percentage of the urban population connected to sewerage and the percentage of wastewater treated increase with income, based on average city-level data from the Global Urban Observatory (UNCHS, 1998).

Livestock production

Estimates for water pollution from livestock wastes, in the form of nitrogen (N) and biochemical oxygen demand (BOD) loading, are also calculated in PoleStar. Loading factors per unit of feed energy requirements are estimated for the base year using data on livestock populations and milk production from the FAOSTAT database of FAO, and the loading factors from WHO (1982). It is assumed that only livestock in feedlots contribute significantly to water pollutant loads. For the policy shocks, relative trends for waste generation by livestock compared to Reference Scenario levels are estimated from relative trends in real transfers of feedgrain to the livestock sector in JOBS, which serves as a proxy for energy needs of lot-fed animals.

Estimated BOD loadings are the sum of BOD that is associated with the wastes directly, as well as oxygen consumed in the oxidation of nitrogen (NBOD). The loading for NBOD is calculated from the nitrogen loading using the relationship:

$$\text{NBOD} = 4.57 \times \text{N}$$

Fertilisers

Following the IPCC guidelines (IPCC/OECD/IEA, 1996), 30% of the nitrogen applied as fertiliser is assumed to be lost to leaching and runoff. Note that this value can vary over a broad range, from as little as 10% to as much as 80%. Nitrogen in manufactured fertilisers is typically in the form of ammonium or nitrate salts. Since nitrate is already oxidised, it does not contribute to oxygen demand. For this analysis it is assumed that fertilisers do not contribute to oxygen demand (in contrast to the nitrogen contained in livestock wastes).

Industry

Estimation of the biochemical oxygen demand contributed by manufacturing is based on a study by the World Bank's Development Research Group (Hettige *et al.*, 1997), which examined data from 12 developing and industrialised countries. Consistent with the approach of Hettige *et al.*, pollution generation is expressed as the product of an activity level – value-added for the industrial sector – and an intensity. An abatement factor is applied to the total generation to estimate the final load:

$$\text{Total Industry Pollution} = \text{Industrial Value Added} \times \text{Pollution Intensity} \times \left(1 - \frac{\%\ \text{Abatment}}{100}\right)$$

The data of Hettige *et al.* suggest that aggregate manufacturing pollution intensities decline steadily from around 5.5 kg/million US$/day at an annual per capita income level of US$500 to around 4.0 kg/million US$/day at an income level of US$5 000 and remain relatively steady, showing only a slight increase, for incomes above that level. For this study it is assumed that aggregate BOD intensities are 5.5 kg/million US$/day for GDP/capita below US$500, 4.0 kg/million US$/day for GDP/capita above US$5 000 and decline steadily for GDP/capita between US$500 and US$5 000. Abatement levels are also assumed to vary with income, following the tabulated values reported in Hettige *et al.* (1997).

4.5. Timber production

Three categories of forest products are considered: fuelwood, paper and pulp, and an aggregate category containing all other wood products. Requirements for fuelwood are driven by requirements for biomass developed in the energy analysis. Requirements for paper and pulp and for other wood products are driven by value-added in the paper and pulp and the wood products sub-sectors in JOBS.

Wood products are designated as being either primary or secondary. Primary products are different kinds of roundwood, that is, felled trees from which the twigs, branches and leaves have been removed. Primary products (such as sawlogs, veneer logs, pulpwood, chips) are used to produce secondary wood products (such as sawnlogs, veneer, paper, pulp and plywood). Wood products are grouped into three categories: paper and pulp, fuelwood and all other wood products. The last category represents all remaining secondary wood products, such as sawnwood, plywood, veneer, etc. Base year data on production and trade are taken from the online FAOSTAT database (FAO, 2000).

The analysis proceeds as follows. First, requirements and production of secondary wood products are developed. Second, the production of secondary products is used to determine the requirements for primary roundwood products, by first converting secondary wood production to primary fibre input requirements, and then taking account of the supply of fibre from sources other than roundwood (*e.g.* non-wood fibre and recycled fibre). It is assumed that for each 1% increase in pulp production, there is a -0.2% decrease in the fraction of fibre that comes from roundwood. After first accounting for trade, regional production of roundwood is then determined. In the Reference Scenario, production is assumed to expand into a mix of plantation and forested land. A fixed fraction of new plantation area is assumed to come from natural forest. Data on land-use conversion rates are scarce. For this study, data for three aggregate tropical regions– Africa, Latin America and Asia– reported in FAO (1996) were examined. Based on these figures, a fraction of 85% for all regions except the Middle East was assumed. For the Middle East, where natural forest area is scarce, the fraction was assumed to be 10%.

For most regions, it is assumed that importers of secondary wood products in the base year maintain their self-sufficiency ratio in the Reference Scenario, defined as the ratio of production to domestic requirements. Most regions that are exporters in the base year maintain their share of total exports in the Scenario. However, Western Europe, an exporter in the base year, has no net exports in 2020, consistent with the results from JOBS.

The fraction from roundwood for paper and pulp is given by the ratio of "pulp for paper" to the total of "pulp for paper", "recovered paper" and "other fibre" from FAOSTAT. For fuelwood, it is assumed that the fraction from roundwood is 0% for developing regions, and 50% otherwise, on the basis that in developing regions fuelwood production is generally obtained by gathering branches and other excess material, rather than by felling trees.

Wood products are produced from either commercially exploitable forest or plantations. Data on the area of commercially exploitable forest and plantations are taken from the FAO-commissioned *Global Fibre Supply Study* (FAO, 1998). The annual increase in biomass on exploitable forest and current plantations are taken from the same source. For an assessment of potential production, the reported value should be reduced, to take into account the amount of the increment that is left in the forest and uncertainties in the reported increment figures. For this analysis, increments are reduced by 25% to estimate the increase in roundwood area.

4.6. Agricultural demand and supply

Base year data on agricultural demand and supply are taken from the FAOSTAT database (FAO, 1996). For this study, crop commodities are classified as Wheat and Coarse Grains, Rice and Other Crops. The livestock commodities are classified as Beef, etc., Other Meat, Milk and Fish. The "Beef, etc." category includes animals that are suitable for grazing, such as cattle, sheep and goats. The "Other Meat" category includes animals that are not grazed, such as pigs and poultry. "Other Meat" also includes eggs and animal by-products. The "Milk" category includes milk products.

Food demand in the analysis is expressed as per capita caloric intake multiplied by population. Total caloric intake is then split into consumption of crop and animal commodities. In the Reference Scenario, dietary patterns adjust with rising incomes per capita. Shifts in diet include greater food consumption in developing regions, with an increasing share of calories derived from animal products. In North America and Western Europe, it is expected that intake levels off, or declines, due to saturation effects and health concerns, *e.g.,* with respect to high-fat diets. In the other regions, as their average incomes grow, diets tend towards the pattern typical of Western Europe today.

Patterns of trade in animal products are based on self-sufficiency ratios (SSRs) for the base year. The SSR is defined as the amount of food (in tonnes) available from domestic sources, divided by requirements. In general, SSRs change little over the period of the Reference Scenario, increasing for regions that were net exporters in the base year and staying at base-year levels for importers. North America, Australia & New Zealand and the Middle East depart from this pattern. In North America and Australia & New Zealand, exports increase significantly over the period of the Reference Scenario to meet part of the rapidly growing world demand. In the Middle East, the SSR decreases over time, in response to increasing pressures placed by livestock on land resources.

To facilitate allocation of feeds to different kinds of livestock in the base year, livestock were separated into two groups. The first includes animals that can make use of grazing resources, such as beef and dairy cattle, sheep and goats. Feed requirements in the Reference Scenario are expressed as changes in feed energy requirements per tonne of meat or milk. Values for OECD regions were based on historical changes in OECD countries as a whole. Assumptions for developing regions were based on scenarios developed by Bouwman (1997).

Assumptions are made for the demand and productivity of crop products. The main variables affecting productivity are the yields, cropping intensities and irrigated area. Changes in these variables are set based on values assumed in Leach (1995) and in the *Bending the Curve* study (Raskin *et al.*, 1998; Heaps *et al.*, 1998). Limits on production are assumed to be determined principally by limits on rainfed crop area. Regional assumptions for domestic production and trade are constrained by the assumption of continuity in trade patterns: *i.e.* SSRs are consistent with historical trends.

Trade assumptions are developed in several steps. First, the self-sufficiency ratios are assumed to be as in the base year in every region. Second, the amount of rainfed crop area is determined which would be required to meet this level of production. This area is then compared to the area of potential cropland (see below). For importing regions, if the required area is below the limit, then the self-sufficiency ratio remains as in the base year. Otherwise, it is lower than in the base year by the amount required to stay within the land constraints. Third, total net import requirements are determined. Finally, production in exporting regions is set so that global demand is met. In general, regions that are importers in the base year remain importers in the Reference Scenario, and the same is true for exporters. The main exceptions are Central & Eastern Europe and the FSU, as they recover from a recent drop in production and as yields in the FSU, which today are well below those of Western Europe and North America, increase.

Estimates of potential productivity of rainfed land for 91 developing countries were prepared by the FAO and the International Institute for Applied Systems Analysis in Vienna (Alexandratos, 1995; Fischer, 1993). The extent of potentially cultivable land and its distribution in different land-use categories was calculated based on these estimates for the study described in Leach (1995), and were used in the study *Bending the Curve* (Raskin *et al.*, 1998; Heaps *et al.*, 1998). For this study, estimated areas of potential cropland in each region were developed by assuming the same distribution of potential cropland under different land uses as in comparable regions in *Bending the Curve*.

4.7. Agricultural inputs

Data for 1995 on water withdrawals and available supply of water are taken from OECD *Environmental Data: Compendium 1999* (OECD, 1999) for OECD countries, and from Shiklomanov (1999) for other countries. For the Reference Scenario, agricultural yield assumptions determine trends in agricultural water use, as described below. For the policy shocks, relative trends in value added for the agriculture sectors in JOBS were used to calculate deviations from the Reference Scenario, to capture the effect of changes in water prices. Real deliveries from the water services sector in JOBS to the crop production sectors were used as a proxy for physical water deliveries. The ratio of the trend in this indicator for a given policy shock scenario and the Reference Scenario was applied to the estimated baseline withdrawals, to give an estimate of water withdrawals in the policy shocks.

For this analysis, all agricultural withdrawals are assigned to irrigation. In the Reference Scenario, irrigation intensities are decomposed into two factors: crop per drop and water use efficiency. The crop per drop factor is assumed to increase with yields, based on empirical yield-response curves (Hexem and Heady, 1978). Efficiencies are assumed not to change in the Reference Scenario. As a result, water intensities increase gradually over the course of the Reference Scenario as yields increase, but at a slower rate than the increase in yields. Irrigation withdrawals in the Scenario are determined by multiplying the intensities by the irrigated harvested area.

The amount of nitrogen fertiliser applied per hectare (the intensity) is calculated in terms of yields, so that the trends in fertiliser use are consistent with the yield assumptions, based on crop production data from FAOSTAT and fertiliser application data from FAO's Land and Water Division (FAO, 1999). Application rates increase with yields, using log-linear regression curves estimated from cross-sectional data. For the policy shocks, the ratio of real deliveries of output from the chemical sector to the agriculture sector in JOBS in the shock to the same indicator in the Reference Scenario were applied to the estimated Reference Scenario figure.

4.8. Land use changes

The principal source for data on land for this analysis is the FAOSTAT database (FAO, 1996), supplemented by information from i.a. the World Conservation Monitoring Centre (WCMC, 1998*a*, *b*).

In the OECD regions, built environment in the Reference Scenario increases at a fixed rate based on historical trends in the US, consistent with the similar economic growth in those regions. In the US, between 1982 and 1992, developed land per capita increased at a rate of about 0.6% per year on average. It is assumed that built environment per capita increases in all OECD regions at half this rate, 0.3% per year, throughout the Reference Scenario. Built environment per capita in Africa, Mexico and Central & South America converges towards the OECD average with increasing income, while China, South Asia, East Asia, the Middle East, Central & Eastern Europe, FSU, ROE and ARW all converge towards the more compact value for Western Europe.

In the Reference Scenario, cropland in the developing regions expands into pasture land and unprotected forest areas in proportion to the amount of potentially cultivable land under those uses, after losses of potential cropland to degradation and conversion have been taken into account. In every region, grazing land expands into unprotected forest and "other" land in proportion to the area of non-cultivable land under these uses. Forest plantations expand largely into natural forest, with the remainder expanding into "other" land.

4.9. Municipal wastes

In the base year, waste generation rates in OECD regions are based on data given in OECD (1997*b*; 1999). In the remaining regions, waste generation rates in rural and urban areas are based on the default regional generation rates given in Doorn and Barlaz (1995). In the Reference Scenario, high-income OECD generation rates are, in accordance with developments over the last decade, assumed to increase at a slightly lower rate than GDP, while other regions converge toward the average rate in high-income OECD regions as incomes increase.

For the OECD regions, the waste processing fractions were calculated using data from OECD (1997*b*; 1999). For other regions the processing fractions are based on the regional aggregate figures in the Global Urban Indicators Database (UNCHS, 1998), which provides city-level data and regional averages. Processing fractions are applied to the total population in high-income OECD and transitional regions, but to the urban population alone in developing regions. In the Reference Scenario, recycling trends in North America to 2020 are based on anticipated trends in the US to 2020 (Tellus Institute, 1998). Trends in Western Europe are based on anticipated trends in the EU. Recycling rates in other regions, which are very low compared to North American and European levels, rise gradually over the period of the Scenario, by 10 percentage points between 1995 and 2020. Landfill fractions decline by the same amount.

Methane emission factors in the Reference Scenario are decomposed into a product of three factors: the degradable organic carbon content (DOC), the fraction of the DOC that is dissimulated to gas, and the fraction of the gas emitted that is methane, following the methodology adopted for SEI's Greenhouse Gas Scenario System (SEI, 1993). The degradable organic content converges gradually in all regions toward the base year value for North America. This convergence reflects the increasing importance of paper, which has a high degradable organic content, in the waste streams of all regions.

Methane recovery in North America, Western Europe and Japan & Korea is close to 10% in 1995. In the Reference Scenario, methane recovery in these regions increases to 60% by 2020, based on anticipated developments in the US (Tellus Institute, 1998). Methane recovery in the Australia & New Zealand region is negligible in the base year, and is assumed to reach 40% by 2020. Recovery rates in all other regions are assumed to reach 20% by 2020, based on the assumptions in Tellus Institute (1998).

4.10. Domestic and industrial water use

Data for 1995 on water withdrawals and available supply of water are taken from OECD *Environmental Data: Compendium 1999* (OECD, 1999) for OECD countries, and from Shiklomanov (1999) for other countries. These sources provide water withdrawal data for agricultural, domestic and industrial uses. Industrial withdrawals include both manufacturing withdrawals and withdrawals for cooling thermal power plants. Separate data on manufacturing and energy withdrawals are available in OECD (1999), but not in Shiklomanov (1999).

4.11. Domestic water use

Domestic water use includes water withdrawals for the household and service sectors. The measure of activity for domestic withdrawals is regional population. In the Reference Scenario, in North America and Australia & New Zealand, domestic water intensities are assumed to decrease by 20% between 1995 and 2050. In Western Europe and Japan & Korea, where current domestic intensities are much less than those in North America and Australia & New Zealand, domestic water intensities are assumed to remain at base year levels. In non-OECD regions, intensities converge toward average OECD patterns. For the policy shocks, relative trends for the household and service sector in JOBS were used to calculate deviations from the Reference Scenario, to capture the effect of changes in water prices. Real deliveries from the water services sector in JOBS to the household and service sectors were used as a proxy for physical water deliveries. The ratio of the trend in this indicator for a given policy shock scenario and the Reference Scenario was applied to the estimated baseline withdrawals, to give an estimate of water withdrawals in the policy shocks.

4.12. Industrial water use

In the analysis, industrial water use is disaggregated into withdrawals for cooling thermal power plants and for manufacturing. Water withdrawals for manufacturing are further disaggregated by subsector. The measure of activity for manufacturing withdrawals is subsectoral value-added, with the water intensity defined as withdrawals per unit value added. For thermal cooling, electricity output from thermal power plants is the measure of activity.

For OECD regions, OECD data on thermal electric cooling were used to allocate total industrial withdrawals to manufacturing and energy in OECD regions (OECD, 1999). Data on withdrawals by manufacturing subsector are scarce. US data on industrial water withdrawals in 1983 (Gleick, 1993) were used to assign relative intensities to individual manufacturing subsectors, using subsectoral value added data for 1983 from UNIDO (1997). The same relative intensities are used for all regions. To allocate total industrial withdrawals to manufacturing and energy in non-OECD regions, the manufacturing and energy intensities for the North American region were scaled by the same factor to recover the total industrial withdrawals for the region.

For the Reference Scenario, trends in water intensity are based on trends in the US, where disaggregated time-series data were available (Gleick, 1993). Assuming a slowing of the historical trend, intensities in OECD regions are assumed to drop by 10% between 1995 and 2025, and by 20% between 1995 and 2050 in all sectors. Manufacturing intensities in the non-OECD regions are assumed to converge toward the average for the OECD regions as incomes rise. For the policy shocks, relative trends in value added for the manufacturing and energy sectors in JOBS were used to calculate deviations from the Reference Scenario, to capture the effect of changes in water prices. Real deliveries from the water services sector in JOBS to the manufacturing and energy sectors were used as a proxy for physical water deliveries. The ratio of the trend in this indicator for a given policy shock scenario and the Reference Scenario was applied to the estimated baseline withdrawals, to give an estimate of water withdrawals in the policy shocks.

5. The MOVE II model

The MOVE II model is a system of spreadsheets that calculates emissions of CO, NMHC, NO_x, N_2O, CH_4 and PM from motor vehicles, as used in Chapters 14 and 15 of this *Outlook*. The system is based on a calculation of the number of kilometres driven in a given region for each technology and each vehicle type in a given calendar year, multiplied by the emissions factor for that vehicle technology/type.

For a given region and vehicle type, the age distribution of the vehicles was estimated. Based on this age distribution, and knowledge of the emissions standards adopted or expected to be adopted in each country, a table was created which determined the technology type for each model year, the number of vehicles of that technology type in a given calendar year, and the number of kilometres driven by vehicles using that technology. Vehicle categories included light duty gasoline vehicles (passenger cars), light duty diesel vehicles, light duty gasoline trucks (including sport utility vehicles), light duty diesel trucks, heavy duty gasoline trucks and buses, heavy duty diesel trucks and buses and motorcycles (including scooters). Emissions of each pollutant were then combined for each vehicle type for calendar years between 1990 and 2030.

Ideally one would calculate emissions for each country of the world separately, recognising the reality that every country is different at least in some important respects. However, such an approach would be prohibitively burdensome and time consuming. Instead the world was divided into regions where the similarities between them were broad enough to provide reasonable emissions estimates. Major factors which were considered were emissions regulations, vehicle types, economic conditions and growth forecasts, etc.

Emissions factors are a function of many variables: vehicle technology, driving characteristics, quality of and frequency of maintenance, fuel composition, meteorology and climate, and traffic patterns to name just a few. Again a precise inventory would require a careful accounting of each of these items on a day-by-day basis, if not hour-by-hour, in any given location. Such an approach is not feasible for most countries, much less for the entire world, due not only to the overall burden of calculations, but also to the lack of reliable data, and hence regional groupings have been used.

Countries around the world are rapidly moving toward more and more stringent vehicle emissions standards. However, three major sets of regulations have evolved – in the European Union, Japan and the US – each with their own driving cycles and test procedures. Other countries have tended to adopt the same standards, although on a different schedule and sometimes with minor modifications.

Production of motor vehicles since the end of World War II has risen from about 5 million motor vehicles per year to almost 50 million. Growth in production has been at a rate of 1 million additional vehicles each year. Beyond cars and trucks, motorcycle production has also grown rapidly, especially in Asia. Since motor vehicle production is increasing at a more rapid rate than vehicle scrappage, worldwide vehicle registrations are growing rapidly and are accelerating. Europe (including Eastern Europe and the former Soviet Union) and North America each have about 35% of the world's motor vehicle population with the remainder now divided among Asia, Latin America, Africa, and Oceania, in that order.

In terms of per capita motor vehicle registration, the US, Japan, and Europe also account for much the largest share of the ownership and use of motor vehicles. Non-OECD countries in Africa, Asia and Latin America – home to more than four-fifths of the world's population – account for only one-fifth of world motor vehicle registrations. It is clear that future growth is likely to occur in Asia and Latin America.

The three primary drivers leading to increases in the world's vehicle fleets are population growth, increased urbanisation and economic improvement. The calculations used here are based on similar assumptions concerning these drivers as were used in the JOBS model and the PoleStar framework.

REFERENCES

Alexandratos, N. (ed.) (1995), *World Agriculture: Towards 2010*, John Wiley and Sons, Chichester, UK.

Bouwman, A.F (1997), *Long-Term Scenarios of Livestock-Crop-Land Use Interactions in Developing Countries*, FAO Land and Water Bulletin 6, Food and Agriculture Organization, Rome.

Burniaux, J-M. (2000), A *Multi-Gas Assessment of the Kyoto Protocol*, Economics Department Working Papers No. 270. OECD, Paris.

Chapra, S. (1997), *Surface Water-Quality Modelling*, McGraw-Hill Companies, Inc.

Doorn, M.R.J. and M. A. Barlaz (1995) *Estimate of Global Methane Emissions from Landfills and Open Dumps*, EPA-600/R-95-019, prepared for US Environmental Protection Agency, Office of Research and Development. US Environmental Protection Agency, Washington, D.C.

FAO (Food and Agriculture Organization of the United Nations) (1996), *Forest Resources Assessment* 1990: *Survey of Tropical Forest Cover and Study of Change Processes*, FAO Forestry Paper 130, FAO, Rome.

FAO (1998), *Global Fibre Supply Model*, FAO Forestry Department, Forest Products Division, Wood and Non-Wood Products Utilisation Branch, FAO, Rome. *www.fao.org/forestry/*FOP/FOPW/GFSM/.

FAO (1999), "Global Forest Products Consumption, Production, Trade and Prices: Global Forest Products Model Projections to 2010" Working Paper GFPOS/WP/01, FAO, Rome.

FAO (1999), "Fertilizer Use by Crop 1999", FAO, Rome. Available from *www.fao.org*.

FAO (2000), FAOSTAT·PC, FAO, Rome.

Fischer, G. (1993), *Personal communication*.

Gleick, P. H. (ed.) (1993), *Water in Crisis: A Guide to the World's Fresh Water Resources*, Oxford University Press, New York.

Heaps, C., E. Kemp-Benedict and P. Raskin (1998), *Conventional Worlds: Technical Description of* "Bending the Curve" Scenarios, PoleStar Series Report No. 9, Stockholm Environment Institute-Boston, Boston.

Hettige, H., M. Mani, and D. Wheeler (1997), "Industrial Pollution in Economic Development: Kuznets Revisited", Development Research Group, December, 1997, World Bank, Washington D.C.

Hexem, R. W. and E. O. Heady (1978), *Water Production Functions for Irrigated Agriculture*, The Iowa State University Press, Ames, Iowa.

IEA (International Energy Agency) (1999), *Energy Statistics* 1960-1997, IEA/OECD, Paris.

IEA (2000), *World Energy Outlook* 2000. IEA/OECD, Paris.

IPCC (Intergovernmental Panel on Climate Change) (1995), IPCC *Greenhouse Gas Inventory Reference Manual*: IPCC *Guidelines for National Greenhouse Gas Inventories*, Vol. 3, IPCC, Bracknell, UK.

IPCC/OECD/IEA (Intergovernmental Panel on Climate Change/Organisation for Economic co-operation and Development / International Energy Agency) (1996), *Revised* 1996 IPCC *Guidelines for National Greenhouse Gas Inventories*, Volume 3, J.T. Houghton, L.G. Meira Filho, B. Lim, K. Treanton, I. Mamaty, Y. Bonduki, D.J. Griggs and B.A. Callender (eds.), UK Meteorological Office, Bracknell.

Kuylenstierna, J. (1998), *Personal communication*.

Leach, G. (1995), *Global Land and Food in the* 21st *Century: Trends and Issues for Sustainability*, PoleStar Series Report No. 5, Stockholm Environment Institute, Stockholm.

Martin, W. and D. Mitra (1999), *Productivity Growth in Agriculture and Manufacturing*, Working Paper No. 2171, World Bank, Washington D.C.

OECD (1995), *Technological Change and Structural Adjustment in* OECD *Agriculture*, OECD, Paris.

OECD (1997*a*), *The World in* 2020: *Towards a New Global Age*, OECD, Paris.

OECD (1997*b*), OECD *Environmental Data: Compendium* 1997, OECD, Paris.

OECD (1999), OECD *Environmental Data: Compendium* 1999, OECD, Paris.

Posch, M., J-P. Hettelingh, J. Alcamo and M. Krol (1996), "Integrated Scenarios of Acidification and Climate Change in Asia and Europe" , in *Global Environmental Change* 6(4), 375-394.

Raskin, P., G. Gallopin, P. Gutman, A. Hammond and R. Swart (1998), *Bending the Curve: Toward Global Sustainability*, PoleStar Series Report No. 8.Stockholm Environment Institute-Boston, Boston.

SEI (Stockholm Environment Institute) (1993), G2S2: *Greenhouse Gas Scenario System*, Database and User Guide, Stockholm Environment Institute-Boston, Boston.

Shiklomanov, I. (1999), *Personal communication*.

Tellus Institute (1998), *Climate and Employment Benefits from Organic Waste Diversion*, prepared for the US EPA Office of Solid Waste.

UNCHS (United Nations Center for Human Settlements) (1998), *Global Urban Indicators Database*, The Global Urban Observatory, Habitat, Nairobi. *www.urbanobservatory.org/indicators/database/*.

UNIDO (United Nations Industrial Development Organization) (1997), UNIDO *Industrial Statistics Database* 1997 (diskette), United Nations Industrial Development Organization, Vienna.

WCMC (World Conservation Monitoring Centre) (1998*a*), *Global Forest Information Service World Tables*, WCMC, Cambridge. *www.wcmc.org.uk*.

WCMC (1998*b*), *Global Protected Areas Summary Statistics*, WCMC, Cambridge. *www.wcmc.org.uk*.

WHO (World Health Organization) (1982), "Rapid Assessment of Sources of Air, Water, and Land Pollution", WHO Offset Publication No. 62, World Health Organization, Geneva.

OECD PUBLICATIONS, 2, rue André-Pascal, 75775 PARIS CEDEX 16
PRINTED IN FRANCE
(97 2001 01 1 P) ISBN 92-64-18615-8 – No. 51591 2001